Hydrogen Transportation and Storage

The success of hydrogen energy markets depends on developing efficient hydrogen storage and transportation methods. Hydrogen may be stored in various ways, including compression, liquefaction, adsorption, hydrides, and reformed fuels. Hydrogen's application, transport method, storage time, and other factors all have an impact on the technology choices available for its long-term storage. This book comprehensively reviews hydrogen storage and transportation technologies along with related safety hazards and challenges.

- Introduces hydrogen storage and transportation materials and standards
- Includes miscellaneous hydrogen storage methods
- Covers different hydrogen transportation technologies
- Comprehensively describes hydrogen storage and transportation safety considerations
- Provides economic assessments and environmental challenges related to hydrogen storage and transportation

Part of the multivolume Handbook of Hydrogen Production and Applications, this stand-alone book guides researchers and academics in chemical, environmental, energy, and related areas of engineering interested in development and implementation of hydrogen production technologies.

Mohammad Reza Rahimpour is a Distinguished Professor of Chemical Engineering at Shiraz University, Iran. He earned a PhD in chemical engineering at Shiraz University, in collaboration with the University of Sydney, Australia, in 1988. He leads a pioneering research group specializing in fuel processing technology, with a particular focus on the catalytic conversion of both fossil fuels, such as natural gas, and renewable fuels, like bio-oils derived from lignin, into valuable energy sources. Throughout his academic career, Dr. Rahimpour has specialized in hydrogen production technologies, including steam methane reforming, water electrolysis, photocatalytic hydrogen production, and hydrogen production from both renewable sources and fossil fuels. His extensive expertise and innovative research significantly contribute to advancements in hydrogen production and its practical applications.

Mohammad Amin Makarem earned a PhD in chemical engineering at Shiraz University. His research interests include gas separation and purification, nanofluids, microfluidics, catalyst synthesis, reactor design, and green energy.

Parvin Kiani earned a degree in chemical engineering at Shiraz University. Her research has focused on gas separation, clean energy, and catalyst synthesis.

Hydrogen Transportation and Storage

Edited by
Mohammad Reza Rahimpour,
Mohammad Amin Makarem and Parvin Kiani

CRC Press
Taylor & Francis Group
Boca Raton London New York

CRC Press is an imprint of the
Taylor & Francis Group, an **informa** business

Designed cover image: © Shutterstock, Macrovector

First edition published 2025
by CRC Press
2385 NW Executive Center Drive, Suite 320, Boca Raton FL 33431

and by CRC Press
4 Park Square, Milton Park, Abingdon, Oxon, OX14 4RN

CRC Press is an imprint of Taylor & Francis Group, LLC

ISBN: 978-1-032-46610-1 (hbk)
ISBN: 978-1-032-46611-8 (pbk)
ISBN: 978-1-003-38255-3 (ebk)

DOI: 10.1201/9781003382553

Typeset in Times
by codeMantra

Contents

SECTION I An Overview of Hydrogen Storage and Transportation Technologies

Mohammad Zarei-Jelyani, Fatemeh Salahi, Fatemeh Zarei-Jelyani, and Mohammad Reza Rahimpour

Dongmin Xi, Hongfang Lu, Shaohua Dong, Zhao-Dong Xu, and Bohong Wang

SECTION II Hydrogen Storage and Transportation Methods

Ameen A. Al-Muntaser, Abdolreza Farhadian, Ismail Khelil, Muneer A. Suwaid, Richard Djimasbe, Mikhail A. Varfolomeev, and Danis K. Nurgaliev

Maneesh Kumar and Sachidananda Sen

John Owolabi, Abdurrazzaq Ahmad, and Chinenye Azie

SECTION III   *Hydrogen Storage and Transportation Safety Considerations*

Preface

As the world intensifies its focus on sustainable energy solutions, hydrogen emerges as a key player due to its versatility and eco-friendly attributes. The *Handbook of Hydrogen Production and Applications* represents a comprehensive and authoritative exploration of the pivotal role hydrogen plays in addressing the global energy landscape and environmental challenges. This ambitious seven-volume project delves into every facet of hydrogen, ranging from its production using both nonrenewable and renewable resources, purification, and separation processes, to its storage, transportation, and various applications across diverse technologies. With an emphasis on fostering a deeper understanding of hydrogen utilization, particularly in fuel cells, this handbook aims to serve as an invaluable resource for researchers, engineers, and policymakers striving to advance the forefront of clean energy technology. Each volume within this compendium is dedicated to a specific aspect, collectively forming a comprehensive guide that illuminates the multifaceted dimensions of hydrogen production and its myriad applications.

Volume 4, *Hydrogen Transportation and Storage*, delves into the critical processes involved in efficiently moving and storing hydrogen, essential for its widespread adoption as a clean energy carrier. In Section I, *An Overview of Hydrogen Storage and Transportation Technologies*, readers are introduced to the fundamental concepts and challenges associated with hydrogen transportation and storage. Engineering properties of hydrogen storage materials are examined, providing insights into the characteristics that influence storage efficiency and safety.

Section II, *Hydrogen Storage and Transportation Methods*, explores a variety of approaches to transporting and storing hydrogen. Liquid hydrogen carriers, pressurized gaseous hydrogen storage, and low-temperature liquefaction methods are discussed, highlighting the diverse technologies available for storing and transporting hydrogen in different states. Additionally, carbonaceous materials, glass microspheres, metal-organic frameworks (MOFs), and pipelines are explored as alternative storage and transportation mediums, showcasing the breadth of options in this field.

Safety considerations take precedence in Section III, where *Hydrogen Storage and Transportation Safety Considerations* are thoroughly examined. Strategies for preventing hydrogen pipeline cracking and leakage, mitigating delayed ignition and explosion risks, and managing the release of liquid hydrogen from pressurized and nonpressurized tanks are discussed in detail. By addressing these safety concerns, stakeholders can ensure the safe and reliable transportation and storage of hydrogen, essential for its widespread adoption as a clean energy solution.

This volume aims to serve as a comprehensive guide for researchers, engineers, and policymakers involved in hydrogen transportation and storage. By offering insights into a diverse array of transportation and storage methods, along with safety considerations, this volume provides a thorough understanding of the challenges and opportunities in effectively managing hydrogen as an energy carrier. We extend our sincere appreciation to all contributors who have shared their expertise to make this volume possible, and we invite readers to delve into the intricacies of hydrogen transportation and storage, paving the way for a sustainable and transformative future in hydrogen utilization.

Mohammad Reza Rahimpour
Mohammad Amin Makarem
Parvin Kiani

Contributors

Abdurrazzaq Ahmad
University of Hull
Hull, England, UK

Ameen A. Al-Muntaser
Kazan Federal University
Kazan, Russia

Chinenye Azie
University of Hull
Hull, England, UK

Moises Bastos-Neto
Federal University of Ceara
Fortaleza, Ceara, Brazil

Hisham Khaled Ben Mahmud
Curtin University Malaysia
Miri, Malaysia

Christian Benedict
Covenant University
Ota, Ogun State, Nigeria

Diana Cristina Silva de Azevedo
Federal University of Ceara
Fortaleza, Ceara, Brazil

Richard Djimasbe
Kazan Federal University
Kazan, Russia

Shaohua Dong
Southeast University
Nanjing, China
and
Southwest Petroleum University
Chengdu, Sichuan, China

Bianca Ferreira dos Santos
Federal University of Ceara
Fortaleza, Ceara, Brazil

Abdolreza Farhadian
Kazan Federal University
Kazan, Russia

Xinmeng Jiang
Southwest Petroleum University
Chengdu, Sichuan, China

Ismail Khelil
Kazan Federal University
Kazan, Russia

Maneesh Kumar
Indian Institute of Technology
Roorkee, India

Ruiling Li
Southwest Petroleum University
Chengdu, Sichuan, China

Minkang Liu
Natural Resources Canada
CanmetMATERIALS
Hamilton, Ontario, Canada

Siming Liu
Southwest Petroleum University
Chengdu, Sichuan, China

Hongfang Lu
School of Civil Engineering
Southeast University
Nanjing, 210096, China

Beatriz Oliveira Nascimento
Federal University of Ceara
Fortaleza, Ceara, Brazil

Danis K. Nurgaliev
Kazan Federal University
Kazan, Russia

Emeka Emmanuel Okoro
University of Port Harcourt
Choba, Rivers State, Nigeria

Babalola Aisosa Oni
North Dakota University
Grand Forks, North Dakota, USA

John Owolabi
University of Hull
Hull, England, UK

Guojin Qin
Southwest Petroleum University
Chengdu, Sichuan, China

Mohammad Reza Rahimpour
Shiraz University
Shiraz, Fars, Iran

Unwana Udoh Robert
Covenant University
Ota, Ogun State, Nigeria

Fatemeh Salahi
Shiraz University
Shiraz, Fars, Iran

Samuel Eshorame Sanni
Covenant University
Ota, Ogun State, Nigeria

Sachidananda Sen
SR University
Warangal, India

Mian Umer Shafiq
UCSI University
Kuala Lumpur, Malaysia

Muneer A. Suwaid
Kazan Federal University
Kazan, Russia

Mikhail A. Varfolomeev
Kazan Federal University
Kazan, Russia

Bohong Wang
Southeast University
Nanjing, China
and
Southwest Petroleum University
Chengdu, Sichuan, China

Yihuan Wang
Southwest Petroleum University
Chengdu, Sichuan, China

Dongmin Xi
Southeast University
Nanjing, China

Ailin Xia
Southwest Petroleum University
Chengdu, Sichuan, China

Zhao-Dong Xu
Southeast University
Nanjing, China
and
Southwest Petroleum University
Chengdu, Sichuan, China

Fatemeh Zarei-Jelyani
Shiraz University
Shiraz, Fars, Iran

Mohammad Zarei-Jelyani
Shiraz University
Shiraz, Fars, Iran

Yimin Zeng
Natural Resources Canada
CanmetMATERIALS
Hamilton, Ontario, Canada

Zhenwei Zhang
Southwest Petroleum University
Chengdu, Sichuan, China

Reviewer Acknowledgments

The editors feel obliged to appreciate the dedicated reviewers (listed below) who were involved in reviewing and commenting on the submitted chapters and whose cooperation and insightful comments were very helpful in improving the quality of the chapters and books in this series.

Dr. Fatemeh Haghighatjoo
Department of Chemical Engineering
Shiraz University
Shiraz, Iran

Dr. Maryam Meshksar
Department of Chemical Engineering
Shiraz University
Shiraz, Iran

Ms. Soheila Zandi Lak
Department of Chemical Engineering
Shiraz University
Shiraz, Iran

Mr. Sina Mosallanejad
Department of Chemical Engineering
Shiraz University
Shiraz, Iran

Section I

An Overview of Hydrogen Storage and Transportation Technologies

1 Introduction to Hydrogen Storage, Transportation, and Distribution Technologies and Challenges

Mohammad Zarei-Jelyani, Fatemeh Salahi, Fatemeh Zarei-Jelyani, and Mohammad Reza Rahimpour

1.1 INTRODUCTION

One of the key challenges posed by climate change is the need to facilitate the transition of energy systems toward technologies that enable a reduction in greenhouse gas (GHG) emissions (Zarei-Jelyani et al., 2023a,b). Hydrogen is now being seen as a prospective participant in both national and worldwide initiatives, with its use expanding across several sectors ranging from industry to transportation (Noussan et al., 2020). However, the overarching concern of environmental sustainability still needs more investigation and clarification. Hydrogen, a molecule devoid of carbon, has the maximum energy density per unit of mass. Consequently, it is considered a pivotal energy carrier in forthcoming low-carbon scenarios, which in turn promotes the growth of hydrogen-based economies (Valente et al., 2017). Hydrogen is a gas that lacks color, odor, and toxicity. It has a propensity for easy ignition and burns with a faint blue flame that is almost imperceptible, rendering it a lighter gas than atmospheric vapors. It has a significant level of danger due to its extensive explosive and flammable region (4%–74% in air), and its ignition requires just a small amount of energy (Park et al., 2022). The combination of these factors, in addition to extremely low temperatures, poses a major challenge in the secure management of hydrogen (Rhodes, 2011). During the process of oxidation, hydrogen loses its two electrons, resulting in the formation of water exclusively. When compared to other transportation fuels, hydrogen's volumetric energy density is low, even when it's a liquid, despite its high gravimetric energy density.

The use of hydrogen has the potential to have an important effect on various industries such as electricity generation, heat production, transportation, and storage of energy within a low-carbon emissions energy framework. This potential is reliant upon the generation of hydrogen using renewable and waste material sources of energy (Staffell et al., 2019). The use of hydrogen may be categorized into three major classifications. Initially, it may be employed as a reagent in hydrogenation operations for producing compounds with reduced molecular weight or to fully saturate them. At present, about 65% of global hydrogen is allocated for the purpose of manufacturing various chemicals, including but not limited to hydrochloric acid, hydrogen peroxide, ammonium, and methanol, among others (Hren et al., 2023). Furthermore, approximately 25% of the global hydrogen supply is utilized as a reagent in various process industries, including petroleum refining, production of petrochemicals, hydrogenation of oils and fats to generate saturated fats, fertilizer production, metallurgical uses such as ore reduction, atomic hydrogen welding, and applications within the field of electronics. A fraction of 10% of the global hydrogen supply is utilized for various purposes, including its use as an agent for removing O_2, as a fuel source for rocket engines and internal combustion engines, and in the role of a coolant in electricity-generating and weather balloons (Ozcanli et al., 2018).

The growing use of hydrogen across various industries has led to a growth in demand for hydrogen production technology. There are differences among hydrogen production systems in terms of their developmental stage, the feedstocks and resources they need, and the corresponding GHG emissions. The color names gray, blue, turquoise, and green are often used to denote conventional, low-CO_2, CO_2-free, and carbon-emissions-free production approaches, respectively (Hermesmann and Müller, 2022). At present, approximately 70 million tons per year of pure hydrogen are produced on a worldwide scale. The predominant source of this hydrogen is derived from fossil fuels, with 76% being obtained through steam methane reforming for natural gas and 22% through gasification of coal. On the other hand, only around 2% of hydrogen production utilizes electrolysis of water, which presents a significant challenge in terms of sustainability. The use of fossil-fuel-based technologies for production is known to significantly contribute to elevated levels of GHG emissions, particularly carbon dioxide. Moreover, these technologies have been shown to have different adverse effects on the environment and are often associated with a gray color.

The integration of low-CO_2 feedstocks and sources of renewable energy should be implemented for the production of hydrogen in order to mitigate its environmental consequences (Agreement, 2015). In addition to the use of natural gas by steam reforming to produce hydrogen, other sources of renewable and waste feedstocks may be employed with the same method. These include alcoholic waste, biogas, and glycerol. Other techniques for producing hydrogen via syngas include auto thermal reforming (ATR) for biogas, lignocellulosic biomass gasification, and a modified Claus process known as acid gas to syngas (AG2S), which employs acid gas as a raw material (Bassani et al., 2019). Furthermore, it is worth noting that acid gas, particularly hydrogen sulfide (H_2S), may be converted into H_2 using photocatalysis (Preethi and Kanmani, 2013). Residual crops without syngas as an intermediary may also undergo hydrogen conversion via iron-based chemical looping (Heng et al., 2018) and dark fermentation (Xiao et al., 2014). In addition to feedstocks requiring pretreatment, like biomass, some feedstocks are available that do not need energy or expensive pretreatment procedures. As an example, aluminum has the capability to produce hydrogen directly during the process of burning (Wang et al., 2009), as well as water via the process of electrolysis (Koj et al., 2015). Technological advancements have been seen in several domains, including water electrolysis. Notably, the field of water electrolysis has experienced significant progress in recent years, mostly attributed to the development of novel electrolyzer membranes and the application of renewable energies to fulfill power requirements (Koj et al., 2017).

After hydrogen has been produced, it needs storage and transportation to reach its intended users. Nonetheless, the limited volumetric density of hydrogen presents a challenge to its worldwide mobility (Qyyum et al., 2021). To overcome this limitation, it is essential to develop large-scale hydrogen storage capabilities and construct an efficient supply-chain infrastructure (Barrett, 2005).

1.2   HYDROGEN STORAGE TECHNOLOGIES

The storage of hydrogen plays a crucial role in hydrogen energy systems, particularly in the context of the widespread utilization of hydrogen on a large scale. In order to effectively meet the current and projected requirements of the hydrogen energy industry, it is essential to have an efficient and reliable storage mechanism for any specific application. The application summarization of hydrogen storage within the framework of the hydrogen industry is shown in Figure 1.1. Hydrogen storage applications may be classified into two distinct categories: stationary and mobile uses. The primary purpose of stationary storage technologies is to provide on-site storage at either the place of production or use, while also to support stationary electrical power generation. Mobile applications may serve two main purposes: the transportation of stored H_2 to a designated storage or use location or its utilization in a vehicle.

The energy density of hydrogen is 9.9 MJ/m^3 LHV (lower heating value), which is lower than that of fossil fuels (Abdin et al., 2020). Consequently, the storage of hydrogen may necessitate the use of quite large vessels. In order to provide an adequate storage capacity for hydrogen, it is necessary to include at least one of the three

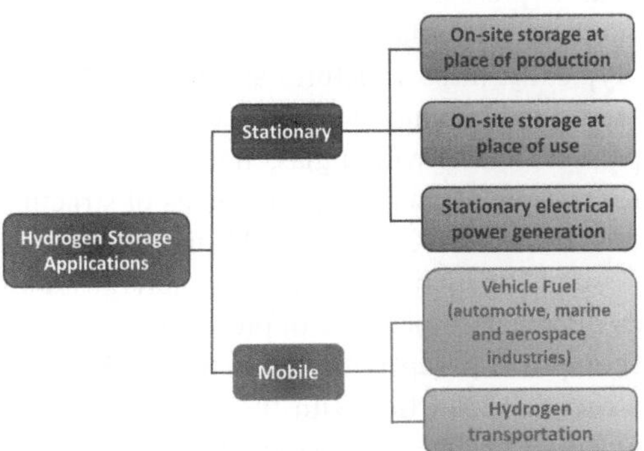

FIGURE 1.1   Hydrogen storage applications: Stationary storage technologies that provide on-site storage at either the place of production or use, as well as support stationary electrical power generation. Mobile applications may serve for the transportation of stored H_2 to a designated storage or use location or its utilization in a vehicle. Information from Moradi and Groth (2019).

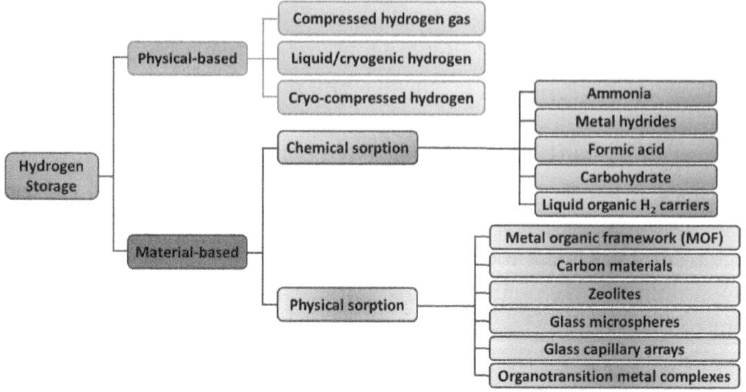

FIGURE 1.2 Hydrogen storage approaches. Information from Moradi and Groth (2019).

characteristics that follow: low storage temperature, high storage pressure, or use of a material with a strong attraction for hydrogen molecules.

The categorization of hydrogen storage approaches may be classified into two primary categories: physical-based and material-based, as shown in Figure 1.2. The first category contains three methods of hydrogen storage: compressed gas, cold/cryo-compressed, and liquid hydrogen storage. There are two main groups under material-based methods of storage, namely chemical sorption (chemisorption) and physical sorption (physisorption) (Akhtar et al., 2021).

1.2.1 PHYSICAL-BASED HYDROGEN STORAGE

1.2.1.1 Compressed Hydrogen Storage

The following are four different kinds of pressure vessels that are suitable for the storage of hydrogen (Niermann et al., 2021).

Type I: Fully metallic pressure vessels. This particular vessel type is considered to be the most economically affordable and has the highest weight density, estimated at roughly 3 lb/L. Typically, these structures are fabricated using aluminum or steel materials and can withstand maximum pressure of 50 MPa.

Type II: Steel pressure vessel with a glass fiber composite overwrap. Both composite and steel materials withstand similar amounts of structural load. The cost of making Type II pressure vessels is around 50% more than that of Type I vessels, while exhibiting a weight reduction of approximately 30%–40%. This particular kind of vessel holds the greatest tolerance of pressure.

Type III: Full composite wrap with metal liner. The primary responsibility for carrying the structural load lies with the composite structure, which is the carbon fiber composite. On the other hand, the liner, composed of aluminum, serves the purpose of providing a seal. Within this particular configuration of a pressure vessel, it is seen that the metal liner supports approximately 5% of the total mechanical stress. The mentioned pressure vessel design has shown its reliability when subjected to a working pressure of 45 MPa. However, it continues to have challenges with successfully passing the aging tests conducted at a higher pressure of 70 MPa (Niermann et al., 2021). Type III has a weight range of 0.75–1 lb/L, almost half that of Type II. However, the cost of Type III is double that of Type II.

Type IV: Fully composite. Typically, in the context of liner materials, high-density polyethylene (HDPE) is often employed, whereas carbon-glass or carbon fiber composites are utilized to withstand the load of structure. This particular kind of vessel has a very low weight, yet it is important to note that it remains comparatively expensive. Type IV pressure vessels have the ability to tolerate pressures as high as 100 MPa.

Additionally, a fully composite, linerless pressure vessel often referred to as a Type V vessel is now in the pre-commercial stage. Composites Technology Development Inc. was the first to produce the aforementioned kind in 2010. The first constructed pressure vessel with an operating pressure of 1.37 MPa exhibited a weight reduction of 20% compared to other vessels of Type IV (Teichmann et al., 2012). However, it is important to note that this pressure level is far below the thresholds necessary for storing considerable amounts of hydrogen beyond laboratory applications.

1.2.1.2 Liquid/Cryogenic Hydrogen Storage

The process of liquefying hydrogen involves subjecting it to very low temperatures, often about −250°C. The primary difficulty in cryogenic hydrogen storage lies in the task of preserving hydrogen at such a low temperature. The process of liquefaction is known to be a time-consuming as well as energy-intensive process that results in a potential loss of up to 40% of the energy content. In comparison, compressed hydrogen storage exhibits a lower energy loss of around 10% (Barthélémy et al., 2017). Therefore, this storage technique is often used for the purpose of storing and transporting massive amounts, including truck delivery or intercontinental hydrogen transportation. In general, cryogenic tankers have the capability to transport around 5,000 kg of hydrogen, a quantity that exceeds the limit of compressed hydrogen gas tube trailers by a factor of approximately five.

In terms of safety considerations, cryogenic vessels are equipped with an extra protective layer known as a vacuum jacket, which serves as an extra layer of protection in the event of accidents. Additionally, it is noteworthy that hydrogen exhibits minimal adiabatic expansion energy when subjected to cryogenic temperatures (Petitpas and Aceves, 2013). Therefore, a significant explosion would not occur in the event of a leak or tanker rupture unless the gas was in some way ignited. The low temperature of the leaking H_2 gas has the potential to cause harm to nearby pressure relief valves and devices that are not designed for strictly low temperatures. This may result in the devices' inability to operate properly. In the year 2016, an incident occurred inside a laboratory specializing in cryogenic hydrogen. The pressure relief valve failed to function at its designated set point due to its placement in a region that was not anticipated to experience cryogenic temperatures, resulting in an inadequate rating (Burgess et al., 2017).

1.2.1.3 Cryo-compressed Hydrogen Storage

This approach to hydrogen storage was initially developed by Aceves et al. (2010). Cryo-compressed H_2 is a supercritical cryogenic gas. At about −233°C, hydrogen gas will be compressed without liquefaction occurring. It has been shown to be effective in terms of both safety and storage. Cryo-compressed storage presents an increase in storage density compared to cryogenic storage, with a density of 80 g/L, surpassing cryogenic storage by about 10 g/L. Additionally, it enables rapid and effective refueling while ensuring a high level of safety using a vacuum enclosure (Stolten et al., 2016). Ahluwalia and coworkers performed a comprehensive technical assessment of this storage technique and determined that it had the capability to fulfill the end goal set by the Department of Energy (DOE) in terms of system volumetric capacity, system gravimetric capacity, and hydrogen loss during periods of inactivity (Ahluwalia et al., 2010). Regarding its potential for broad application, it is essential to include this technology in research and development projects. Nevertheless, the primary difficulties that remain are the accessibility and financial implications associated with infrastructure.

1.2.2 MATERIAL-BASED HYDROGEN STORAGE

In the context of both chemical and physical sorption storage, it is worth noting that some base materials are originally present in a powder state, while others may exist in a liquid form, for instance, liquid organic H_2 carriers. Heat may be generated or absorbed during hydrogen charge and discharge processes, and material in powder form is not an effective type of material when it comes to heat transmission. Hence, the base materials undergo several pre-processing procedures as outlined by Ren et al. (2017), including templating, casting, coating, foaming, and uniaxial pressing. Subsequently, the treated substance will be suitably placed into a containment structure. Typically, the integration of a heat exchanger within the containment design is observed for the goal of thermal management. According to Lototskyy et al. (2015), this also involves a set of connections that enable the regulation of hydrogen flow and the filtering of both input and output hydrogen gas. Jehan and Fruchart (2013) have presented a design idea that demonstrates the possibility of using hydrogen storage at a fuel station size, serving as a proof of concept. However, as previously said, the implementation of these approaches for commercial purposes is improbable in the near future. This is due to many factors that will be elaborated upon in the subsequent discussion (Figure 1.3).

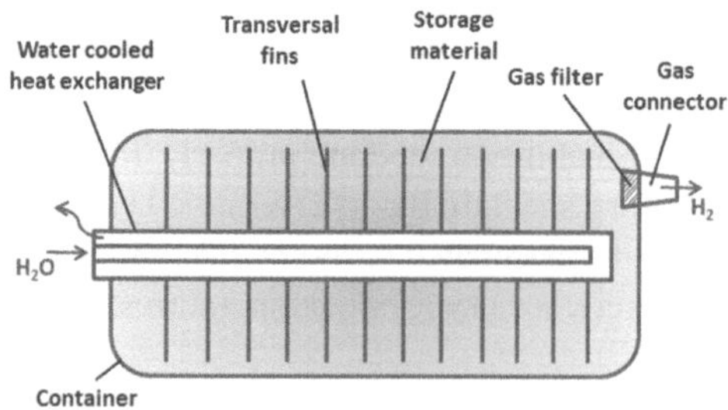

FIGURE 1.3 A proposed design for a material-based storage. Reproduced from Lototskyy et al. (2015).

1.2.2.1 Chemical Sorption

Chemical sorption involves the dissociation of hydrogen molecules into individual atoms, which are then incorporated into the chemical structure of the substance. Metal hydrides are well recognized as a prominent category of materials suitable for chemical sorption. Extensive data and an exhaustive compilation of relevant scholarly sources regarding metal hydrides may be accessed in Ren et al. (2017). The major remaining issues in the field of chemical sorption materials include decreasing overall expenses, weight, and temperature of operation; advances in charge-discharge kinetics; and effectively managing the generation of undesirable gases during desorption. It is noteworthy to note that liquid organic hydrogen carriers (LOHCs) provide a very promising alternative as well. Chemical bonding with hydrogen-lean molecules stores hydrogen in the LOHC storage system, and a catalytic dehydrogenation process releases it (Preuster et al., 2017). The attraction of these storage systems is in their simplicity of management under ambient conditions, their carbon-free store and release operations, and the ability to reuse the carrier liquid without consumption. These carriers do not possess corrosive or toxic properties, and their storage pressure is rather low. The primary concern is the limited hydrogen storage capacity of LOHCs, with a maximum value of 7.2% wt. This puts limitations on the applications of LOHCs (Teichmann et al., 2012).

1.2.2.2 Physical Sorption

Porous material-based storage systems have the ability to serve as an effective means to make storage units with high capacity and reliability. Porous carbon materials and metal organic frameworks (MOFs) have been identified as very promising (Xia et al., 2013). The use of this approach yields many advantages, including a substantial increase in surface area, a decrease in hydrogen binding energy, enhanced kinetics during charge-discharge operations, and inexpensive resources. Additionally, the possibility of physical absorption may help reduce thermal management challenges that arise during the charging and discharging processes of the storage unit. However, there are some concerns associated with this approach, including the weight of the carrier materials, requiring high pressures and low temperatures, and the comparatively low gravimetric and volumetric densities of hydrogen (Zhang et al., 2016). Currently, the use of physical sorption technologies remains limited due to their lack of widespread adoption. This is mainly attributed to the fact that existing investigations have been done on a small scale, leading to unsatisfactory specifications of performance such as gravimetric density of hydrogen, volumetric density, temperature, and pressure.

1.3 HYDROGEN TRANSPORTATION AND DISTRIBUTION TECHNOLOGIES

The delivery of hydrogen has a crucial impact on determining the expenses, consumption of energy, and pollution caused by hydrogen systems. In the context of centralized H_2 generation, the process of delivering hydrogen to end consumers may be divided into two primary phases: transportation, which involves transporting hydrogen from production plants to city gates, and distribution, which includes carrying hydrogen from city gates to fueling stations or final consumers. There are three primary routes to transportation and distribution (T&D), the selection of which is mostly reliant upon the mode of storage:

   i. Gaseous hydrogen T&D
  ii. Liquid hydrogen T&D
 iii. Material-based hydrogen carriers

The selection of the T&D mechanism will be dependent upon certain geographical and market features, such as the target population and their consumption patterns, population density, the scale of refueling stations, and the level to which fuel cell vehicles as well as other hydrogen-consuming systems have penetrated the market.

1.3.1 GASEOUS HYDROGEN T&D

Gas pipelines and compressed H_2 pressure vessels organized in tube trailers (as discussed in the section on hydrogen storage technologies) are two common ways to transport gaseous hydrogen.

1.3.1.1 Transportation of Gaseous Hydrogen with Pipelines

Approximately 2,600 km of hydrogen pipelines are now accessible in the United States, mostly concentrated in areas close to major hydrogen users such as refineries and ammonia plants (Moradi and Groth, 2019). In order to facilitate the transportation of hydrogen throughout the United States, a substantial network of specialized hydrogen pipelines covering hundreds of thousands of kilometers is required (Moradi and Groth, 2019). Consequently, policymakers and researchers have begun investigations into the use of the existing infrastructure of natural gas pipelines for the distribution of hydrogen throughout the country.

1.3.1.2 Transportation of Gaseous Hydrogen with Tube Trailers

Hydrogen gas transportation using tube trailers has attracted significant interest from the DOE and its various collaborators for a number of years. The development of a high-pressure tube trailer, named TITAN, is outlined in a study authored by HEXAGON Lincoln (Baldwin, 2017). These tanks possess an operational pressure of 250 bar. A TITAN, in particular, has a maximum capacity of 616 kg of hydrogen. It is worth noting that the mass of hydrogen contained in a TITAN is roughly 7% of the TITAN weight.

The use of tube trailers is of considerable interest due to their inherent simplicity in terms of infrastructural needs. Furthermore, the historical background regarding the storage and transportation of other gaseous substances has provided us with an extensive comprehension of the physics involved. A further benefit of tube trailers is the low loss of hydrogen and cost-effective compression at fueling stations. Elgowainy et al. (2014) have shown that these costs may be further reduced by 60% compared to the transportation of liquid hydrogen, which will be described in the subsequent section. In order to ensure the safety of the tube trailer, a series of tests are conducted, including a hydrostatic burst test, penetration testing, leak before burst, as well as temperature and pressure cycle tests. A number of challenges can be identified with these products. These include liner blistering caused by saturation and decompression, which has been observed empirically at a test pressure of 350 bar. Additionally, the manufacturing cost of the COPV vessels accounts for 70% of the overall expense. Furthermore, the storage capacity of the tube trailer still remains low, and there are constraints on the dimensions and maximum pressure of tanks due to transportation standards.

1.3.2 LIQUID HYDROGEN T&D

Although it was previously noted that the energy loss is greater, this approach is regarded as cost-effective for substantial demands (exceeding 500 kg/day) and moderate distances (Moradi and Groth, 2019). According to a report published by the California Air Resources Board (CARB), it is predicted that hydrogen fueling stations will be supplied with liquid hydrogen, mostly owing to its much greater storage capacity. The process of cryogenic hydrogen delivery has three primary phases: liquefaction, storage, and transportation using cryogenic tanks to reach the target consumers. North America now has a total of eight liquefaction plants, each of which has a production capacity ranging from 5 to 10 metric tons per day. To effectively meet the needs of the upcoming hydrogen market, more liquefaction plants need to be built with higher production rates, lower specific energy consumption (with a target value of 6 kWh/kg of liquid hydrogen, which means a 40% reduction), lower capital costs, and better efficiency. This will facilitate the adoption of liquid hydrogen as the preferred delivery method for hydrogen in the future. Cardella et al. conducted a study to explore the most effective methods for achieving optimum large-scale and economically feasible liquefaction methods (2017a,b). There are also current attempts underway to advance the efficiency of methods. Asadnia and Mehrpooya (2017) have presented a novel approach for large-scale liquefaction, obtaining an energy consumption rate of 7.69 $kWh/(kgLH_2)$. This is much lower than the energy consumption range of existing plants, which typically ranges between 12.5 and 15 $kWh/(kgLH_2)$.

1.3.3 MATERIAL-BASED HYDROGEN CARRIERS

Material-based hydrogen delivery systems have the potential to provide enhanced safety features owing to their controllable characteristics under ambient conditions, low storage pressure, and acceptable gravimetric density in comparison to gaseous storage (for instance, in a tube trailer, the weight of hydrogen constitutes only 7% of the total tank weight). However, it should be noted that these material-based hydrogen carriers may not be suitable for high-demand uses.

1.4 CONCLUSION

Hydrogen is emerging as a viable and eco-friendly option for transportation and energy storage, owing to its abundant availability and numerous production sources. Considerable governmental funding is being allocated toward research in this field, and there is increasing public talk about hydrogen's potential as a future fuel source. The hydrogen economy encompasses the fundamental elements of hydrogen generation, storage, transportation, and distribution. This chapter has provided a general introduction to the various applications of hydrogen and the primary techniques used for hydrogen production. The chapter has discussed hydrogen storage systems, which may be generally classified into two categories: physical-based and chemical-based approaches. The section on physical approaches has covered three distinct methods: liquid/cryogenic hydrogen, compressed hydrogen gas, and cryo-compressed hydrogen. Chemical sorption and physical sorption are discussed as subcategories of material-based storage methods for hydrogen. This chapter also addresses the technology involved in transporting and distributing hydrogen. Each related section has also talked about the associated difficulties in the domains of hydrogen production, storage, transportation, and distribution.

ABBREVIATIONS AND SYMBOLS

AG2S	Acid Gas to Syngas
ATR	Auto Thermal Reforming
CARB	California Air Resources Board
COPV	Composite Overwrapped Pressure Vessel
DOE	Department of Energy
GHG	Greenhouse Gas
HDPE	High-Density Polyethylene
LHV	Lower Heating Value
LOHCs	Liquid Organic Hydrogen Carriers
MOFs	Metal Organic Frameworks
T&D	Transportation and Distribution

REFERENCES

Abdin, Z., Zafaranloo, A., Rafiee, A., Merida, W., Lipiński, W. & Khalilpour, K. R. 2020. Hydrogen as an energy vector. *Renewable and Sustainable Energy Reviews*, 120, 109620.

Aceves, S. M., Espinosa- Loza, F., Ledesma- Orozco, E., Ross, T. O., Weisberg, A. H., Brunner, T. C. & Kircher, O. 2010. High-density automotive hydrogen storage with cryogenic capable pressure vessels. *International Journal of Hydrogen Energy*, 35, 1219–1226.

Agreement, P. 2015. Adoption of the Paris Agreement Proposal by the President, United Nations Framework Convention on Climate Change. FCCC/CP/2015/L. 9 https://unfccc.int/resource/docs/2015/cop21/eng/l09.pdf.

Ahluwalia, R., Hua, T., Peng, J.-K., Lasher, S., Mckenney, K., Sinha, J. & Gardiner, M. 2010. Technical assessment of cryo-compressed hydrogen storage tank systems for automotive applications. *International Journal of Hydrogen Energy*, 35, 4171–4184.

Akhtar, M. S., Dickson, R. & Liu, J. J. 2021. Life cycle assessment of inland green hydrogen supply chain networks with current challenges and future prospects. *ACS Sustainable Chemistry & Engineering*, 9, 17152–17163.

Asadnia, M. & Mehrpooya, M. 2017. A novel hydrogen liquefaction process configuration with combined mixed refrigerant systems. *International Journal of Hydrogen Energy*, 42, 15564–15585.

Baldwin, D. 2017. *Development of High Pressure Hydrogen Storage Tank for Storage and Gaseous Truck Delivery*. Hexagon Lincoln LLC: Lincoln, NE.

Barrett, S. 2005. The European hydrogen and fuel cell strategic research agenda and deployment strategy. *Fuel Cells Bulletin*, 2005, 12–19.

Barthelemy, H., Weber, M. & Barbier, F. 2017. Hydrogen storage: recent improvements and industrial perspectives. *International Journal of Hydrogen Energy*, 42, 7254–7262.

Bassani, A., Van Dijk, H., Cobden, P., Spigno, G., Manzolini, G. & Manenti, F. 2019. Sorption enhanced water gas shift for H2 production using sour gases as feedstock. *International Journal of Hydrogen energy*, 44, 16132–16143.

Burgess, R. M., Post, M. B., Buttner, W. J. & Rivkin, C. H. 2017. *High Pressure Hydrogen Pressure Relief Devices: Accelerated Life Testing and Application Best Practices*. National Renewable Energy Lab (NREL): Golden, CO.

Cardella, U., Decker, L. & Klein, H. 2017a. Roadmap to economically viable hydrogen liquefaction. *International Journal of Hydrogen Energy*, 42, 13329–13338.

Cardella, U., Decker, L., Sundberg, J. & Klein, H. 2017b. Process optimization for large-scale hydrogen liquefaction. *International Journal of Hydrogen Energy*, 42, 12339–12354.

Elgowainy, A., Reddi, K., Sutherland, E. & Joseck, F. 2014. Tube-trailer consolidation strategy for reducing hydrogen refueling station costs. *International Journal of Hydrogen Energy*, 39, 20197–20206.

Heng, L., Xiao, R. & Zhang, H. 2018. Life cycle assessment of hydrogen production via iron-based chemical-looping process using non-aqueous phase bio-oil as fuel. *International Journal of Greenhouse Gas Control*, 76, 78–84.

Hermesmann, M. & Muller, T. 2022. Green, turquoise, blue, or grey? Environmentally friendly hydrogen production in transforming energy systems. *Progress in Energy and Combustion Science*, 90, 100996.

Hren, R., Vujanović, A., Van Fan, Y., Klemeš, J. J., Krajnc, D. & Čuček, L. 2023. Hydrogen production, storage and transport for renewable energy and chemicals: an environmental footprint assessment. *Renewable and Sustainable Energy Reviews*, 173, 113113.

Jehan, M. & Fruchart, D. 2013. McPhy-Energy's proposal for solid state hydrogen storage materials and systems. *Journal of Alloys and Compounds*, 580, S343–S348.

Koj, J. C., Schreiber, A., Zapp, P. & Marcuello, P. 2015. Life cycle assessment of improved high pressure alkaline electrolysis. *Energy Procedia*, 75, 2871–2877.

Koj, J. C., Wulf, C., Schreiber, A. & Zapp, P. 2017. Site-dependent environmental impacts of industrial hydrogen production by alkaline water electrolysis. *Energies*, 10, 860.

Lototskyy, M. V., Davids, M. W., Tolj, I., Klochko, Y. V., Sekhar, B. S., Chidziva, S., Smith, F., Swanepoel, D. & Pollet, B. G. 2015. Metal hydride systems for hydrogen storage and supply for stationary and automotive low temperature PEM fuel cell power modules. *International Journal of Hydrogen Energy*, 40, 11491–11497.

Moradi, R. & Groth, K. M. 2019. Hydrogen storage and delivery: review of the state of the art technologies and risk and reliability analysis. *International Journal of Hydrogen Energy*, 44, 12254–12269.

Niermann, M., Timmerberg, S., Drunert, S. & Kaltschmitt, M. 2021. Liquid organic hydrogen carriers and alternatives for international transport of renewable hydrogen. *Renewable and Sustainable Energy Reviews*, 135, 110171.

Noussan, M., Raimondi, P., Scita, R. & Hafner, M. 2020. The role of green and blue hydrogen in the energy transition-a technological and geopolitical perspective. *Sustainability*, 13, 298.

Ozcanli, M., Bas, O., Akar, M. A., Yildizhan, S. & Serin, H. 2018. Recent studies on hydrogen usage in Wankel SI engine. *International Journal of Hydrogen Energy*, 43, 18037–18045.

Park, S. W., Kim, J. H. & Seo, J. K. 2022. Explosion characteristics of hydrogen gas in varying ship ventilation tunnel geometries: an experimental study. *Journal of Marine Science and Engineering*, 10, 532.

Petitpas, G. & Aceves, S. M. 2013. Modeling of sudden hydrogen expansion from cryogenic pressure vessel failure. *International Journal of Hydrogen Energy*, 38, 8190–8198.

Preethi, V. & Kanmani, S. 2013. Photocatalytic hydrogen production. *Materials Science in Semiconductor Processing*, 16, 561–575.

Preuster, P., Papp, C. & Wasserscheid, P. 2017. Liquid organic hydrogen carriers (LOHCs): toward a hydrogen-free hydrogen economy. *Accounts of Chemical Research*, 50, 74–85.

Qyyum, M. A., Dickson, R., Shah, S. F. A., Niaz, H., Khan, A., Liu, J. J. & Lee, M. 2021. Availability, versatility, and viability of feedstocks for hydrogen production: product space perspective. *Renewable and Sustainable Energy Reviews*, 145, 110843.

Ren, J., Musyoka, N. M., Langmi, H. W., Mathe, M. & Liao, S. 2017. Current research trends and perspectives on materials-based hydrogen storage solutions: a critical review. *International Journal of Hydrogen Energy*, 42, 289–311.

Rhodes, R. 2011. Explosive lessons in hydrogen safety. *Ask Magazine*, 41, 46–50.

Staffell, I., Scamman, D., Abad, A. V., Balcombe, P., Dodds, P. E., Ekins, P., Shah, N. & Ward, K. R. 2019. The role of hydrogen and fuel cells in the global energy system. *Energy & Environmental Science*, 12, 463–491.

Stolten, D., Samsun, R. C. & Garland, N. 2016. *Fuel Cells: Data, Facts, and Figures*. John Wiley & Sons: Berlin.

Teichmann, D., Arlt, W. & Wasserscheid, P. 2012. Liquid organic hydrogen carriers as an efficient vector for the transport and storage of renewable energy. *International Journal of Hydrogen Energy*, 37, 18118–18132.

Valente, A., Iribarren, D. & Dufour, J. 2017. Harmonised life-cycle global warming impact of renewable hydrogen. *Journal of Cleaner Production*, 149, 762–772.

Wang, H., Leung, D. Y., Leung, M. & Ni, M. 2009. A review on hydrogen production using aluminum and aluminum alloys. *Renewable and Sustainable Energy Reviews*, 13, 845–853.

Xia, Y., Yang, Z. & Zhu, Y. 2013. Porous carbon-based materials for hydrogen storage: advancement and challenges. *Journal of Materials Chemistry A*, 1, 9365–9381.

Xiao, R., Zhang, S., Peng, S., Shen, D. & Liu, K. 2014. Use of heavy fraction of bio-oil as fuel for hydrogen production in iron-based chemical looping process. *International Journal of Hydrogen Energy*, 39, 19955–19969.

Zarei-Jelyani, M., Babaiee, M., Baktashian, S., Eqra, R. & Shirani- Faradonbeh, H. 2023a. Assessment of the formation process effect on the lithium-ion battery performance at low temperatures. *Journal of Materials Science: Materials in Electronics*, 34, 1890.

Zarei-Jelyani, M., Loghavi, M. M., Babaiee, M. & Eqra, R. 2023b. The significance of charge and discharge current densities in the performance of vanadium redox flow battery. *Electrochimica Acta*, 443, 141922.

Zhang, F., Zhao, P., Niu, M. & Maddy, J. 2016. The survey of key technologies in hydrogen energy storage. *International Journal of Hydrogen Energy*, 41, 14535–14552.

2 Engineering Properties of Hydrogen Storage Materials

Dongmin Xi, Hongfang Lu, Shaohua Dong,
Zhao-Dong Xu, and Bohong Wang

2.1 INTRODUCTION

In the contemporary era characterized by the escalating concerns surrounding global warming and the consequential extreme climate change, there has been a profound global consensus on the urgent need to curtail energy consumption and reduce carbon footprint. Within this context, hydrogen energy has emerged as a recognized and pivotal clean energy source, thereby propelling the advancement of a low-carbon hydrogen economy. Distinguished by its renewability, hydrogen energy can be generated through diverse methodologies that harness other existing energy sources.

Notably, the progress of hydrogen energy is critically reliant on the establishment of efficient hydrogen storage mechanisms. Currently, three primary forms of hydrogen storage prevail, namely high-pressure gaseous hydrogen storage, low-temperature liquid hydrogen storage, and solid-state hydrogen storage. Among these alternatives, solid-state hydrogen storage exhibits a range of advantages over gaseous and liquid-state storage techniques, including higher volumetric hydrogen density, lower working pressure requirements, and superior safety performance.

The deployment of solid-state hydrogen storage systems predominantly relies on the utilization of carrier materials such as metal hydrides, complex hydrides, or carbon-based materials. These materials function as repositories for hydrogen storage and achieve this objective through either physisorption or chemisorption mechanisms. This intricate process facilitates the secure and efficient storage of hydrogen within the solid-state framework, ensuring its availability for subsequent utilization when required.

In summary, in the current era defined by the imperatives of global warming and extreme climate change, the international community has unequivocally acknowledged the significance of reducing energy consumption and minimizing carbon emissions. Consequently, hydrogen energy has emerged as a prominent and environmentally friendly energy source, bolstering the development of a low-carbon hydrogen economy. The progress of hydrogen energy is intricately linked to the establishment of effective hydrogen storage systems. Solid-state hydrogen storage, with its notable advantages of high volumetric density, low working pressure, and enhanced safety performance, represents a compelling avenue for realizing efficient hydrogen storage.

DOI: 10.1201/9781003382553-3

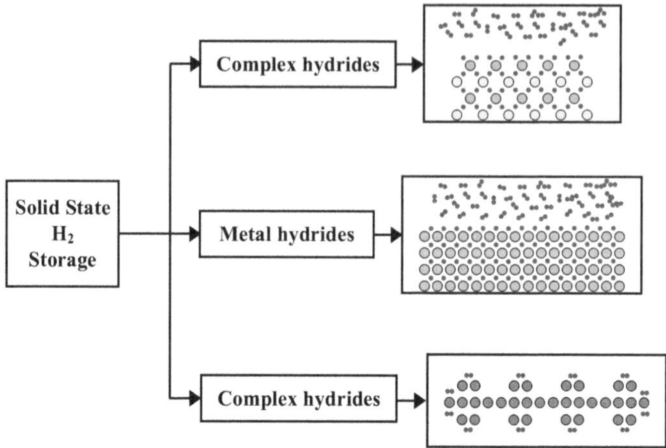

FIGURE 2.1 Different solid-state hydrogen storage options (Mohan et al., 2019).

By employing carrier materials like metal hydrides, complex hydrides, or carbon-based substances, solid-state hydrogen storage achieves the crucial task of securely storing hydrogen through physisorption or chemisorption mechanisms, ensuring its accessibility for subsequent utilization (Zhang et al., 2019). There are various types of solid-state hydrogen storage options available, as illustrated in Figure 2.1.

This chapter presents an introduction to the properties of solid-state hydrogen storage materials, including metal hydrides, complex hydrides, and carbon materials. The properties of these materials are explored from various aspects such as physical and chemical properties, adsorption isotherms, thermodynamic properties, microstructure, preparation, and characterization.

2.2 MECHANISM OF HYDROGEN ADSORPTION

2.2.1 PHYSISORPTION

Physisorption is a process that relies on the existence of van der Waals forces between gas molecules and a solid surface. This process results in gaseous molecules being enriched on the surface. The interaction energy E between the solid surface and hydrogen molecules can be defined as (Mohan et al., 2019):

$$E = \frac{\alpha_{H_2} \alpha_{\text{substrate}}}{D^6} \tag{2.1}$$

where α is the polarizability of hydrogen which is fixed and D is the interaction distance. To increase the interaction energy, the selected materials should be high polarizable. The average interaction energy generally ranges from 4 to 5 kJ/mol. This is a weak interaction because hydrogen is bound to the material in molecular form (Mohan et al., 2019).

According to calculations, adsorbing one mole of hydrogen requires a minimum surface area of $85.917\,m^2{\cdot}mol^{-1}$. Furthermore, a single graphene sheet has a surface area of $1{,}315\,m^2{\cdot}mol^{-1}$, which means that graphite has the maximum hydrogen storage capacity of 3 wt%. To calculate the theoretical hydrogen storage capacity (wt%), one can multiply the specific surface area (SSA) by 2.27×10^{-3} (Mohan et al., 2019). In summary, hydrogen storage materials that rely on physisorption typically exhibit high SSA.

2.2.2 CHEMISORPTION

The adsorption of adsorbent molecules and solid surface atoms (or molecules) involves electron transfer, exchange, or sharing, forming adsorption chemical bonds that result in the ability to combine atoms or ions. In other words, it is the result of the chemical bonding between the solid surface and the adsorbed material. The influence of the chemical bonding in chemisorption is stronger compared to the weak interaction of physisorption (Liu et al., 2022). This is why chemisorption is typically irreversible. However, it is worth noting that physisorption and chemisorption are not completely separate, and they may occasionally coexist.

2.3 CARBON-BASED HYDROGEN STORAGE MATERIALS

2.3.1 PROPERTIES OF CARBON-BASED MATERIALS

Solid-state hydrogen storage materials commonly include carbon-based materials with different internal morphological structures, inorganic porous materials, metal-organic frameworks (MOFs), covalent organic frameworks (COFs), metal hydrides, chemical hydrides, and more (Wu et al., 2013; Rao et al., 2013; Ding and Yazaydin, 2013; Li and Yang, 2006). Among them, carbon-based materials are favored by researchers due to their high hydrogen absorption, light weight, ease of desorption, and recyclability. Additionally, these materials have a strong hydrogen storage capacity due to their porous microstructure with SSA and low mass density (SUSlick, 1998; Sharma and Anil Kumar, 2017).

2.3.2 HYDROGEN STORAGE MECHANISM OF CARBON-BASED MATERIALS

Carbon-based materials rely on the principle that carbon-based adsorbent materials can adsorb and store hydrogen by physisorption at low temperatures and desorb hydrogen at high temperatures to achieve hydrogen storage and utilization (Zhang et al., 2022). It is widely accepted that the mechanism of hydrogen storage in carbon-based materials is based on physisorption. In practical applications, studies on the hydrogen storage properties of carbon-based materials are typically carried out at room temperature and relatively high-pressure conditions (Li et al., 2022).

Graphite is an example of a carbon-based material. Through the use of ultraviolet photon spectroscopy and scanning tunneling microscopy, the chemisorption of atomic hydrogen on graphite was experimentally observed (Neumann et al., 1992; Ruffieux et al., 2000). The chemisorption energy can then be estimated theoretically using *ab initio* quantum chemical methodologies (Jeloaica and Sidis, 1999; Sha and Jackson, 2002; Zecho et al., 2002). Experimental results have shown that the energy barrier for the chemisorption of hydrogen atoms on graphite is 0.15–0.2 eV at room temperature, while the diffusion energy barrier and desorption energy barrier for chemisorption of hydrogen atoms are 1.4 and 0.8 eV, respectively (Psofogiannakis and Froudakis, 2011). Due to

TABLE 2.1

Hydrogen Storage Properties of Four Carbon-Based Materials (Qu et al., 2014)

Category	Abbreviations	Temperature (K)	Pressure (MPa)	Mass Hydrogen Storage Density (%)
Activated carbon	AC	77	2–4	5.3–7.4
		93	6	9.8
		Room temperature	7.04	3.8
Graphite nanofiber	GNF	25	12	67
		Room temperature	11	12
Carbon nanofiber	CNF	Room temperature	10–12	10
Carbon nanotube	CNT	80	12	8.25
		Room temperature	0.05	6.5

the high energy barrier, the possibility of chemisorption between hydrogen atoms and carbon atoms of graphite is low. Therefore, in theory, carbon-based materials predominantly rely on physisorption for hydrogen storage. Carbon materials commonly used for hydrogen storage include activated carbon (AC), graphite fibers, nanofibers, and carbon nanotubes (CNTs), as shown in Table 2.1 (Qu et al., 2014).

2.3.3 ACTIVATED CARBON

AC is a black powdered, granular, or columnar porous carbon material with an amorphous microstructure and a large SSA. AC hydrogen storage relies on the principle of supercritical gas adsorption, with a particular focus on the hydrogen storage performance of AC with ultra-high SSA and a well-developed pore structure at low temperatures (Zhang et al., 2022).

2.3.3.1 Preparation Methods

Currently, there are two primary techniques utilized for the production of AC with a high SSA: chemical activation and combined chemical-physical activation (Zhan et al., 2002a).

2.3.3.1.1 Chemical Activation Method

The chemical activation method can vary depending on the type of raw materials used, and its general process flow is illustrated in Figure 2.2.

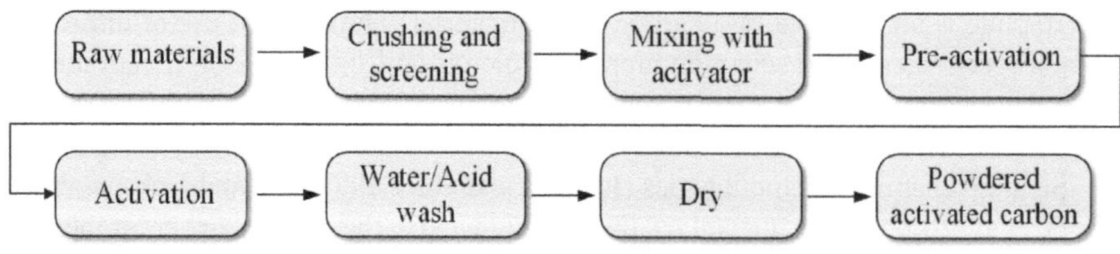

FIGURE 2.2 Process flow for preparation of high specific surface area AC by chemical activation method (Xu, 2006).

The chemical activation method typically employs activators such as alkali metals, alkaline earth metal hydroxides, inorganic salts, and certain acids. Among these, KOH is recognized as the most effective active agent for producing AC with superior performance (Xu, 2006).

Zhan et al. (2002a) conducted an orthogonal experimental study using high-sulfur coke as a raw material and industrial KOH as the activator to prepare a series of super-ACs with SSAs ranging from 2,332 to 3,886 m^2/g. The hydrogen storage performance of these carbons was also investigated. The results revealed that the hydrogen storage capacity of the AC decreased as the temperature increased. Specifically, the hydrogen storage capacity was found to be 9.8 wt% at 93 K and 6 MPa, and 1.9 wt% at 293 K and 5 MPa (Zhang et al., 2022).

2.3.3.1.2 *Combined Chemical-Physical Activation Method*

Various factors in the chemical activation method can affect the pore structure and performance of the resulting AC, including the amount of activator used, activation temperature, activation time, and type of raw material. In the production of AC with a high SSA, it is crucial to carefully control the pore size, with particular attention given to inhibiting the formation of mesopores and macropores in order to achieve optimal adsorption performance (Xu, 2006).

The combined chemical-physical activation method typically involves a process where chemical activation is followed by physical activation (Xu, 2006). The physical activation method, also referred to as gas activation, involves introducing a gas activator such as carbon dioxide, water vapor, oxygen (air), nitrogen, or oxygen during the activation process.

2.3.3.2 **Thermomechanical Properties**

2.3.3.2.1 *The Adsorption Isotherm*

Zhan et al. (2002a) investigated the hydrogen storage performance of super-AC, and observed that the adsorption isotherm of hydrogen on super-AC with a SSA of 3,886 m^2/g displayed the characteristics of a type-I isotherm (Figure 2.3).

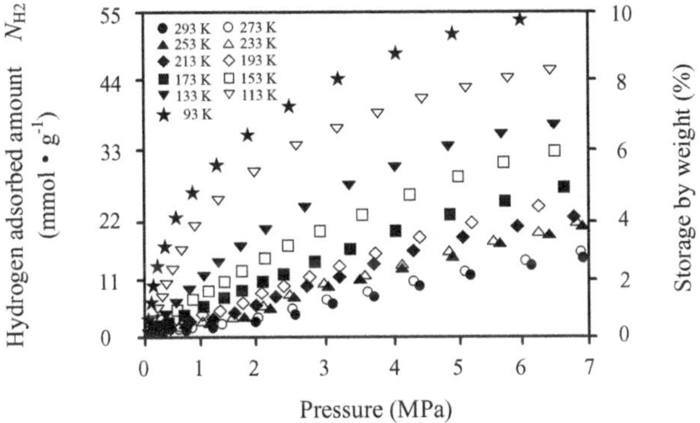

FIGURE 2.3 Adsorption isotherm of hydrogen on super AC with 3,886 m^2/g (Zhan et al., 2002a).

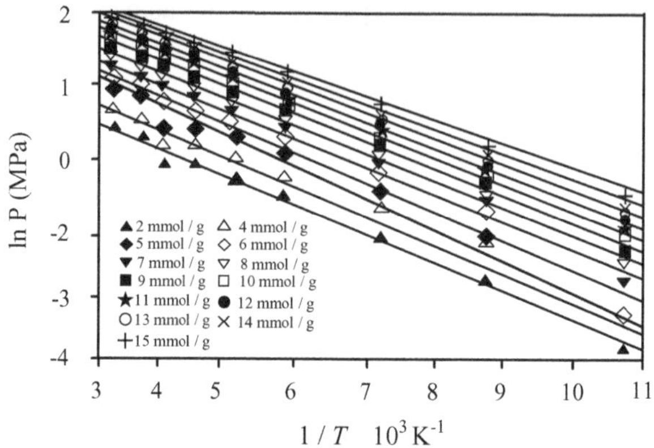

FIGURE 2.4   Adsorption isometric lines of hydrogen on super AC (Zhan et al., 2002a).

The graph illustrates that the adsorption of hydrogen on super-AC is more favorable at lower temperatures. In particular, in the low-pressure region of 0–6 MPa, the hydrogen adsorption significantly increases with increasing gas pressure. Additionally, in the higher-pressure region, the SSA of AC has a more pronounced effect on the hydrogen adsorption capacity. The adsorption and desorption rates of hydrogen on super-AC are fast, and the resulting hydrogen storage mass fraction is high. Specifically, the hydrogen storage capacity has reached 1.9 wt% and 9.8 wt% at 293 K/5 MPa and 93 K/6 MPa, respectively (Zhan et al., 2002a).

2.3.3.2.1   The Heat of Adsorption Effect

The heat of adsorption effect is commonly referred to as the equivalent heat of adsorption (q_{st}). This parameter is defined by the Clausius-Clapeyron equation:

$$q_{st} = \Delta H = -R\left[\frac{d \ln f}{d\left(\frac{1}{T}\right)}\right]_n \tag{2.2}$$

where f is the gas fugacity which can be replaced by the pressure P and R is the universal gas constant. The above equation illustrates that a set of equivalent adsorption lines is obtained as a straight line of $\ln P$ to $1/T$ based on the adsorption isotherm data (Figure 2.4). The slope of these equivariant adsorption lines is the value of q_{st}, as shown in Figure 2.5. As seen in Figure 2.5, the equivalent heat of adsorption is small and predominantly concentrated in the range of 4.8–6.5 kJ/mol (Zhan et al., 2002a).

2.3.4   GRAPHITE NANOFIBERS (GNFs)

GNFs, or graphene nanofibers, are a type of nanofiber formed by multiple layers of graphite flakes without any significant hollow structure. They are typically produced

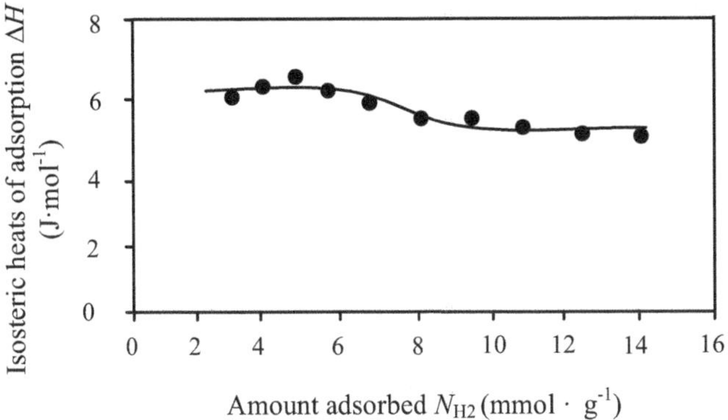

FIGURE 2.5 Isosteric heat of adsorption (Zhan et al., 2002a).

through the pyrolysis of a mixture of olefin, hydrogen, and carbon monoxide on a catalyst surface at high temperatures (700–900 K), using a specific metal or alloy as the catalyst. The microstructure of GNFs comprises many tiny graphite flake layers with a width of 30–50 nm, which are regularly stacked together with a spacing of 0.34 nm (Zhang et al., 2022). Depending on the angle between the graphite layer and fiber axis, as well as whether they are hollow or not, the microstructure of GNFs can be classified into four types: flat, herringbone, tubular, and cup-like (Zhan et al., 2002b). Among these, the herringbone type has demonstrated the best adsorption performance (Zhang et al., 2022).

Chambers et al. (1998) produced three types of GNFs in flat, herringbone, and tubular shapes through catalytic cracking. The resulting hydrogen storage mass fractions for these three structures were 53%, 67%, and 11% at room temperature and 11.2 MPa, respectively (Wei et al., 2014). However, some researchers are skeptical of these results due to the difficulty in verifying them. Studies have shown that the hydrogen storage mass fraction for GNFs can range from 1% to 15%, depending on the experimental methods, sample preparation and processing conditions, and testing methods used (Browning et al., 2002; Gupta and Srivastava, 2001; Fan et al., 1999; Lueking et al., 2005; Wu et al., 2010).

Kim et al. (2008) utilized a chemical reduction method to introduce Pt nanoparticles onto the surface of GNFs. The study investigated the effect of Pt doping on the hydrogen storage capacity of GNFs, for Pt content ranging from 1.3% to 7.5% (mass fraction). The microstructure of the Pt/porous GNFs was characterized using X-ray diffraction and transmission electron microscopy, while the hydrogen storage behavior was evaluated using a PCT apparatus at 298 K and 10 MPa. The results indicated that the amount of hydrogen stored increased with increasing Pt content up to 3.4% (mass fraction), after which it decreased. These findings suggest that the hydrogen storage mass fraction of porous GNFs is related to their metal content and dispersion rate.

2.3.5 Carbon Nanofibers (CNFs)

CNFs are fibrous carbon nanomaterials that are composed of convoluted multilayer graphite flakes. They typically have a diameter of 10–500 nm and a length distribution of 0.5–100 µm (Marella & Tomaselli, 2006). CNFs are considered to be one-dimensional carbon materials that exist between nanotubes and ordinary carbon fibers.

2.3.5.1 Preparation Methods

There are several methods for synthesizing CNFs, including the following:

2.3.5.1.1 Chemical Vapor Deposition (CVD) Method

This method involves synthesizing CNFs by thermally decomposing hydrocarbon compounds on metal catalysts at a certain temperature (500°C–1,000°C). Baldé et al. (2006) utilized Ni/SiO_2 with a 5% mass fraction as a catalyst to prepare a carbon matrix using CVD, which was then loaded with $NaAlH_4$ via impregnation and drying techniques. The hydrogen storage performance of the loaded CNFs was tested using programmed temperature desorption, with the mass fraction of hydrogen desorption reaching 4.8% at 300°C (Wei et al., 2014).

2.3.5.1.2 Electrospinning Method

The electrospinning method involves spraying and stretching a polymer solution/melt under the influence of high-voltage electrostatic field force, which is then cured by solvent evaporation to obtain nanofibers (Kang et al., 2017). Kim et al. (2011) combined coaxial electrostatic PAN-based fibers, prepared using the coaxial electrospinning technique mixed with palladium salts, and carbonized them at 800°C in $Ar + H_2O$ to obtain PCNFs with a nanopore structure. Surface chemical composition and degree of carbonization were confirmed using X-ray photoelectron spectroscopy (XPS) and Raman spectroscopy. The test results displayed that the reversible hydrogen adsorption capacity was 0.35 wt% at 298 K and 0.1 MPa.

2.3.5.1.3 Solid Phase Synthesis Method

This method is distinct from the previous single synthesis methods, as it utilizes solid carbon sources as raw materials to prepare CNFs.

2.3.5.2 Effects of SSA and Micropore Volume on Hydrogen Storage Performance

Zhang et al. (2007) utilized N_2 adsorption isotherms at 77 K to determine the SSA and micropore volume of CNFs treated by different post-treatment methods (QB2, QB5, QB7, and QB9). Based on the obtained data, the effects of SSA and micropore volume on CNFs were analyzed. The correlation between the SSA of CNFs and their hydrogen adsorption capacity is linear, as shown in Figure 2.6. The effect of SSA on the hydrogen adsorption capacity is greater at higher pressures, as evidenced by the larger slope of the correlation line. On the other hand, the correlation between the micropore volume of CNFs and their hydrogen adsorption capacity is parabolic, as illustrated in Figure 2.7. The hydrogen adsorption capacity increases rapidly with the

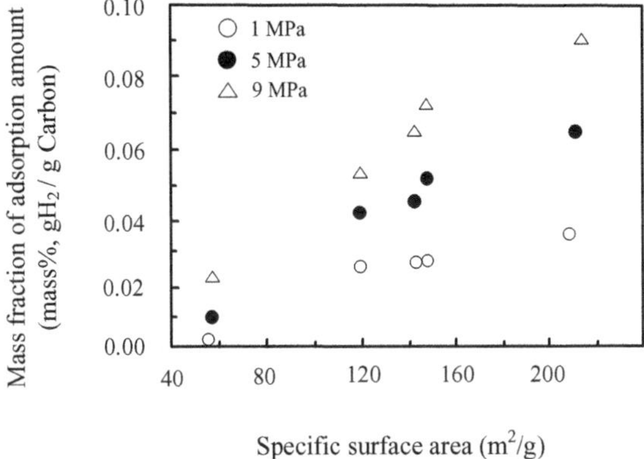

FIGURE 2.6 The correlation curves between the SSA of CNFs and the hydrogen adsorption capacity (Zhang et al., 2007).

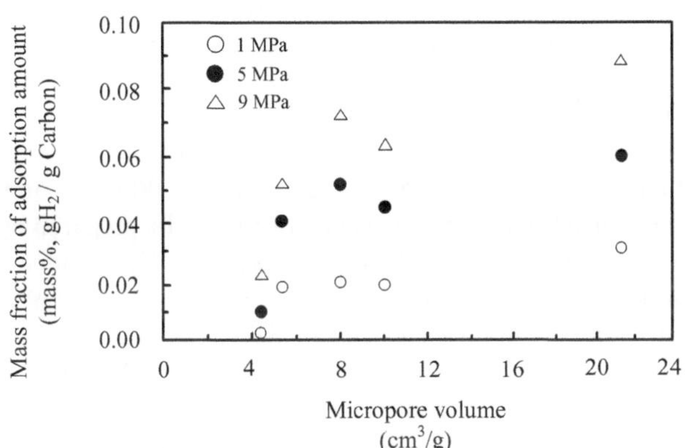

FIGURE 2.7 The correlation curves between the micropore volume of CNFs and the hydrogen adsorption capacity (Zhang et al., 2007).

increase of micropore volume when the micropore volume is small. The increasing trend becomes more obvious at higher pressures.

2.3.6 Carbon Nanotubes

CNTs are a type of carbon isotope that can be described as cylindrical tubes made up of graphene layers wrapped around a core with diameters ranging from a few nanometers to tens of nanometers. The tube walls are extremely thin, forming a hollow cylinder that resembles barbed wire. CNTs can be classified into two types, single-walled CNTs (SWCNTs) and multi-walled CNTs (MWCNTs), based on the number of graphene layers present (Guo et al., 2010).

2.3.6.1 Preparation Methods

The preparation of CNTs usually involves dissociating the carbon source into atomic or ionic form using various applied energies to achieve the desired one-dimensional carbon structure. The most commonly used methods are the electric arc method (Zhang et al., 2002), laser evaporation method, catalytic cracking, and CVD method. While CNTs prepared using the electric arc and laser evaporation method are highly crystalline and straight, the yield is often low and mixed with impurities such as graphite fragments, amorphous carbon, and carbon nanoparticles. Acid or alkali treatment is usually employed to remove impurities and purify the CNTs. On the other hand, catalytic cracking and CVD methods involve decomposing carbon-containing gases (such as carbon monoxide, methane, and ethylene) using catalysts to crack carbon-containing compounds into carbon atoms at high temperatures. Carbon atoms then attach to the surface of catalyst micro-particles, forming CNTs, particularly when transition metals are used as catalysts (Liu and Su, 2006).

Cheng et al. (2000) developed a hydrogen arc method, which utilizes a growth promoter to produce high-purity SWCNTs in large quantities. Their research demonstrated that after pretreatment such as HCl pickling and vacuum 773 K heat treatment, the hydrogen storage capacity of SWCNTs can reach 4.2–4.7 wt%, which is approximately 2–3 times greater than that of metal hydride.

2.3.6.2 Hydrogen Storage Capacity of CNTs

CNTs are considered one of the best materials for hydrogen storage due to their larger SSA and the presence of micropores (Liu and Su, 2006). Liu and Cheng (2001) measured the mass hydrogen storage capacity of SWCNTs, prepared by the hydrogen arc method, at room temperature and moderate pressure. By introducing high-purity hydrogen with a pressure of about 10 MPa and a purity of 99.999% into the system, the initial hydrogen storage capacity of SWCNTs after preliminary pretreatment was up to 4.2 wt%. Since the mass fraction of SWCNTs in the sample is about 50 wt%, it can be inferred that the mass hydrogen storage capacity of pure SWCNTs should be around 8 wt% (Guo et al., 2010). Darkrim and Levesque (2000) conducted Monte Carlo molecular simulation of hydrogen storage in SWCNTs and found that the SWCNTs with a diameter of 1.174 nm and a wall spacing of 0.7 nm exhibited the best hydrogen storage performance at 293 K and 10 MPa. In addition to having a certain lumen and thin wall, improving the surface characteristics of CNTs is also a significant factor for further improving the hydrogen storage capacity.

Zhu et al. (2000) investigated the differences in hydrogen storage performance of CNTs subjected to two different surface treatments. In the first treatment process, the pristine CNTs underwent ball-milling for 30 minutes, followed by soaking the milled products in concentrated nitric acid (about 65%) for 72 hours (CNT1) or boiling (CNT2). In the second treatment process, the pristine CNTs were heated in a 3 mol/L NaOH solution for 1 hour, and the resulting black powder was filtered, washed with deionized water, dried in an oven at 373 K for 2 hours, roasted in a crucible at 823 K for 1 hour, and finally soaked in 15% dilute sulfuric acid solution for 1.5 hours (or boiled for 10 minutes). The sample was then washed with deionized water and dried at 373 K to obtain CNT3. The results showed

TABLE 2.2

Adsorption Data of CNTs Processed by Different Methods (Zhu et al., 2000)

Samples	Mass (g)	Mass Fraction of Adsorbed Hydrogen (%)
CNT1	0.3504	2.67
CNT2	0.7990	1.16
CNT3	0.1640	5.15

TABLE 2.3

Hydrogen Storage Capacity of MWCNT Samples Ball-Milled for Different Times at 30°C and 10 MPa (Yao et al., 2006)

Samples	Time for Ball-Milling (h)	The Maximum Adsorption Capacity (%)
CNT1	0	1.6
CNT2	6	2.1
CNT3	12	2.55
CNT4	24	1.75

that the surface treatment of CNTs using concentrated nitric acid and sodium hydroxide solution improves the SSA and surface activity, resulting in a hydrogen adsorption rate of 5%, as shown in Table 2.2.

Yao et al. (2006) suggest that ball-milling treatment can increase the SSA and surface activity of MWCNTs. The maximum hydrogen adsorption capacity of samples ball-milled for different times at 30°C and 10 MPa is presented in Table 2.3. As the milling time increases, the hydrogen storage capacity of MWCNTs initially increases and then decreases. After ball-milling for 12 hours, the maximum adsorption capacity of MWCNTs has increased from 1.60 to 2.55 wt%, indicating that the hydrogen adsorption capacity of MWCNTs can be significantly improved through ball-milling modification.

2.4 METAL HYDRIDE

2.4.1 Hydrogen Storage Mechanism of Metal Hydride

Hydrogen storage alloys can produce metal solid solutions MH_x and metal hydrides MH_y through a reversible reaction with hydrogen under certain temperature and pressure conditions, where hydrogen can be re-emitted through diffusion, phase change, and chemical combination under opposite conditions (Gao et al., 2022). Typically composed of two types of metals, hydrogen storage alloys include a metal element with negative hydrogen binding energy (known as A element, containing Mg, Ti, V, La, Zr, etc.), which controls hydrogen storage, and a metal element with positive hydrogen binding energy (known as B element, containing Cr, Mn, Fe, Co, Ni, Cu, Zn, Al, etc.), which controls the reversibility of hydrogen absorption and discharge, while also regulating the heat of generation and decomposition pressure (Li et al., 2022; Xu et al., 2006).

The hydrogen storage process of metal hydride can be broadly classified into three stages (Gao et al., 2022; Xu et al., 2006):

Step 1: The hydrogen storage alloys are exposed to hydrogen, resulting in the formation of solid solutions MH_x (α phase) with a small amount of hydrogen at a specific temperature and pressure. During this step, the structure of the hydrogen storage alloys remains unchanged. The relationship between the solubility of hydrogen $[H]_M$ and the equilibrium hydrogen pressure of solid solution P_{H_2} can be expressed as follows:

$$P_{H_2}^{\frac{1}{2}} \propto [H]_M \tag{2.3}$$

Step 2: The formed solid solution continues to react with hydrogen and undergoes a phase transition, leading to a change in the structure of the hydrogen storage alloy. This change allows the metal to absorb more hydrogen and form a metal hydride MH_y (β phase). The chemical reaction can be represented as follows:

$$\frac{2}{y-x}MH_x + H_2 \leftrightarrow \frac{2}{y-x}MH_y + Q \tag{2.4}$$

where x represents the equilibrium concentration of hydrogen in the solid solution, and y represents the concentration of hydrogen in metal hydride.
Step 3: As the hydrogen pressure continues to increase, the hydrogen content in the alloy will continue to increase slightly.

Therefore, hydrogen storage alloy materials can continuously and reversibly absorb large amounts of hydrogen. The forward reaction involves absorbing hydrogen and releasing heat, while the reverse reaction involves desorbing hydrogen and absorbing heat (Gao et al., 2022; Xu et al., 2006).

2.4.2 Thermomechanical Properties

The thermodynamics of hydrogen absorption and desorption of metal hydrides are influenced by system temperature, pressure, and alloy composition. The pressure-composition-temperature (PCT) curve can be used to characterize the hydrogen storage performance, reflecting the reversible hydrogen storage capacity, equilibrium hydrogen pressure, platform slope, and hysteresis effects (Figure 2.8) (Li et al., 2022).

The OA section corresponds to the stage of α phase formation, while the AB section corresponds to the stage where α phase is transformed to β phase. In the AB section, the equilibrium hydrogen pressure remains approximately constant, thus the AB section is also commonly referred to as the plateau region, which indicates the

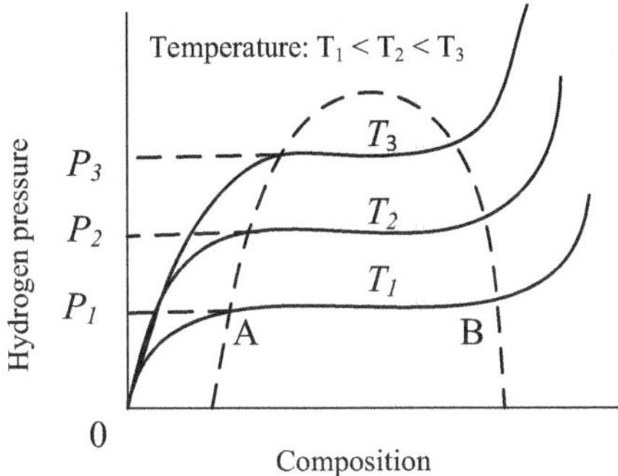

FIGURE 2.8 PCT curves of hydrogen storage alloys (Li et al., 2022).

effective hydrogen storage capacity. After the B point, the α phase is completely transformed into β phase. With increasing temperature, it can be observed from Figure 2.8 that the plateau pressure zone becomes shorter, indicating that high temperature is not conducive to the hydrogen absorption reaction (Ma et al., 2018).

The Van't Hoff equation, shown in Equation 2.5, is a useful tool for calculating the enthalpy and entropy changes of hydrogen absorption and desorption reactions, which are important thermodynamic performance parameters for hydrogen storage materials.

$$\ln P_{H_2} = \frac{\Delta H}{RT} - \frac{\Delta S}{R} \tag{2.5}$$

where P_{H_2} is the equilibrium hydrogen pressure, MPa; T is the temperature, K; ΔS is the entropy change, J/(K·mol); ΔH is the standard enthalpy change, kJ/mol; R is the gas constant. The entropy value that does not vary significantly in different metal hydrogen systems is approximately 130 J/(K·mol). Therefore, enthalpy becomes the first criterion for characterizing the thermal stability of hydrides. The more negative the enthalpy is, the more stable the hydrides are. As a general requirement, the reaction enthalpy should be between 30 and 40 kJ/mol at room temperature and normal pressure (Li et al., 2022).

2.5 COMPLEX HYDRIDE

Complex hydrides are a type of hydrogen storage material that differs from metal hydrides as they are salts. In complex hydrides, the hydrogen atoms are covalently bonded to the central atom of the complex anion, which then combines with metal ions through ionic bonds to form hydrides (Kim et al., 2020). These hydrides have the general chemical formula $A_xMe_yH_z$, where A is usually a first or second group element in the periodic table and Me is typically B, Al, or N. The theoretical mass density of hydrogen storage for complex hydrides ranges from 5.5% to 21%. Hydrogen release from complex hydrides can be achieved through hydrolysis or pyrolysis.

Complex hydrides can be further divided into three categories based on the ligands present. The first category consists of metal hydrides with [AlH$_4$]-groups as ligands, with notable examples including LiAlH$_4$, Ca(AlH$_4$)$_2$, and NaAlH$_4$. The second type of metal hydride features a [NH$_2$]-group as the ligand, with representative metal amides including Mg(NH$_2$)$_2$, NaNH$_2$, and LiNH$_2$. The third category is metal hydrides with [BH$_4$]-groups as ligands, which includes typical metal borohydrides like LiBH$_4$, NaBH$_4$, KBH$_4$, and Mg (BH$_4$)$_2$ (Li et al., 2022).

2.5.1 ALUMINUM HYDRIDE

2.5.1.1 Preparation Methods

Various studies have investigated the preparation of NaAlH$_4$, which is a type of aluminum hydride. According to current research (Zaluski et al., 1999; Wang et al., 2006; Bogdanović et al., 2009), three primary methods have been developed for the preparation of NaAlH$_4$, including (i) treatment of NaAlH$_4$ and dopants using mechanical ball-milling or wet synthesis, (ii) treatment of NaH/Al and dopants using mechanical ball-milling or wet synthesis followed by hydrogenation, and (iii) synthesis of NaAlH$_4$ from NaH/Al and dopants in hydrogen by mechanical ball-milling.

2.5.1.2 The Decomposition of NaAlH$_4$

The decomposition of NaAlH$_4$ occurs in two steps (Liu, 2012):

$$NaAlH_4 \rightarrow \frac{1}{3} Na_3AlH_6 + \frac{2}{3} Al + H_2 \tag{2.6}$$

$$Na_3AlH_6 \rightarrow 3NaH + Al + \frac{2}{3} H_2 \tag{2.7}$$

The first step decomposition reaction releases 3.7 wt% hydrogen at 185°C–230°C. The second step releases about 1.9 wt% hydrogen at 260°C. The total hydrogen release amount theoretically reaches 5.6 wt%. However, the decomposition reaction of NaH cannot occur until 425°C (Liu, 2012). Therefore, it has no practical significance for applications.

2.5.2 BOROHYDRIDE

2.5.2.1 Preparation Methods

NaBH$_4$ is one of the most studied borohydrides for hydrogen storage. There are two main industrial methods for producing NaBH$_4$: the Schlesinger method and the Bayer method (Zhou et al. 2014).

2.5.2.1.1 Schlesinger Method

$$4NaH + (CH_3)_3 BO_3 \rightarrow NaBH_4 + 3CH_3ONa \tag{2.8}$$

2.5.2.1.2 Bayer Method

$$Na_2B_4O_7 + 16Na + 8H_2 + 7SiO_2 \rightarrow 4NaBH_4 + 7Na_2SiO_3 \tag{2.9}$$

2.5.2.2   The Decomposition of NaBH$_4$

The decomposition of NaBH$_4$ is carried out in the following reaction equation (Zhou et al., 2014):

$$NaBH_4 + 2H_2O \rightarrow 4H_2 + NaBO_2 \tag{2.10}$$

It should be noted that although the theoretical hydrogen storage capacity of NaBH$_4$ is high, the practical hydrogen storage capacity is lower due to the high activation energy and slow kinetics of the dehydrogenation reaction. Therefore, various methods, such as catalyst doping and ball-milling, have been developed to improve the hydrogen storage performance of NaBH$_4$ (Ouyang et al., 2017). Additionally, the hydrolysis reaction of NaBH$_4$ has some limitations, including the production of borate and the requirement for an alkaline environment, which can lead to the corrosion of storage and transportation equipment (Liu, 2012). Nevertheless, the hydrolysis of NaBH$_4$ still shows great potential in hydrogen storage and fuel cell applications due to its high hydrogen content and easy controllability.

2.6   CONCLUSION

Hydrogen energy, often regarded as a promising new form of clean energy, has faced challenges in achieving large-scale application primarily due to limitations in hydrogen storage technology. The development of hydrogen storage materials with exceptional performance is crucial to enable the widespread utilization of hydrogen energy, meeting the demands of society while mitigating environmental degradation.

Metal hydrides have emerged as one of the most extensively studied and widely applicable hydrogen storage materials, owing to their high bulk hydrogen storage density and thermodynamic stability. These materials exhibit the ability to absorb and release hydrogen through reversible reactions, providing an efficient and reliable storage solution for hydrogen energy systems.

The high bulk hydrogen storage density of metal hydrides refers to their capacity to store a large amount of hydrogen per unit volume or mass. This characteristic is essential for achieving practical energy storage, as it allows for the storage of a significant amount of hydrogen within a limited space. Metal hydrides can store hydrogen through various mechanisms, such as chemisorption or physisorption, depending on the specific properties of the metal hydride system.

Furthermore, the thermodynamic stability of metal hydrides ensures that hydrogen is securely stored and can be readily released when needed. The stability of a metal hydride is determined by factors such as the strength of metal-hydrogen bonds and the energy required for hydrogen desorption. Metal hydrides can exhibit high stability, maintaining their structural integrity and hydrogen storage capacity over prolonged periods.

Considering the current research landscape, it is evident that metal hydrides hold great potential as hydrogen storage materials, and their further exploration and development should be prioritized. Efforts should be made to enhance the understanding of metal hydride systems, including their fundamental properties, hydrogenation and dehydrogenation kinetics, and stability under various conditions. Additionally, researchers should focus on optimizing the design and synthesis of metal hydrides to improve their storage capacity, kinetics, and overall performance.

By emphasizing the importance of metal hydrides in the future of hydrogen storage technology, adequate attention can be directed toward advancing this field. Continued research and development in metal hydrides will contribute significantly to realizing the large-scale application of hydrogen energy, meeting societal demands for clean and sustainable energy sources, and mitigating environmental deterioration.

ABBREVIATIONS

AC	Activated Carbon
CNF	Carbon Nanofiber
CNT	Carbon Nanotube
COF	Covalent Organic Framework
CVD	Chemical Vapor Deposition
GNF	Graphite Nanofiber
MOF	Metal-Organic Framework
MWCNT	Multi-walled Carbon Nanotube
PCT	Pressure-Composition-Temperature
SEM	Scanning Electron Microscope
SSA	Specific Surface Area
SWCNT	Single-Walled Carbon Nanotube
XPS	X-ray Photoelectron Spectroscopy

NOMENCLATURE

α	the polarizability of hydrogen
D	the distance between solid surface and hydrogen molecules
qst	the equivalent heat of adsorption
R	the universal gas constant
f	the gas fugacity
T	the absolute temperature
HM	the solubility of hydrogen
PH_2	the equilibrium hydrogen pressure of the solid solution
MH_x	the metal solid solution
MH_y	the metal hydride
x	the equilibrium concentration of hydrogen in the solid solution
y	the concentration of hydrogen in metal hydride
ΔS	the entropy change
ΔH	the standard enthalpy change

ACKNOWLEDGMENT

This work is funded by the Natural Science Foundation of Jiangsu Province (Grant No. BK20220848).

REFERENCES

Baldé, C.P., Hereijgers, B.P., Bitter, J.H. and de Jong, K.P., 2006. Facilitated hydrogen storage in NaAlH4 supported on carbon nanofibers. *Angewandte Chemie*, 118(21), pp. 3581–3583.

Bogdanović, B., Felderhoff, M., Pommerin, A., Schüth, F., Spielkamp, N. and Stark, A., 2009. Cycling properties of Sc-and Ce-doped NaAlH4 hydrogen storage materials prepared by the one-step direct synthesis method. *Journal of Alloys and Compounds*, 471(1–2), pp. 383–386.

Browning, D.J., Gerrard, M.L., Lakeman, J.B., Mellor, I.M., Mortimer, R.J. and Turpin, M.C., 2002. Studies into the storage of hydrogen in carbon nanofibers: proposal of a possible reaction mechanism. *Nano Letters*, 2(3), pp. 201–205.

Chambers, A., Park, C., Baker, R.T.K., and Rodriguez, N.M., 1998. Hydrogen storage in graphite nanofibers. *The Journal of Physical Chemistry B*, 102(22), pp. 4253–4256.

Cheng, H.M., Liu, C., and Cong, H.T., 2000. The synthesis and hydrogen uptake of high-purity single-walled carbon nanotubes. *Physics Beijing*, 29(8), pp. 449–450.

Darkrim, F. and Levesque, D., 2000. High adsorptive property of opened carbon nanotubes at 77 K. *The Journal of Physical Chemistry B*, 104(29), pp. 6773–6776.

Ding, L. and Yazaydin, A.O., 2013. Hydrogen and methane storage in ultrahigh surface area metal-organic frameworks. *Microporous and Mesoporous Materials*, 182, pp. 185–190.

Fan, Y.Y., Liao, B., Liu, M., Wei, Y.L., Lu, M.Q., and Cheng, H.M., 1999. Hydrogen uptake in vapor-grown carbon nanofibers. *Carbon*, 37, pp. 1649–1652.

Gao, Y., Yu, G., Zhang, L., and Huang, D., 2022. Talking about metal hydride hydrogen storage and common metal hydrogen storage materials. *Journal of Applied Chemical Industry*, 51(10), pp. 2975–2978.

Guo, L., Li, H., and Guo, H., 2010. Research into hydrogen storage in carbon nanotubes. *Hebei Journal of Industrial Science and Technology*, 27(3), pp. 139–142.

Gupta, B.K. and Srivastava, O.N., 2001. Further studies on microstructural characterization and hydrogenation behaviour of graphitic nanofibres. *International Journal of Hydrogen Energy*, 26(8), pp. 857–862.

Jeloaica, L. and Sidis, V., 1999. DFT investigation of the adsorption of atomic hydrogen on a cluster-model graphite surface. *Chemical Physics Letters*, 300(1–2), pp. 157–162.

Kang, W., Fan, L., Deng, N., He, H., Ju, J., and Cheng, B, 2017. Research progress in preparation and application of electrospinning porous carbon nanofibers. *Journal of Textile Research*, 38(11), pp. 168–176.

Kim, B.J., Lee, Y.S., and Park, S.J., 2008. Preparation of platinum-decorated porous graphite nanofibers, and their hydrogen storage behaviors. *Journal of Colloid and Interface Science*, 318(2), pp. 530–533.

Kim, H., Lee, D., and Moon, J., 2011. Co-electrospun Pd-coated porous carbon nanofibers for hydrogen storage applications. *International Journal of Hydrogen Energy*, 36(5), pp. 3566–3573.

Kim, J.B., Han, G., Kwon, Y., Bae, J., Cho, E., Cho, S., and Lee, B.J., 2020. Thermal design of a hydrogen storage system using La (Ce) Ni5. *International Journal of Hydrogen Energy*, 45(15), pp. 8742–8749.

Li, Y. and Yang, R.T., 2006. Hydrogen storage in metal-organic frameworks by bridged hydrogen spillover. *Journal of the American Chemical Society*, 128(25), pp. 8136–8137.

Li, J., Ren, C., Luo, C., and Chen, H., 2022. Technological progress on the research and development of solid hydrogen storage materials. *Journal of Petroleum and New Energy*, 34 (5), pp. 14–20.

Liu, Y., 2012. Hydrogen storage properties of complex hydride materials. *Journal of Yunyang Teachers College*, 32(6), pp. 39–44.

Liu, C. and Cheng, H., 2001. Synthesis and hydrogen storage properties of single-walled carbon nanotubes. *Journal of Bulletin of Chinese Academy of Sciences*, 33(1), pp. 32–35.

Liu, S. and Su, D., 2006. Research situation in preparation and hydrogen storage of carbon nanotube. *Journal of Chemical Propellants & Polymeric Materials*, 56(1), pp. 31–34.

Liu, S., Liu, J., Liu, X., Shang, J.X., Yu, R., and Shui, J., 2022. Non-classical hydrogen storage mechanisms other than chemisorption and physisorption. *Applied Physics Reviews*, 9(2), p. 021315.

Lueking, A.D., Pan, L., Narayanan, D.L., and Clifford, C.E., 2005. Effect of expanded graphite lattice in exfoliated graphite nanofibers on hydrogen storage. *The Journal of Physical Chemistry B*, 109(26), pp. 12710–12717.

Ma, T., Gao, L., Hu, M., Hu, L., Wen, L., and Hu, M., 2018. Research progress of solid hydrogen storage materials. *Journal of Functional Materials*, 49(4), pp. 4001–4006.

Marella, M. and Tomaselli, M., 2006. Synthesis of carbon nanofibers and measurements of hydrogen storage. *Carbon*, 44(8), pp. 1404–1413.

Mohan, M., Sharma, V.K., Kumar, E.A., and Gayathri, V., 2019. Hydrogen storage in carbon materials-a review. *Energy Storage*, 1(2), p. e35.

Neumann, D., Meister, G., Kürpick, U., Goldmann, A., Roth, J., and Dose, V., 1992. Interaction of atomic hydrogen with the graphite single-crystal surface: I. Angle-resolved photo-emission studies. *Applied Physics A*, 55, pp. 489–492.

Ouyang, L., Chen, W., Liu, J., Felderhoff, M., Wang, H., and Zhu, M., 2017. Enhancing the regeneration process of consumed NaBH4 for hydrogen storage. *Advanced Energy Materials*, 7(19), p. 1700299.

Psofogiannakis, G.M. and Froudakis, G.E., 2011. Fundamental studies and perceptions on the spillover mechanism for hydrogen storage. *Chemical Communications*, 47(28), pp. 7933–7943.

Qu, H., Lou, Y., Du, J., Pu, C., Huang, T., Li, Z., and Wu, Z., 2014. Research progress of carbon-based hydrogen storage materials. *Journal of Materials Reports*, 28 (13), pp. 69–71+77.

Rao, D., Lu, R., Meng, Z., Xu, G., Kan, E., Liu, Y., Xiao, C., and Deng, K., 2013. Influences of lithium doping and fullerene impregnation on hydrogen storage in metal organic frameworks. *Molecular Simulation*, 39(12), pp. 968–974.

Ruffieux, P., Gröning, O., Schwaller, P., Schlapbach, L., and Gröning, P., 2000. Hydrogen atoms cause long-range electronic effects on graphite. *Physical Review Letters*, 84(21), p. 4910.

Sha, X. and Jackson, B., 2002. First-principles study of the structural and energetic properties of H atoms on a graphite (0001) surface. *Surface Science*, 496(3), pp. 318–330.

Sharma, V.K. and Anil Kumar, E., 2017. Metal hydrides for energy applications-classification, PCI characterisation and simulation. *International Journal of Energy Research*, 41(7), pp. 901–923.

SUSlick, K.S., 1998. *Kirk-Othmer Encyclopedia of Chemical Technology*. Wiley & Sons: New York, Vol. 26, pp. 517–541.

Wang, P., Kang, X.D., and Cheng, H.M., 2006. Dependence of H-storage performance on preparation conditions in TiF3 doped NaAlH4. *Journal of Alloys and Compounds*, 421(1–2), pp. 217–222.

Wei, W., Huang, B., Qian, Q., Liu, X., Xiao, L., and Chen, Q., 2014. Progress of research on nanostructured. *Journal of Advances in Fine Petrochemicals*, 15(1), pp. 39–41.

Wu, H.C., Li, Y.Y., and Sakoda, A., 2010. Synthesis and hydrogen storage capacity of exfoliated turbostratic carbon nanofibers. *International Journal of Hydrogen Energy*, 35(9), pp. 4123–4130.

Wu, X.J., Zheng, J., Li, J., and Cai, W.Q., 2013. Molecular simulation on hydrogen storage capacities of porous metal organic frameworks. *Acta Physico-Chimica Sinica*, 29(10), pp. 2207–2214.

Xu, C., 2006. Preparation of high specific surface area active carbon and research progress of its storing hydrogen performance. *Chinese Journal of Chemical Education*, 23(9), pp. 8–10+14.

Xu, W., Tao, Z., and Chen, J., 2006. Progress of research on hydrogen storage. *Journal of Progress in Chemistry*, Z1, pp. 200–210.

Yao, Y.J., Zhang, S.P., and Yan, Y.J., 2006. Ball milling process and its effect on hydrogen adsorption storage of MWNTS. *Chinese Journal of Process Engineering*, 6(5), p. 837.

Zaluski, L., Zaluska, A., and Ström-Olsen, J.O., 1999. Hydrogenation properties of complex alkali metal hydrides fabricated by mechano-chemical synthesis. *Journal of Alloys and Compounds*, 290(1–2), pp. 71–78.

Zecho, T., Güttler, A., Sha, X., Jackson, B., and Küppers, J., 2002. Adsorption of hydrogen and deuterium atoms on the (0001) graphite surface. *The Journal of Chemical Physics*, 117(18), pp. 8486–8492.

Zhan, L., Li, K., Lv, C., Zhu, X., Song, Y., and Ling, L., 2002a. The preparation of super-activated carbons and its properties for hydrogen storage. *Journal of New Carbon Materials*, 22(1), pp. 31–34.

Zhan, L., Li, K., Zhu, X., Song, Y., Lv, C., and Ling, L., 2002b. The preparation of super-activated carbons and its properties for hydrogen storage. *Materials Science and Engineering*, 20(1), pp. 31–57.

Zhang, C., Bai, J., and Zhou, G., 2007. The performance analysis of hydrogen adsorption storage in carbon nanofibers. *Journal of Refrigeration*, 48(4), pp. 341–344.

Zhang, H.Y., Chen, K.X., Zhu, Y.J., Chen, Y.M., He, Y.Y., Wu, C.Y., Wang, J.H., and Liu, S.H., 2002. Preparation and Raman spectroscopy of single wall carbon nanotubes using CO_2 continuous laser vaporization. *Acta Physica Sinica*, 51(2), pp. 444–448.

Zhang, N., Chen, H., Ma, X., Shen, S., and Wang, G., 2019. Research progress of high density solid-state hydrogen storage materials. *Manned Spaceflight*, 25(1), pp. 116–121.

Zhang, S., Tang, W., Liu, J., Li, G., and Ren, T., 2022. Research progress of hydrogen storage materials. *Journal of Henan University (Natural Science)*, 52(5), pp. 579–586.

Zhou, P., Liu, Q., Sui, J., and Jin, H., 2014. Research progress in chemical hydrogen storage. *Journal of Chemical Industry and Engineering Progress*, 33(8), pp. 2004–2011.

Zhu, H.W., Chen, A., Mao, Z.Q., Xu, C.L., Xiao, X., Wei, B.Q., Liang, J., and Wu, D.H., 2000. The effect of surface treatments on hydrogen storage of carbon nanotubes. *Journal of Materials Science Letters*, 19, pp. 1237–1239.

Section II

Hydrogen Storage and Transportation Methods

3 Liquid Hydrogen Carriers

Ameen A. Al-Muntaser, Abdolreza Farhadian,
Ismail Khelil, Muneer A. Suwaid,
Richard Djimasbe, Mikhail A. Varfolomeev,
and Danis K. Nurgaliev

3.1 INTRODUCTION

Recently, hydrogen energy, as an emerging direction of the global energy system, has emerged as an alternative to traditional energy sectors that use hydrocarbon feedstocks (oil, natural gas, etc.) as the main energy carriers. The transition to hydrogen energy uses hydrogen to accumulate, transport, and generate energy (Staffell et al., 2019). Persistence Market Research has predicted that the global energy hydrogen market will experience an average annual growth rate of 6.1% between 2017 and 2025. They estimate the market will reach \$200 billion by 2025 (in 2017, this indicator was \$130 billion).

From both a technological and ecological standpoint, hydrogen-based energy systems are appealing. Compared to other fuel types, hydrogen has a higher specific energy (119 kJ/g under normal conditions), and hydrogen combustion does not produce greenhouse gases. However, hydrogen energy development depends on creating secure and effective technologies in storing and transportation (Mazloomi and Gomes, 2012; Preuster et al., 2017). This is a challenging task since hydrogen remains in a gaseous state when cooled down to ultra-low temperatures, and its low density limits the amount of energy stored per unit volume. As a result, hydrogen infrastructure costs are relatively high. In recent decades, the storage and transportation of hydrogen for specialized, small-scale applications have been adequately addressed in recent years. However, new technological solutions suitable for large-scale infrastructure and logistics are necessary to establish a global hydrogen energy sector and market. It is important to note that each hydrogen transportation and storage method has energy costs, technological benefits, and safety measures. Therefore, when constructing specific infrastructure projects, choosing the optimal technologies to support the entire supply chain is crucial. The materials and technologies employed for transporting and storing natural gas and other gases are often not applicable to hydrogen due to the "hydrogen embrittlement" (Aakko-Saksa et al., 2018). Consequently, no infrastructure for hydrogen can be built using the existing gas pipes. Hence, developing new energy-efficient and secure technologies is imperative to establish an infrastructure for transporting and storing hydrogen. These technologies are being developed through various research and projects. For instance, since 2013, the hydrogen-based energy storage program of the International Agency of Energy (IAE) has been operating and implemented.

DOI: 10.1201/9781003382553-5

To achieve its research objectives, this program/project aims to develop various materials for storing hydrogens; the research centers on developing new technologies for energy storage and hydrogen systems, including porous materials, liquid organic hydrogen carriers (LOHCs), and complex and intermetallic hydrides, with a particular emphasis on advancements in electrochemical and thermal storage. However, the program/project also aims to explore other emerging trends in research in the near future (Mizuno et al., 2016; Aakko-Saksa et al., 2018; Aziz et al., 2019; Makepeace et al., 2019).

One of the promising directions for transporting and storing hydrogen is via LOHCs. LOHCs are aromatic hydrocarbons or heterocyclic aromatic compounds or mixtures that are stable under various conditions (biphenyl, diphenylmethane, carbazoles, indoles, etc.). Under certain conditions, LOHCs chemically absorb hydrogen (up to 7.2 wt%) and turn into cyclic (aliphatic) compounds, which are also stable. In this form, hydrogen can be easily stored and transported to any place using the infrastructure and vehicles used for liquid fuels at atmospheric pressure. Hydrogen can be obtained from these carriers by heat treatment with a catalyst at 180–240 °C. Using LOHCs involves recycling, which reduces costs. LOHCs have advantages over methanol, ethanol, and ammonia as hydrogen carriers. LOHC systems present an appealing storage and transportation solution for hydrogen generated through electrolysis by utilizing energy produced by sources like sunlight and wind energy plants (Geburtig et al., 2016). Transporting hydrogen across significant distances is made possible by LOHC, which serves as a storage medium for hydrogen. The transportation process is made feasible by following a series of conversion steps, starting with storing hydrogen within a LOHC molecule through an exothermic hydrogenation reaction at the initial point of the supply chain. Then, the filled LOHC can be transported and stored. Hydrogen can be released upon reaching the use site through an endothermic dehydrogenation reaction. The now unloaded LOHC can then be returned to the hydrogen production point. The optimal LOHC for transportation would possess characteristics akin to crude oil-based liquids such as diesel and gasoline and be in a liquid state under normal environmental conditions (Modisha et al., 2019).

This chapter comprehensively reviews current knowledge and the latest research advancements in utilizing LOHCs as an efficient and cost-effective method for storing and transporting hydrogen in its liquid form. To achieve this goal, we begin with a general outline of the primary elements involved in long-distance hydrogen storage and transportation using LOHCs. We further divided the following content into distinct subsections, each focusing on a specific aspect of LOHC technology. These subsections provide supporting evidence for the overarching concept presented in the initial overview. In addition, this section illustrates the different subjects, including a brief history of LOHCs, the creation and advancement of LOHCs and their respective hydrogenation and hydrogen removal catalysts that are currently being designed and developed, challenges and need for further research, classification of LOHC media and catalytic systems, modification of the current LOHC technology catalysts, and a short introduction to risk assessment. After a systematic comparison, this study concludes and offers recommendations. A brief review of the introductory chapter will give the reader a general understanding of hydrogen storage and conveyance studies using liquid hydrogen organic carriers. This will serve a useful purpose by demonstrating the importance of further improvement in the hydrogen energy sector.

3.2 BRIEF HISTORY OF LOHCS

Research shows that using LOHCs for hydrogen storage through hydrogenation/dehydrogenation processes has been explored since the early 1980s (Taube et al., 1983). The toluene/methylcyclohexane (TOL/MCH) system was the most prominent LOHC concept based on (de)hydrogenation processes (Klvana et al., 1988). Later (Pez et al., 2008; Campbell et al., 2010; Sotoodeh and Smith, 2010; Luo, Zakharov and Liu, 2011; Sobota et al., 2011; Sotoodeh, Huber and Smith, 2012), Japanese researchers investigated a cyclohexane/benzene LOHC system in the early 2000s. NEC was suggested as a possible candidate for LOHC by Pez et al. (2008), and other research groups have since continued to study alternative azaborine carrier materials (Campbell et al., 2010; Luo, Zakharov and Liu, 2011). Crabtree (2008) proposed that LOHC materials containing N-containing heterocyclics would be more beneficial, while in a thermodynamic evaluation, Muller et al. suggested that nitrogen-consisting aromatic compounds were ideal for enhanced hydrogen storage (Müller, Völkl and Arlt, 2013). In 2018, Chiyoda Corporation of Japan achieved a significant milestone in the field of LOHC technology by constructing and operating a pilot-plant facility for the large-scale hydrogenation and dehydrogenation of LOHC materials (Okada, Mikuriya and Yasui, 2015). LOHCs have been suggested for practical uses, including decentralized energy storage networks (Teichmann et al., 2012) and combined heat and power systems (Haupt and Müller, 2017). Eypasch et al. have proposed a theoretical foundation for utilizing LOHCs in industrial production plants' energy supply (2017). Meanwhile, Niermann et al. emphasize the practicality of LOHCs as secure hydrogen storage and transportation materials from both technological and economic perspectives (2019). In 2018, a prototype plant facility capable of conducting extensive hydrogenation and dehydrogenation of LOHC elements was constructed (Agency, 2009). Several LOHC candidates have been proposed and applied in real-world scenarios, including decentralized energy storage networks (Teichmann et al., 2012) and systems that generate both heat and electricity (Haupt and Müller, 2017). Moreover, Chiyoda Corporation of Japan has presented a strategy aimed at storing and transporting hydrogen on a large scale, as illustrated in Figure 3.1 (Crabtree, 2008).

The scientific literature has proposed several possible LOHCs, considering their thermodynamic characteristics (Crabtree, 2008; Müller, Völkl and Arlt, 2013). Theoretical methods such as ab initio-DFT calculations and density functional theory (DFT) have proved useful for studying catalytic processes and creating effective LOHCs (Preuster, Papp and Wasserscheid, 2017). Moreover, researchers have investigated essential catalytic aspects of LOHC hydrogenation and dehydrogenation.

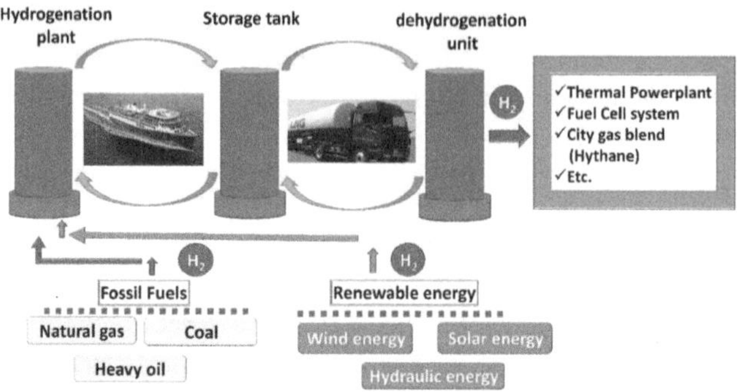

FIGURE 3.1   The demonstration of a large-scale plan for hydrogen transportation and storage, the Chiyoda Corporation's approach (Japan) 1 (Rao and Yoon, 2020).

3.2.1 ENERGY-CARRYING COMPOUNDS

The energy-carrying compounds concept is based on chemical substances in energy-rich and energy-poor states. LOHC system is formed when a liquid organic compound undergoes hydrogenation and dehydrogenation to achieve these states. The LOHCs are charged with hydrogen obtained from renewable sources via catalytic hydrogenation at an energy-rich location and time, then transported and stored until needed. To release the energy, a catalytic dehydrogenation reaction is initiated to produce hydrogen, which can be used for energy production through processes such as fuel cells or hydrogen combustion. The carrier material can be recycled as an energy-poor compound, allowing it to travel from the energy source to the user's location without being consumed. LOHCs offer the potential for separating energy production and usage spatially and temporally. Scientists, researchers, industry professionals, and politicians consider hydrogen a promising energy transport option (Sartbaeva et al., 2011; McQueen et al., 2020). Hydrogen can be generated from renewable and non-renewable energy sources and possesses high gravimetric energy content (Scherer and Newson, 1998; Zhao, Oyama and Naeemi, 2010). However, because of its low density, storing and handling large quantities of hydrogen requires high pressures or cryogenic temperatures, which is a challenge (Schneider, 2015). LOHCs offer a more practical solution to hydrogen storage, and several potential LOHC candidates exist. In this context, we will focus on heterocyclic aromatic hydrocarbons, specifically N-ethyl carbazole, which can store up to 5.8 wt% of hydrogen when fully hydrogenated. Usually, the hydrogenation process in this LOHC setup involves the use of a noble metal catalyst, such as ruthenium on aluminum oxide, which is supported on a substrate. The reaction takes place at a temperature range between 130 and 160 °C and a pressure of 7 MPa. On the other hand, the dehydrogenation reaction is conducted at slightly above ambient pressures and temperatures ranging from 200 to 230 °C, using supported Pt or Pd catalysts. Using LOHCs for energy storage allows for high energy density and simple, leak-free storage associated with the storage method (Cooper et al., 2010; Teichmann et al., 2011). The DoE-funded project has developed and evaluated multiple dehydrogenation reactors: "Reversible Liquid

Carriers for Integrated Hydrogen Production, Storage and Delivery" (Eblagon et al., 2010; Sotoodeh, Huber and Smith, 2012).

3.2.2 ELECTRICAL NETWORKS STABILITY—LOADING OF LOHC FROM RENEWABLE ENERGIES AND THEIR USE IN POWER GENERATION

Utility companies and electricity network operators face challenges transitioning to renewable energy sources. An obstacle that needs to be addressed is related to the pricing of electric energy, which is traded on the European Energy Exchange. In order to improve pricing, negative prices have been introduced since September 2008. Negative prices were observed in approximately 1 TWh of the total trade volume of 135 TWh on the EPEX spot market in 2009. The distribution of positive and adverse spot prices at the EEX (day-ahead auction) is shown in Figure 3.2, which indicates that negative prices are a frequent occurrence. Wind power is Germany's primary source of sustainable electricity production, accounting for 38 TWh out of 94 TWh in 2009. Several studies show a significant correlation between negative pricing incidents at the EEX and wind energy facility input levels. This transition poses significant challenges for utility companies and network operators. An analysis conducted by Brueckl and Koch revealed that during negative pricing periods, wind plants operated at 25% higher capacity than those hours with positive pricing (Andor et al., 2010; Hoisl, 2014).

According to researchers, as the share of renewable energy sources in the power system increases, negative prices are becoming more common, which happens when there is excess electricity supply and insufficient demand. These prices are crucial in signaling the need for more flexibility in the power generation and distribution market and increased investment in energy storage and grid infrastructure. Additionally, the authors mention that certain market players, such as power traders and operators of hydroelectric storage facilities, have adapted to this situation and are taking advantage of negative prices by buying surplus electricity and selling it back to the grid during times of high demand (Wang, Zhou and Ouyang, 2016). However, most commentators do not see this as a long-term solution and believe that a more comprehensive approach, including energy storage and grid expansion, is needed to address the challenges posed by the increasing share of renewable energy. It is necessary to transition toward a more flexible

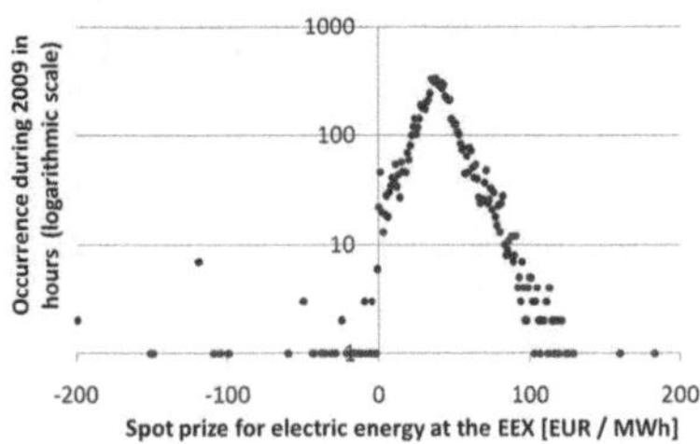

FIGURE 3.2 Frequency of negative prices at the EEX (Teichmann et al., 2011).

energy system, which can be achieved by storing excess energy from intermittent sources. However, this requires a significant increase in compensation capacity, which is predicted to reach 28 GW (40 TWh) for only Germany in 2020, according to a study by the Boston Consulting Group (Pieper and Rubel, 2011). The current capacity is 7 GW only, mostly through pumped hydroelectric storage sites. Another challenge is significant investments in the electric grid's intermediate and low-voltage lines and high-voltage parts. The German Energy Agency's analysis (Hernandez-Santoyo and Sanchez-Cifuentes, 2010) suggests that an additional 3,600 km of power lines will be needed to connect offshore wind farms to large industrial centers. The expansion of the power grid is needed to cope with high-voltage lines and due to the growing number of power producers connected to the medium and low-voltage sections. These include photovoltaic house systems, medium-sized solar or wind parks, and other power producers. As a result, the power network needs to be strengthened and expanded to meet the increasing demand. Considering the difficulties, we suggest employing LOHCs to store intermittent renewable energy for the medium to long term. Additionally, we propose utilizing LOHCs to connect this energy to sustainable mobile applications, as elaborated below. During periods of high energy production from sustainable sources, it is recommended to use excess energy for on-site hydrogen production through water electrolysis. The hydrogen can then be stored in LOHCs in its energy-rich form, providing good handling qualities and high energy densities. The energy carrier is simple to transport and can be used for different purposes such as heating and transportation. Moreover, the energy stored can be converted back to electricity when required or can be kept in local storage. To implement this system, electrolysis devices, chemical converters, storage tanks, and a gas turbine for electricity generation can be employed as depicted in Figure 3.3. Although large storage facilities of up to 50 MW are possible, the benefit of this concept lies in local energy storage near intermittent sustainable power producers (e.g., 1–5 MW). This not only helps to bridge the discrepancy between fluctuating demand and sporadic supply, but also reduces the need for local grid amplification, saving time and money. As a result, this system is expected to impact electric grids' stability positively.

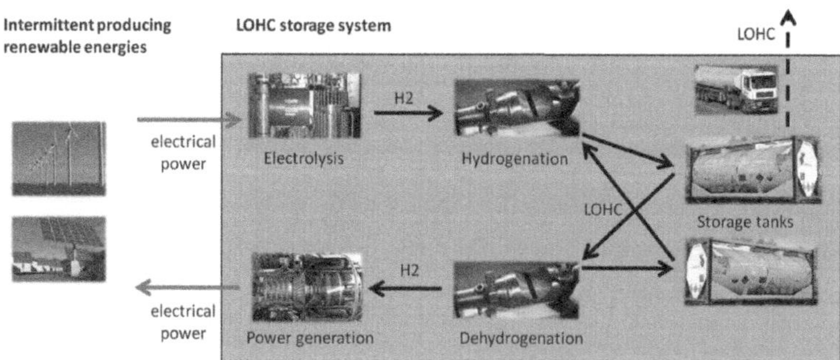

FIGURE 3.3   Concept of the LOHC electricity storage device (Teichmann et al., 2011).

3.2.3 LOHCs as a Mode of Energy Transportation for Mobility Purposes

Transportation relies heavily on fossil fuels. Battery-powered electric vehicles are currently being designed for use in urban areas with short travel distances. In terms of mobile applications, hydrogen is being examined as a potential energy carrier for the future. For more than four decades, BMW, a German automaker, has been working on developing hydrogen car technology. In 2006, it produced a test fleet of 100 hydrogen-powered vehicles, successfully operating on public roads for millions of test kilometers. However, storing sufficient amounts of hydrogen for mobile applications poses challenges due to space and weight limitations and strict safety and usability requirements. To date, two hydrogen storage concepts have been employed in cars: compressed gaseous hydrogen, which is stored at pressures up to 70 MPa, and liquid hydrogen, which is stored cryogenically at temperatures below 253 °C. The technical characteristics, pros, and cons of these storage methods have been described in various publications (von Wild, Freymann and Zenner, 2008). Notably, no other technologies the general public utilizes operate at such high pressures or low temperatures, so implementing these storage methods may not be simple. In addition, there is currently no established delivery system for hydrogen to gas stations. The use of hydrogen pipelines for mobile applications would be costly and vulnerable. Therefore, LOHCs are considered alternatives. LOHCs offer interesting energy densities for storage systems and can be stored in standard gasoline tanks under ambient conditions. To achieve a cruising range of 500 km, approximately 5 kg of onboard hydrogen is needed, corresponding to 100 L of N-ethyl carbazole. The LOHC system's charging and discharge involve chemical reactions that are either exothermic (hydrogenation) or endothermic (dehydrogenation). Charging an exothermic system generates heat rates of around 600 kW within several minutes, which cannot be achieved safely at typical gasoline stations. Thus, LOHCs offer an advantage in that the charging process can take place in large-scale hydrogenation facilities, which allows for the optimization of process efficiency and the recovery and utilization of reaction heat, such as steam production. There are two ways to utilize the produced hydrogen: thermally by using it in a combustion engine or electrochemically to power a fuel cell (as shown in Figure 3.4). In order to maintain the necessary temperature

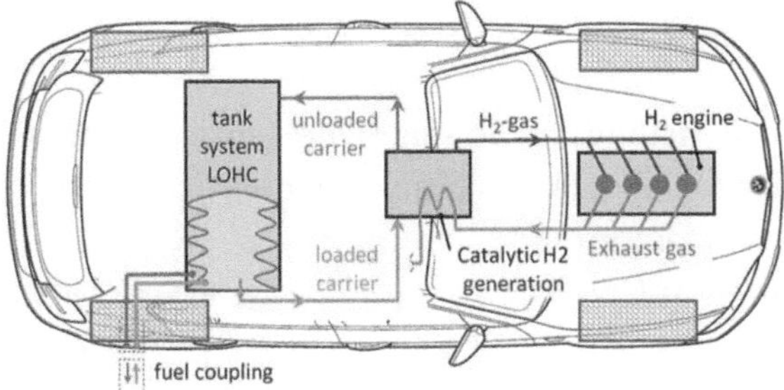

FIGURE 3.4 Liquid organic hydrides in a hydrogen automobile concept device (Teichmann et al., 2011).

for the dehydrogenation process, a catalytic device is used in the hydrogen generation unit, which requires heating. One way to accomplish this is by utilizing the waste heat produced by the hydrogen combustion engine (Taube et al., 1983).

3.2.4 Hydrogen Storage

The classification of storage technology into three main groups is based on how the hydrogen system interacts with the storage environment (Table 3.1).

A range of hydrogen storage methods have been developed and are currently under development, each with unique characteristics such as the quantity of hydrogen held per unit of the storage medium's mass or volume, the time required for hydrogen storage and release, and the cost of transporting hydrogen (Nafchi et al., 2018; Kikuchi et al., 2019; Müller, 2019). These characteristics determine the suitability of each storage method for different applications. For instance, liquid hydrogen is more appropriate for long-distance transportation due to its transportation costs, while compressed hydrogen gas is ideal for shorter distances and small quantities. On the other hand, pipelines are more suitable for large hydrogen volumes. In order to make hydrogen storage and transportation feasible, it is necessary to compress it as its volumetric energy density is low (under normal conditions, 1 kg of hydrogen takes up more than $11\,m^3$ of space). Typically, there are two conventional methods of storing hydrogen: compressing it into steel gas cylinders (with a maximum 20 MPa pressure and a temperature of 300 K) or storing it in liquid form in Dewar vessels, stationary or transport cryogenic containers (with a temperature of 20.4 K). Zhao et al. (Yanxing et al., 2019) performed a thermodynamic assessment of hydrogen storage methods that operate at room temperature, comprising compressing hydrogen in its gaseous form, transforming hydrogen into a liquid state, and partially converting hydrogen into a liquid by pressurizing it. According to the study's results, pressurized partial liquefaction could produce a high density of hydrogen storage while consuming less energy. It is recommended to maintain a specific temperature range of 35–110 K and a pressure range of 5–70 MPa to achieve these favorable conditions. This leads to a high ortho-hydrogen conversion to para-hydrogen and a hydrogen density of $71.5\,kg/m^3$. Moreover, the ratio of hydrogen density to energy consumption for compression falls within the range of $1.50–2.30\ kg/m^3/kW$ under these conditions.

TABLE 3.1

Categorization of Methods for Storing Hydrogen according to Its Interplay with the Storage Medium (Andersson and Grönkvist, 2019)

The Nature of the Interaction of Hydrogen with the Storage Medium	Storage Method
Physical	In a compressed gaseous form (in pressurized cylinders)
	In liquid form (in cryogenic containers)
Physicochemical	As metal hydrides (in pressurized cylinders)
Chemical	In the form of LOHCs (in containers at normal conditions)
	In the form of ammonia (in cylinders or cryogenic containers)

Various hydrogen storage methods are used, including absorption in materials with weak van der Waals interactions with hydrogen molecules. Various substances such as activated carbon, zeolites, organometallic frameworks, and carbon nanostructures are examples of such materials. Another option is using metal hydrides as hydrogen carriers, which come in the form of hydrides of metals, alloys, intermetallic compounds, and composites in solid form. Solid metal hydrides and intermetallic compounds can absorb and release several hundred volumes of hydrogen per unit mass upon heating. A literature review of research on metal hydrides as carrier substances in hydrogen storage systems and related studies is provided in the source (Tarasov et al., 2021). Hydrides with light elements such as alkaline or alkali earth metals (ionic hydrides CaH_2, LiH, and NaH) are required to achieve high hydrogen storage capacity. Among these, magnesium hydride (MgH_2) is a promising candidate due to its abundant reserves, low reactivity, and high hydrogen storage capacity (7.6 wt%). However, the hydrogen adsorption/desorption enthalpy of MgH_2 is 76 kJ/mol, which limits its practical application at ambient temperatures. To address this limitation various strategies are being explored to address this limitation, involving MgH_2 in the form of nanoparticles or combining it with additional substances (Mg_2Ni, Mg_2Si). It is possible to store hydrogen in chemical compounds of various compositions covalently. Under specific conditions, these compounds release hydrogen. For example, ammonia can undergo catalytic decomposition at temperatures between 800 and 900 °C, and unsaturated hydrocarbons can undergo reversible hydrogenation to release hydrogen. Organic compounds used for hydrogen storage under ambient conditions exist in liquid states, simplifying transportation and storage. This liquid state facilitates heat and mass transfer during reversible catalytic hydrogenation/dehydrogenation processes. Storing and transporting hydrogen in a chemically bound state has the key advantage of high volumetric (~100 kg/m^3) and mass (17.7 wt%) hydrogen content. However, the main disadvantage is the limitations of multiple reuses. Ammonia, methanol, and acetic acid are among the hydrogen-rich compounds that have established handling and transportation infrastructures, making them suitable for a hydrogen economy. LOHCs are also available. Hydrogen storage systems designed for transportation purposes are examined (Table 3.2). The crucial aspects of such systems during transportation are ensuring that the refueling process is carried out rapidly and safely while maintaining a compact size and low mass for the storage system (Rivard, Trudeau and Zaghib, 2019).

Currently, many of the hydrogen storage systems that have been developed and researched are limited in practical applications due to their low mass and volume density of hydrogen, high-energy thermodynamics, and slow kinetics of the ongoing processes. Thus, efforts to discover novel hydrogen storage systems with enhanced properties persist. The concept of LOHC was first introduced in the early 1980s, and they are materials that can undergo reversible hydrogenation/dehydrogenation reactions while remaining in a liquid state in both forms (Latapí, Davíðsdóttir and Jóhannsdóttir, 2023). LOHCs have several benefits over other hydrogen storage methods, such as storing and releasing hydrogen at high depths, high mass, and volume hydrogen content, and low technological and investment risks as they are compatible with existing transportation infrastructure. Works on LOHCs as efficient hydrogen storage systems are widely discussed (Brayton and Jensen, 2015; Dong et al., 2015; Markiewicz et al., 2015; Mehranfar, Izadyar and Esmaeili, 2015; Aslam et al., 2016;

TABLE 3.2

Classification of Techniques for Storing Hydrogen (Rivard, Trudeau and Zaghib, 2019)

Storage Method	Mass Content Hydrogen (wt%)	Volume Density Energy (MJ/L)	Temperature (K)	Pressure (MPa)
In a compressed gaseous form (in cylinders under pressure)	6.7	4.9	293	70
In liquid form (in cryogenic containers)	7.5	6.4	20	0
Partially compressed under pressure	5.4	4.0	40–80	30
In adsorbed form on organometallic framework structures	4.5	7.2	78	2–10
Adsorbed on carbon nanostructures	2.0	5.0	298	10
In the form of metal hydrides	7.6	13.2	260–425	2
In the form of metal boron hydrides	14.9–18.5	9.8–17.6	130	10.5
In the form of organic compounds (LOHCs)	8.5	7	293	0

Choi et al., 2016; Geburtig et al., 2016; Stark et al., 2016). It should be noted that most storage system studies have been carried out in a laboratory-controlled setting, and the technological advancements have not yet allowed for the development of experimental storage systems. Table 3.3 compares three potential hydrogen storage methods based on Aziz, Oda and Kashiwagi (2019) and Wijayanta et al. (2019). As is evident from the presented data, the LOHC system is the most effective.

Developers of LOHCs aim to create storage systems principles of cost-effective and easily available organic compounds with significant ability to store and release hydrogen, allowing the same catalyst to be used in both hydrogenation and dehydrogenation reactions under mild conditions. Additionally, LOHC systems must be compatible with existing transportation and refueling infrastructure. The storage of hydrogen in such systems is accomplished by subjecting it to a high-pressure (20 atm) and exothermic hydrogenation reaction, while hydrogen release is carried out through an endothermic dehydrogenation reaction performed under low pressure (5 atm) (Modisha et al., 2019). LOHC systems are discussed in detail in various articles (Teichmann et al., 2012; Müller, Völkl and Arlt, 2013; Müller, Stark and Emel'yanenko, 2015; Stark et al., 2015; Müller et al., 2016; Niermann et al., 2019). A storage system diagram presents a possible approach to transporting and storing hydrogen produced by water electrolysis. Water electrolysis produces hydrogen to hydrogenate unsaturated LOHC molecules. The resultant saturated organic

TABLE 3.3

Comparison of Hydrogen Storage Methods in the Hydrogenated Form of Liquid Organic Carriers of Hydrogen, Liquid Hydrogen, and Ammonia (Aziz, Oda and Kashiwagi, 2019)

Characteristic	Liquid Organic Hydrogen Carriers	Liquid H_2	Ammonia (NH_3)
Method of application	Dehydrogenation and purification with subsequent use in fuel cells	Direct use in fuel cells	Dehydrogenation and purification with subsequent use in fuel cells; direct use in fuel cells
Infrastructure	Ability to use existing infrastructure for liquid hydrocarbons (gasoline)	Infrastructure needs to be developed for large-scale applications	Ability to use existing propane infrastructure
Autoignition temperature of the storage phase	~283 °C	535 °C	651 °C
Advantages	Possibility of storage in liquid form (minimum losses during transportation); use of existing infrastructure and storage standards	High purity; no dehydrogenation and purification required	The cheapest carrier of energy; direct use is possible; operation of the existing infrastructure and storage standards
Disadvantages	A heat source is required for dehydrogenation; 30% of H_2 energy is used in dehydrogenation; low number of storage cycles	Very low temperatures (-250 °C) are required, 45% of H_2 energy is used for liquefaction; complexity of long-term storage	Special treatment of process equipment is required due to pungent odor and toxicity; 13% of H_2 energy is absorbed during dehydrogenation and purification.
Development stage	Demonstration stage	Small scale—applied in practice; large scale—development stage	Stage of research work; entering the demo stage
Development prospects	Improving the efficiency of dehydrogenation; new catalysts for dehydrogenation and hydrogenation	Increasing the efficiency of liquefaction; it is necessary to develop and approve standards for transportation and storage	Fuel cells based on NH_3; increasing the energy efficiency of storage system synthesis

molecules can be stored without energy loss under normal conditions for an extended duration. The hydrogen in the form of saturated molecules is then transported to energy consumers through existing pipelines for crude oil, road, rail, or sea, where saturated LOHC molecules undergo dehydrogenation to generate hydrogen gas for energy generation. Studies have extensively examined the prospects of producing and storing hydrogen efficiently onboard various types of vehicles. As renewable energy generation capacities are seasonally dependent, seasonal hydrogen storage, such as a hydrogenated form of LOHCs, is necessary (Reuß et al., 2017). The cost of energy usage and greenhouse gas emissions during hydrogen production from LOHCs for fuel cells onboard vehicles has been calculated in a study, showing that LOHC storage technology is the most economically promising. Recycling LOHC waste is also discussed, which is crucial for commercially attractive full-cycle hydrogen storage technologies.

LOHC systems use aromatic compounds as hydrogen storage carriers. The first LOHCs, such as cyclic alkanes like cyclohexane, methylcyclohexane, and decalin, can release up to 6 wt% hydrogens via dehydrogenation into corresponding aromatic compounds at temperatures ranging from 300 to 350 °C using platinum catalysts (Wu and Wang, 2016; Preuster, Papp and Wasserscheid, 2017). One of the most popular LOHC cycles is the methylcyclohexane-toluene reaction route, which has a hydrogen content of 6.1 wt% by mass and 47.0 kg H_2/m^3 by volume. It is reversible, highly selective, and does not produce carcinogenic products. Most studies on cyclohexane LOHCs focus on lowering the dehydrogenation temperature by modifying the LOHC molecules to reduce the reaction enthalpy. In addition, they focus on finding more active, stable, and economic dehydrogenation catalysts to replace expensive noble metal catalysts. Several alternative carrier substances have been developed for LOHC systems that show promise for practical use. Some examples include benzyltoluene and dibenzyltoluene (Durbin, 2013), borane-based amines (Müller et al., 2012), methylcyclohexane (Alhumaidan, Cresswell and Garforth, 2011), and other compounds. Table 3.4 summarizes the physical properties of some of these studied carrier substances (Wang, Zhou and Ouyang, 2016).

In the course of hydrogenation and dehydrogenation of ethylcarbazole, a range of intermediate compounds with varying hydrogenation levels is produced. Studies have shown that intermediate compounds of varying degrees of hydrogenation are created throughout the hydrogenation and dehydrogenation of ethyl carbazole, and their content in the reaction products depends on both the reaction time and the catalyst properties. Using a LOHC system based on dibenzyltoluene-perhydrodibenzyltoluene as a source of hydrogen in the hydrogenation of toluene has also been investigated as a model for studying the possibility of achieving a full-cycle process (Geburtig et al., 2016). Experiments conducted under strictly controlled conditions

TABLE 3.4

Physical Properties of Some Substance Carriers in the LOHC Composition (Wang, Zhou and Ouyang, 2016)

Carrier	Melting Point (K)	Boiling Point (K)	Mass Content of Hydrogen (%)
Cyclohexane	279.65	353.85	7.2
Methylcyclohexane	146.55	374.15	6.2
cis-Decalin	230.15	466.15	7.3
trans-Decalin	242.75	458.15	7.3
Carbazole	517.95	628.15	6.7
Ethylcarbazole	341.15	463.15	5.8

with a stoichiometric ratio of perhydrodibenzyltoluene and toluene to H_2 showed a conversion of the initial component above 60% at 270 °C. The experiments utilized heterogeneous catalysts containing noble metals such as Pd, Pt, and Ru. The highest toluene conversion rates were achieved with the Pt/Al_2O_3 catalyst, with 89% in 1 hour and 99% in 5 hours, while the ruthenium-based catalyst exhibited the lowest activity. Shi et al. present the results of using dibenzyltoluene-perhydrobenzyltoluene as LOHCs in which the process of adding and removing hydrogen atoms was conducted in the same reactor with the same catalyst, increasing the efficiency of the complete hydrogen storage/release cycle (Shi et al., 2019). The study determined optimal temperatures of 140 °C for hydrogenation and 270 °C for dehydrogenation, with a total hydrogenation time of 35 minutes and dehydrogenation time of 180 and 300 minutes, resulting in 56.5% and 60.1% conversion, respectively. The most efficient platinum catalyst tested on various supports was 3 wt% Pt/Al_2O_3. The cyclic testing of the system demonstrated high thermal stability, with 84.6% hydrogen storage efficiency after five test cycles. The system was also effective in continuous stationary operation. Wijayanta et al. compared various LOHC systems presented in Table 3.5. Despite the low hydrogen mass and volume content, the authors suggest that the methylcyclohexane-toluene system has considerable industrial advantages over other systems (Wijayanta et al., 2019).

Li et al. (2016) explored a novel support called 2-methylindole, which possesses a hydrogen storage capacity of 5.76 wt% under moderate hydrogenation/dehydrogenation conditions, making it a promising candidate for LOHCs (Li et al., 2016). Meanwhile, in another study, researchers investigated 1,2-dimethylindole, which has

---

TABLE 3.5

Comparison of the Potential Advantages of Different Hydrogen Storage Systems Incorporating Liquid Organic Carriers (Wijayanta et al., 2019)

	System			
Characteristic	**Toluene-methylcyclohexane**	**Naphthalene-decalin**	**Benzene-cyclohexane**	**Dibenzyltoluene–perhydrodibenzyltoluene**
Phase state of the system under normal conditions	Liquid	Two-Phase Solid/Liquid System	Liquid	Liquid
Hydrogenation				
Temperature (°C)	200–300	150–250	150–250	180
Pressure (bar)	1–5	2–5	1–5	1–5
Volume density H_2 (kg H_2/m^3)	47.4	65.4	55.9	57
Mass density H_2 (wt%)	6.16	7.29	7.2	6.2
Enthalpy of reaction (kJ/mol)	204.8	319.6 (cis) 332.5 (trans)	205.9	588.5
Dehydrogenation				
Temperature (°C)	250–350	300–350	330	260–320
Pressure (MPa)	0.1–0.5	0.1–0.4	0.1–0.4	0.1–0.5

ability to store hydrogen up to 5.23 wt%; the compound can undergo rapid and easy hydrogenation and dehydrogenation at mild conditions, and possesses a melting point of 55 °C (Dong et al., 2019). A first-order equation with an apparent activation energy of 85.1 kJ/mol can be used to describe the hydrogenation reaction of 1,2-dimethyl-indole. The product of this reaction, octahydro-1,2-dimethylindole, can be dehydro-genated using a 5 wt% Pd/Al_2O_3 catalyst at temperatures between 170 and 200 °C. The apparent activation energy for this process is 111.9 kJ/mol. The utilization of N-ethylindole as a LOHC was investigated through cycles of catalytic hydrogena-tion/dehydrogenation (Dong et al., 2015). At room temperature, both the chemical and its hydrogenation products exist in liquid form, enabling complete hydrogenation and dehydrogenation reactions under relatively mild conditions. The hydrogenation process of N-ethylindole was carried out using a 5 wt% Ru/Al_2O_3 catalyst at a tem-perature range of 160–190 °C and 9 MPa pressure. This process proceeds in a series of steps involving the conversion of N-ethylindole into 2H-N-ethylindole, then to 4H-N-ethylindole, and finally to 8H-N-ethylindole, with the intermediate products being rapidly assimilated. The 5 wt% Pd/Al_2O_3 catalyst was utilized to accomplish the reverse dehydrogenation of octahydro-N-ethylindole within the identical temper-ature range, resulting in complete dehydrogenation and high-purity hydrogen. Based on the promising results obtained, N-ethylindole may be a potential LOHC.

Several studies, both experimental and theoretical, have explored possible car-rier materials used in LOHC systems, such as N-phenylcarbazole (Emel'yanenko et al., 2019), furfuryl alcohol (Verevkin, Siewert and Pimerzin, 2020), and diphenyl ether with aromatic substituents (Verevkin, Pimerzin and Sun, 2020). These carrier substances were analyzed for their thermodynamic characteristics, which include enthalpies of formation, sublimation, and vaporization, in addition to the revers-ible hydrogenation/dehydrogenation energy indices. It is crucial to maintain hydro-gen purity generated from LOHCs to establish practical industrial storage systems (Jorschick et al., 2019). Extensive analysis was conducted (Bulgarin et al., 2020) on the purity of hydrogen produced through perhydrodibenzyltoluene dehydrogenation. The investigation demonstrated that high-purity hydrogen (>99.999%) with less than 0.2 ppm CO content could be achieved if the perhydrodibenzyltoluene utilized in the dehydrogenation reaction was first purified and dried. Additionally, the research indi-cated that even small quantities of water dissolved in perhydrodibenzyltoluene dur-ing catalytic dehydrogenation could generate CO and CO_2 instead of acting as inert additives. According to the research conducted, it was found that in LOHC systems, the majority of impurities that fall under the purview of global hydrogen quality regulations for fuel cells are produced due to the presence of water and oxygen-based impurities.

3.3 CLASSIFICATION OF LOHC MEDIA AND CATALYTIC SYSTEMS

Numerous organic carrier molecules have been reported in the field of LOHCs, and various influential review articles have covered different aspects of this topic (He, Pei and Chen, 2015; Zhu and Xu, 2015; Preuster, Papp and Wasserscheid, 2017; Gianotti et al., 2018). This discussion focuses on current advancements in catalytic systems for hydrogenation and dehydrogenation. Efficient dehydrogenation cata-lysts are widely studied due to high heat demand. As a result, we have identified

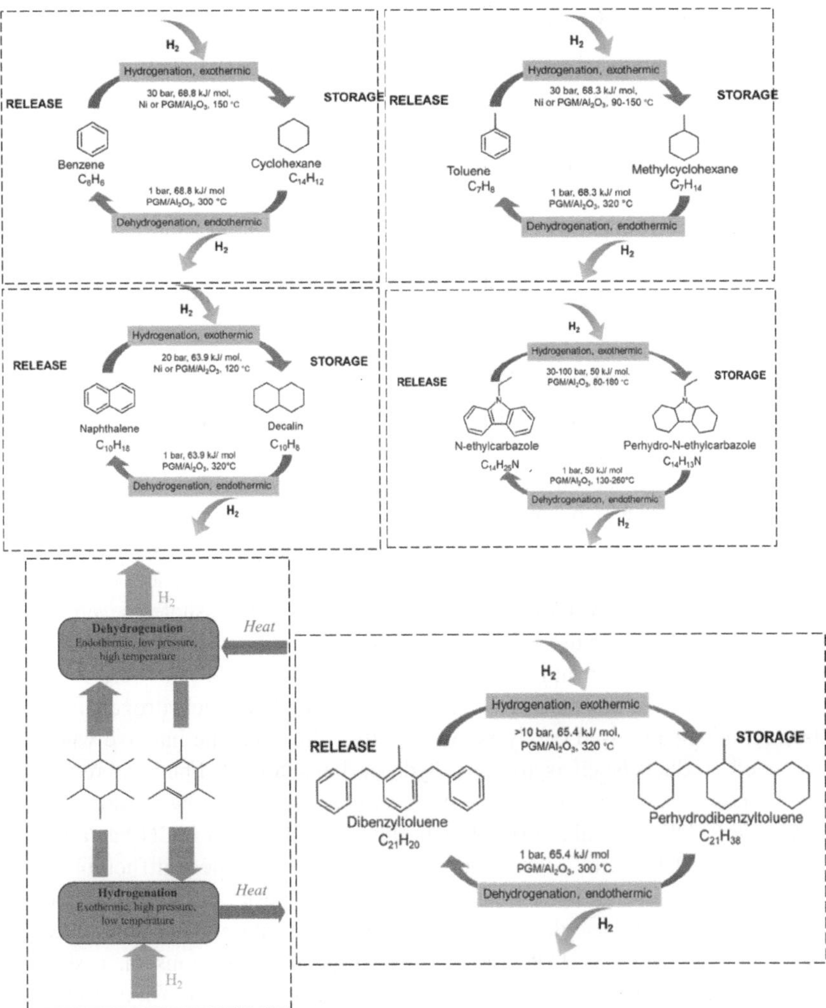

FIGURE 3.5 Schematic representations of various LOHC systems (Modisha et al., 2019).

and selected several promising LOHC systems based on their structural characteristics and intended use. These systems have been categorized into (i) homocyclic and (ii) heterocyclic compounds. In addition, recent theoretical efforts have been made toward developing model dehydrogenation catalysts for some of these compounds. Section 3.1 centers on LOHC molecules, particularly toluene, naphthalene, and monobenzyl toluene LOHC systems. This section highlights the latest progress in heterocyclic compounds, including phenazine, indole derivatives, and 2-(n-methyl benzyl)pyridine. However, we exclude coverage of NEC-derived derivatives since other recent reviews have already covered these particular LOHC systems (Matthias Niermann et al., 2019; Zhou et al., 2020). Figure 3.5 illustrates the hydrogenation and dehydrogenation conditions (temperatures, pressures, and processes) of the most studied LOHCs (Modisha et al., 2019).

3.3.1 Homocyclic Compounds

Precious metal catalysts are commonly used for aromatic LOHC hydrogenation, but dehydrogenation reactions have been less explored. Extensive studies have been conducted on hydrogenating several prospective LOHC systems, including benzyltoluene (Marlotherm LH, BT), naphthalene, toluene, and utilizing effective catalysts to produce MCH, decalin, and perhydrobenzyl toluene, correspondingly. This section will focus on recent advancements in dehydrogenation catalytic systems and their associated physical and chemical characteristics. The ultimate goal is to identify practical ways these catalytic systems can be utilized in liquid hydrogen carrier systems.

3.3.1.1 Methylcyclohexane Dehydrogenation

One of the most well-known LOHC systems in the early stages of hydrogen storage and transport technology was TOL/MCH (Figure 3.6). Various efficient heterogeneous catalysts are used during hydrogenation to convert toluene into liquid-phase methylcyclohexane, which can reach a capacity for storing hydrogen at elevated levels (6.2 wt%) at ambient conditions, satisfying the hydrogen energy target (5.5 wt%). Moreover, MCH's liquid state is advantageous for long-distance transportation and can be employed in numerous applications. MCH necessitates severe thermodynamic conditions because of the significant heat demand (-68.3 kJ/mol H_2) for dehydrogenation. Sinfelt and his team reported that Pt catalysts were recognized as the most effective and selective catalysts for cycloalkane dehydrogenation (1977, 2000). Although Pt-based catalysts are effective, they become inactive when dehydrogenated products (such as aromatic hydrocarbons) are adsorbed and require high temperatures ($\geq 350\,°C$), which can result in coke and other by-products (such as dealkylation) (Okada et al., 2006). Furthermore, the fact that MCH has a low boiling point of $100.9\,°C$ and a low flash point of $-3\,°C$ could pose difficulties regarding its ability to reverse its selectivity and its usefulness in practical applications. Nonetheless, despite these drawbacks, Chiyoda Corporation, the Japanese firm, has effectively manufactured TOL/MCH LOHC system applications on a significant scale, as per the source reference provided (Wolf, 2015; Rao and Yoon, 2020).

As reported in the literature, there are two primary means by which the catalytic efficacy of MCH dehydrogenation can be enhanced: via the synergistic influences

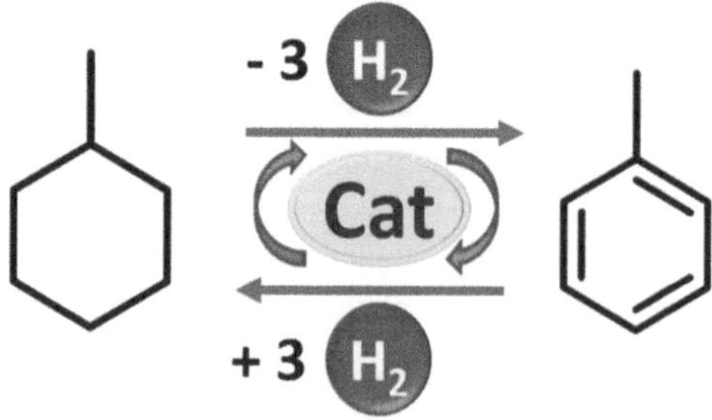

FIGURE 3.6 A diagrammatic representation of the LOHC system of methylcyclohexane and toluene (Modisha et al., 2019).

that result from bimetallic catalysts (for instance, Pt-Mn and Pt-Sn) and through the utilization of catalytic supports (such as Al_2O_3 and SiO_2). Incorporating second metals (typically through bimetallic catalysts like Pt-Sn and Pt-Mn) such as Sn, Re, Mn, and Ni decreases the reliance on expensive metal catalysts or enhances their effectiveness (Al-ShaikhAli et al., 2015). These bimetallic catalysts show greater effectiveness than their monometallic equivalents. Catalytic supports such as metal oxides (e.g., ZrO_2, TiO_2, and MnO_2) and perovskites (including La0.7Y0.3NiO_3) have been found to hold significant potential as catalysts for MCH dehydrogenation (Shukla, Karmakar and Biniwale, 2012).

3.3.1.2 Decalin Dehydrogenation

The decalin/naphthalene (Figure 3.7) LOHC system is another significant aromatic hydrocarbon LOHC system. As is well known, the decalin (DEC) hydrogen storage capacity is 7.3 wt% (Ninomiya et al., 2006). Although it has a higher storage capacity, the dehydrogenated product (naphthalene) is in a solid phase at normal ambient temperature (melting point of 8 °C). In addition, the need for a significant amount of heat (66.3 kJ/mol H_2) to dehydrate, the fact that it cannot be reversed, and the challenges involved in transporting it present obstacles to the implementation of this system in real-world scenarios, which require further research to overcome. Several research studies have highlighted the use of dehydrogenation catalysts to convert decalin to naphthalene (Jiang et al., 2012; Suttisawat et al., 2012; Li et al., 2014).

3.3.1.3 Benzyltoluene Dehydrogenation

According to literature sources (Brückner et al., 2014; Müller et al., 2015; Jorschick et al., 2020), the perhydrobenzyltoluene (H12-BT)/benzyltoluene (H0-BT) system is a highly promising LOHC system with a hydrogen storage capacity of 6.2 wt% (as illustrated in Figure 3.8). H12-BT is available as heat transfer oil and possesses

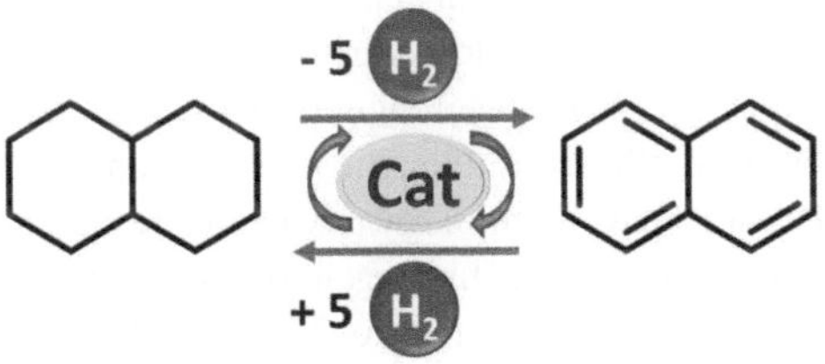

FIGURE 3.7 A diagrammatic depiction of the LOHC system using decalin/naphthalene (Modisha et al., 2019).

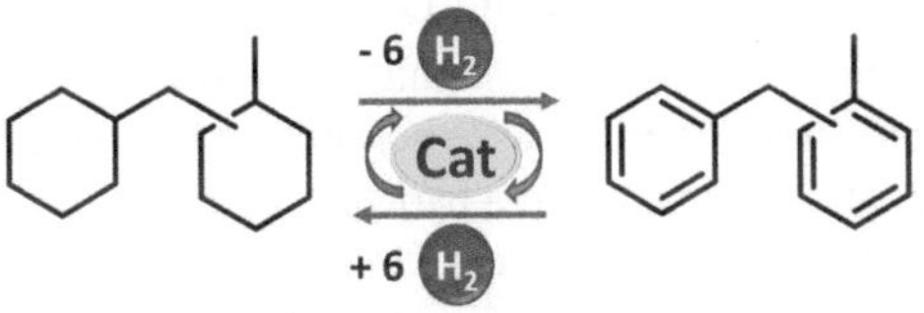

FIGURE 3.8 A diagrammatic representation of the LOHC system involving perhydrobenzyl toluene/benzyl toluene (Rao and Yoon, 2020).

advantageous physicochemical properties for use as LOHCs. While H12-BT dehydrogenation demands less energy per H_2 mole than H18-DBT (63.5 vs 65.4 kJ/mol H_2), its capacity for liquid-phase dehydrogenation is restricted by its low boiling point (270 °C). Despite this, H12-BT has advantages that make it more attractive for low-temperature energy storage applications than H18-DBT. These advantages include faster hydrogen release at 270 °C in the presence of the Pt/Al_2O_3 catalyst, formation of intermediates with simpler structures during hydrogenation/dehydrogenation processes, and at a temperature of 20 °C, H0-BT and H12-BT have lower viscosity values (3.94 cP and 6.97 cP, respectively) compared to H18-DBT, which has much higher viscosity values (49 cP for H0-DBT and 389 cP for H18-DBT), as reported in various research review articles (Leinweber and Müller, 2018).

3.3.2 HETEROCYCLIC COMPOUNDS

The benefits of the research have been emphasized by the Pez and Crabtree groups for N-heterocyclic-based LOHCs over cyclic aromatic hydrocarbons, such as favorable kinetics and thermodynamics, low dehydrogenation temperature, and reversibility (Rao and Yoon, 2020). However, heterocyclics like dibenzofuran and dibenzothiophene are not commonly used due to undesired products during hydrogenolysis (Eblagon and Tsang, 2014). Indole derivatives, similar to carbazole, have been extensively studied in the literature. Phenazine and 2-n-methylbenzyl pyridine are also significant N-based LOHC systems.

3.3.2.1 Hydrogenation/Dehydrogenation of Indole Derivatives

The advantageous kinetic and thermodynamic characteristics of indole and its derivatives suggest they could be promising options for LOHC uses. These substances share resemblances with carbazole derivatives and display properties like low melting points (Bachmann et al., 2019), lower dehydrogenation temperatures, and sufficient storage capacities. Jessop et al. (Cui et al., 2008) proposed that introducing alkyl groups to the pyrrolidine rings of indole derivatives can effectively lower enthalpy and reduce storage capacity. Thus, researchers have investigated several alkyl-substituted indoles that can be synthesized in a safe and environmentally friendly manner for their potential use in LOHC technology. Extensive research has been conducted on the catalytic hydrogenation and dehydrogenation of various LOHC systems such as N-methyl indole/N-methyl perhydroindole, N-ethylindole/N-ethylperhydroindole, 2-methylindole/2-methyl perhydroindole, and 1,2-dimethyl indole/1,2-dimethyl perhydroindole, which have demonstrated significant potential as LOHCs (as illustrated in Figure 3.9) (Dong et al., 2015; Rao and Yoon, 2020).

3.3.2.2 Hydrogenation/Dehydrogenation of 2-(n-Methylbenzyl Pyridine)

In recent years, researchers have investigated a novel N-heterocyclic compound, 2-(n-methylbenzyl pyridine) or MBP, as a possible LOHC (Figure 3.10). According to the work of Suh et al., MBP can hold a hydrogen density of 6.15 wt% and be found in liquid form under normal environmental conditions (Oh et al., 2018). The ease of synthesizing MBP and theoretical calculations indicates that it could overcome thermodynamic obstacles and be readily available for technical use. By analyzing several

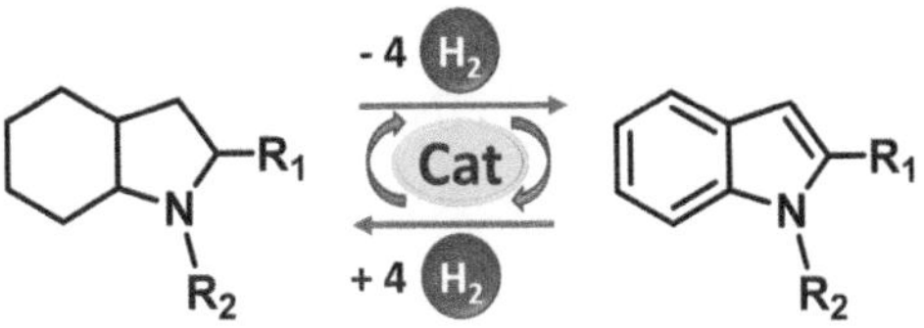

1. R_1 = H, R_2 = CH$_3$; 2. R_1 =CH$_3$, R_2 = H

3. R_1 = CH$_3$, R_2 = CH$_3$; 4. R_1 = C$_2$H$_5$, R_2 = H

FIGURE 3.9 A diagrammatic representation of the LOHC system employing perhydroindole derivatives/indole derivatives (Rao and Yoon, 2020).

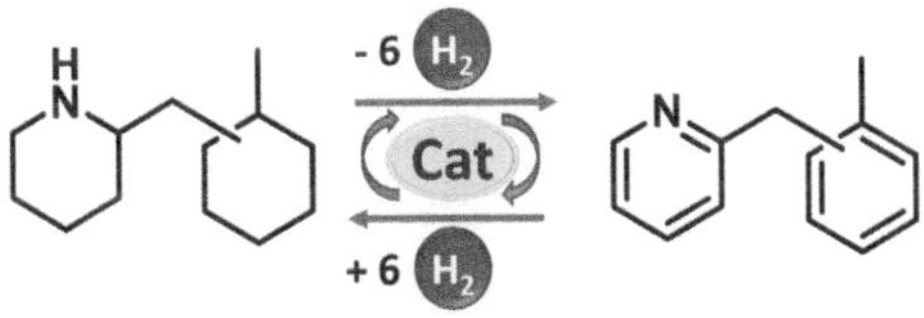

FIGURE 3.10 A diagrammatic representation of the LOHC system involving perhydro-2-(*n*-methylbenzyl pyridine) and 2-(*n*-methylbenzyl pyridine) (Rao and Yoon, 2020).

catalysts that were either synthesized or commercially available, it was determined that Ru/Al$_2$O$_3$ was a proficient catalyst for hydrogenation, while Pd/C was more efficient for dehydrogenation. MBP's low viscosity (5.9 cP at 25 °C) compared to other LOHCs such as dibenzyl toluene (DBT) and ethylene glycol and favorable copper strip corrosion test results make it an ideal option for transportation. Additionally, MBP has demonstrated stability and hydrogen storage effectiveness during three consecutive cycles. Table 3.6 shows the results of H12-MBP dehydrogenation at 250 °C and 4 hours of reaction time.

TABLE 3.6

The Dehydrogenation Performance of H12-MBP Evaluated Using GC Data after 4 hours at 250 °C (Rao and Yoon, 2020)

Catalyst	H12-MBP Conversion (mol%)	Selectivity to H0-MBP (mol%)	H$_2$ Yield (mol%)
Pd/Al$_2$O$_3$	65.3	63.3	53.3
Pd/2.0CCA	72.1	75.4	63.2
Pd/3.3CCA	80.6	76.4	71.1
Pd/4.5CCA	65.5	62.3	53.1
Pd/6.0CCA	60.1	29.2	38.8
Pd/C	59.0	78.2	52.6

3.4 HYDROGENATION/DEHYDROGENATION CATALYSTS FOR LOHC TECHNOLOGIES

3.4.1 Hydrogenation/Dehydrogenation Catalytic Systems Used in LOHC Technologies

Platinum group metal-based monometallic catalysts are being developed as potential LOHC systems. This is due to their superior activity, selectivity, and stability in dehydrogenation reactions (Usman, 2011). A specific investigation focused on catalyzing octahydroindole dehydrogenation. The research revealed that the Pt atoms in the crystallographic plane (111) on the catalytic surface preferred to absorb octahydroindole, indoline, and indole. Moreover, the dehydrogenation process proceeded through distinct pathways, resulting in hydrogen liberation (Ouma, Modisha and Bessarabov, 2019). In a fixed catalyst bed reactor, various platinum catalysts were tested, and it was discovered that a monometallic Pt catalyst displayed the greatest catalytic activity, selectivity, and stability in dehydrogenating methylcyclohexane (Alhumaidan et al., 2013). Additionally, bimetallic (bifunctional) catalysts that activate both hydrogenation and dehydrogenation reactions are being studied for LOHC systems. For instance, a Ru-Pd/Al$_2$O$_3$ bifunctional catalyst was created through direct impregnation. This catalyst exhibited significant catalytic activity, stability, and selectivity in the hydrogenation of N-propyl carbazole and the dehydrogenation of perhydro-N-propyl carbazole (Zhu et al., 2019). Among the studied bimetallic catalysts with Pd–Cu nanoparticles supported on reduced graphene oxide (rGO), the Pd1.2Cu/rGO catalyst demonstrated significantly higher catalytic activity and selectivity in the dehydrogenation of perhydro-N-ethyl carbazole. This result highlights the potential of Pd1.2Cu/rGO as a promising catalyst for the LOHC system. Bimetallic catalysts substitute noble metals for cheaper alternatives. Wet impregnation produced a bimetallic catalyst Pd–Ni/Al$_2$O3 (Pd:Ni = 1:1.5 wt%) that surpassed Pd/Al$_2$O$_3$ in activity during perhydropropyl carbazole dehydrogenation (volumetric dehydrogenation rate up to 285 h^{-1} compared to 176 h^{-1}) while retaining high stability (Wang et al., 2019). A low-cost Pd–Me/SiO$_2$ bimetallic catalyst series was also proposed for the N-ethyl carbazole/dodecahedron-N-ethyl carbazole LOHC system, where Me = Cu, Ni. Among the investigated Pd–Ni bimetallic catalysts supported on SiO$_2$, the Pd3–Ni1/SiO$_2$ catalyst exhibited the highest catalytic activity, achieving complete conversion of N-ethyl carbazole, high selectivity of 91.1% to N-ethyl carbazole, and a hydrogen storage capacity of 5.63 wt% (at 190 °C, 101.325 kPa, 8 hours). This catalyst's activity (in TOF units) increased by 42.4% compared to the Pd/SiO$_2$ monometallic catalyst (Chen et al., 2020). Under microwave activation, the activity of Pd/TiO$_2$ catalysts in the dehydrogenation of perhydro-N-ethyl carbazole can be enhanced by introducing a second metal as a promoter to create bimetallic catalysts Pd–Ru/TiO$_2$ and Pd–Pt/TiO$_2$ (Kustov, Tarasov and Kirichenko, 2017). Research on the hydrogenation of N-ethyl carbazole in the LOHC system utilizing Ru–Ni catalysts supported by various modifications of TiO$_2$ (rutile, anatase) revealed that TiO$_2$'s crystal structure plays a crucial role in hydrogenation. After evaluating various titanium supports, it was found that the commercial support P25 comprises of around 1:4 ratio of anatase and rutile phases, exhibiting the highest effectiveness. This result is significant as it demonstrates the potential of P25 support as a suitable material for catalytic applications, particularly in LOHCs (Yu et al., 2020). A study investigating the degree of dehydrogenation of perhydrodibenzyltoluene in LOHC system found that the presence of 1 wt% Pt/Al_2O_3 catalyst significantly increased the degree of dehydrogenation. There was an increase in the extent of dehydrogenation from 40% to 90% with a temperature rise from 290 °C to 320 °C in mono- and multi-metallic platinum and palladium catalysts (Modisha et al., 2019). The increase in temperature resulted in a higher rate of hydrogen formation in the LOHC system. However, it also decreased hydrogen purity due to by-product formation. The degree of dehydrogenation at 320 °C was 96% with 2 wt% Pt/Al_2O_3 catalyst, while with Pd, 1 wt% Pt, and Pt–Pd catalysts, it was 11%, 82%, and 6%, respectively, which aligns with DFT calculations. The results suggest that Pd is unsuitable for the dehydrogenation of perhydrodibenzyltoluene, whether used in a monometallic or bimetallic system. The dehydrogenation process was observed to follow a first-order reaction kinetics, with the activation energies calculated to be 205 kJ mol^{-1} for 1 wt% Pt/Al_2O_3, 84 kJ mol^{-1} for 1 wt% Pd/Al_2O_3, and 66 kJ mol^{-2} for 1:1 wt% $Pt–Pd/Al_2O_3$. A study investigated a one-pot hydrogen storage system and found that a mixture of methanol and 1,2-diamine in the presence of a bifunctional catalyst, the ruthenium complex $[Ru\{HN(CH_2CH_2Pi\ Pr2)_2\}H(CO)Cl]$, could undergo reversible dehydrogenation/hydrogenation reactions to release H_2. The study showed that the system had a theoretically attainable hydrogen storage capacity of 6.6 wt% (Kothandaraman et al., 2017). Fe and Ir organic complexes serve as catalysts for the reversible hydrogenation reaction of alcohols, resulting in the production of carbonyl compounds (Bernskoetter and Hazari, 2018). According to Yu et al. (2021), catalysts utilizing pincer-type ligands with manganese and iron have demonstrated high efficiency in hydrogenation and dehydrogenation processes. Additionally, these catalysts can undergo reversible reactions for hydrogen uptake and release, making them suitable for applications such as hydrogen storage and fuel cells in different industries. These catalysts show potential for rechargeable LOHC systems (Yu et al., 2021). It has been revealed that the conventional $LaNi_5$ alloy with $LaNi_{5+x}$ particles, measuring around 100 nm in size, exhibits excellent efficiency and stability in facilitating the reversible binding and release of hydrogen in N-ethyl carbazole-based systems (Peng et al., 2019).

Catalytic processes are typically employed for the addition and removal of hydrogen from a molecule of LOHCs during hydrogen storage and release, with catalysts that are supported by metals from the platinum group being commonly used (Goebel et al., 1984; Baykara and Bilgen, 1989; Teichmann et al., 2012). The efficiency of reversible hydrogenation/dehydrogenation cycles varies, with hydrogenation often achieving high yields of end products (98%). Carrying out the dehydrogenation process, specifically the hydrogen evolution stage, is a challenging task when attempting to achieve high conversion rates under unfavorable conditions. This is largely due to the difficulty of developing dehydrogenation catalysts that can maintain high activity and stability during the process. The selection of base metal supports and catalyst preparation techniques is crucial to optimize dehydrogenation performance. In the industry, platinum or palladium catalysts supported on different materials, such as activated carbon, Al_2O_3, or SiO_2, are commonly used to dehydrate aromatic

TABLE 3.7

Characteristics of Hydrogen Evolution on Dehydrogenation Catalysts (Hamayun et al., 2020)

	Catalyst		
Indicator	**Pt/γ-Al$_2$O$_3$**	**Pt/TiO$_2$ rutile anatase**	**Pt/TiO$_2$ anatase**
Substrate structure	γ-Al$_2$O$_3$ 100%	Anatase 86% Rutile 14%	Anatase 100%
Pt content (wt%)	0.9 ± 0.1	0.9 ± 0.1	0.9 ± 0.1
Hydrogen evolution in 45 min (% of theoretical maximum)	55 ± 4	65 ± 5	32 ± 5

compounds. In addition, various catalysts were evaluated for their efficacy in dehydrogenating perhydrodibenzyltoluene, with Pt/C and Pt/Al$_2$O$_3$ catalysts demonstrating the highest activity, while Pd/C and Pd/Al$_2$O$_3$ catalysts exhibited lower activity. The Pt/Al$_2$O$_3$ catalyst led to over 95% conversion of perhydrodibenzyltoluene in 3.5 hours at 290 °C, with the maximum rate of hydrogen production observed at 310 °C (Acar and Dincer, 2016). Activated carbon-based carriers are cost-effective but lack mechanical strength. The use of metal oxide-supported catalysts, specifically those containing titanium and lanthanum, resulted in greater efficacy in the dehydrogenation of methylcyclohexane compared to catalysts supported on Al$_2$O$_3$. Agrafiotis et al. reported that Pt/TiO$_2$ra catalyst was more active than Pt/TiO$_2$ana or Pt/Al$_2$O$_3$, with Pt/TiO$_2$ra approaching that of Pt/C on activated carbon (Agrafiotis et al., 2014). The study established that the catalyst activity differences were unrelated to carriers' surface area or particle size. However, platinum particles' characteristics, such as size and Pt(II) content, significantly impacted their activity. Table 3.7 illustrates hydrogen evolution characteristics using various studied dehydrogenation catalysts (Hamayun et al., 2020).

Several studies state that in the removal of hydrogen from methylcyclohexane, conventional catalysts such as Pt/TiO$_2$, Pt/Al$_2$O$_3$, Pt/V$_2$O$_5$, Pt/Y$_2$O$_3$, and Pt–Re/Al$_2$O$_3$ are commonly used. In addition, various catalyst supports, including perovskite, La$_2$O$_3$, ZrO$_2$, CeO$_2$, TiO$_2$, and MnO$_2$, have been investigated in several studies. Among them, La$_2$O$_3$ has been found to have the most profound impact on the activity of the catalyst. To further enhance the catalyst activity, several promoters, such as Mo, Re, Cu, and Mn, have been incorporated (Eblagon et al., 2010; Teichmann et al., 2011). Various catalysts have been investigated for the dehydrogenation of methylcyclohexane, including Pt/Al$_2$O$_3$ (Andor et al., 2010) and Pt/Ce–Mg–Al–OPt/Al$_2$O$_3$ (Hernandez-Santoyo and Sanchez-Cifuentes, 2010). The studies focused on how reaction conditions affect side product production and also the hydrogenation of polyaromatic compounds that vary in their level of condensation, such as benzene, biphenyl, and diphenylbenzene. In addition, the resulting polycyclic naphthenes that can be dehydrogenated afterward (cyclohexane, bi-cyclohexane, and perhydrodiphenylbenzene) were investigated in the LOHC system using a platinum catalyst on the original subunit support (Energiespeicher, 2009). The catalytic dehydrogenation of every substrate was examined at a space feed rate of approximately 1 h^{-1} and a temperature range of 260–340 °C, emphasizing the effect of molecular structure on conversion in the dehydrogenation reaction and the impact of isomerization on the process kinetics. The efficacy of isomeric terphenyls as LOHCs has been investigated, and it has been demonstrated that the catalytic system of 5% Pt/SiO_2 is highly effective and stable in both hydrogenation (at 160°C) and dehydrogenation (at 320°C) reactions of terphenyls (Neupert et al., no date). von Wild, Freymann and Zenner (2008) have reported on the potential of mesoporous MoO_3 as catalyst support for N-ethyl carbazole hydrogenation. The study found that at a temperature of 220 °C for 6 hours, a significant amount of tetrahydro-N-ethyl carbazole and perhydro-N-ethyl carbazole were produced, resulting in hydrogen absorption of 0.97 wt%. These results indicate that mesoporous MoO_3 may act as a catalyst for hydrogenation reactions. The catalytic activity of the 0.5 wt% Pd/MoO_3 catalyst was superior to that of conventional catalysts like 0.5 wt% Pd/Al_2O_3 and 0.5 wt% Ru/Al_2O_3, particularly in converting octahedron-N-ethyl carbazole to perhydro-N-ethyl carbazole. The method of hydrogenation utilizing heterogeneous catalysts comprised of noble metals supported on MoO_3 differs (von Wild, Freymann and Zenner, 2008) from that of conventional heterogeneous hydrogenation catalysts. Added minute amounts of Pd to the surface of MoO_3 speeds up the breaking of the H-H bond and augments the hydrogen concentration in H_xMoO_3, which consequently amplifies the catalytic efficacy of the catalyst supported on MoO_3. Moreover, elevating the temperature also enhances the hydrogen concentration in $HxMoO_3$ (von Wild, Freymann and Zenner, 2008). The researchers have synthesized a catalyst, $Ru-Pd/Al_2O_3$, which exhibits remarkable performance in hydrogenating N-propyl carbazole and dehydrogenating perhydro-N-propyl carbazole, demonstrating high selectivity and stability (von Wild et al., 2010). This newly developed catalyst performs better than the commercial Pd/Al_2O_3 catalyst, which is attributed to a synergistic effect between the ruthenium and palladium nanoparticles, leading to enhanced activity. Even after three hydrogenation and dehydrogenation cycles, the catalyst continued to function. The authors suggest this catalyst can be utilized in industrial hydrogen storage using LOHCs.

Metal complex catalysts with pincer ligands have shown promising results in hydrogenation and dehydrogenation reactions. One example is developing a chemical hydrogen storage system using a soluble ruthenium complex with an organic pincer ligand, which achieved a theoretical mass capacity of 6.52 wt%, a high value for LOHCs. The dehydrogenation yield was close to 100% equilibrium, and the system was rechargeable. Additionally, metal complex catalysts with pincer ligands have enabled the use of base metals like iron and manganese in reversible hydrogenation/dehydrogenation reactions, which were previously impossible due to their low activity. This has allowed for replacing expensive noble metal catalysts with more affordable base metal catalytic systems (Taube et al., 1983; Klvana et al., 1988; Itoh et al., 2000). Chiyoda Corporation, a Japanese corporation, has developed the SPERA hydrogen system, designed for hydrogen mass storage using a toluene/methylcyclohexane LOHC pair. The catalyst used in this system is a partially sulfided nanosized Pt cluster on Al_2O_3. During laboratory tests involving the dehydrogenation of methylcyclohexane, the catalyst demonstrated a lifetime of more than 10,000 hours, with a conversion rate of methylcyclohexane exceeding 95% and a selectivity for toluene of over 99%. It is suggested that sulfur in this catalyst prevents methylcyclohexane decomposition in the platinum cluster. Platinum catalysts can be more effective in hydrogen evolution from LOHC systems by sulfidation, according to Kariya et al. (2002). They modified the preparation procedure for Pt/Al_2O_3 catalysts by incorporating sulfur-containing additives, resulting in increased catalyst activity and decreased by-products. The modification process with sulfur was investigated using IR spectroscopy, which indicated that sulfur compounds selectively blocked Pt atom sites with small coordination defects and were responsible for side reactions. Moreover, the addition of sulfur affected the electronic environment of Pt atoms, promoting the aromatic dehydrogenation products' desorption (Kariya, Fukuoka and Ichikawa, 2002). The Russian Academy of Sciences' Institute of Problems in Chemical Physics focuses on finding efficient catalysts for aromatic chemical hydrogenation. These catalysts are based on platinum (Pt) and palladium (Pd) and are supported by materials such as activated carbon, Al_2O_3, and glass fabrics. Using glass fabrics as support allows for a controlled degree of catalyst permeability, which is useful for continuous hydrogenation of aromatic LOHCs. These catalysts hydrogenate exhaustively under mild conditions.

3.4.2 ACTIVE METALS IN CATALYSTS

Developing cost-effective catalytic systems with high activity, selectivity, stability, and long operating life for hydrogenation/dehydrogenation in modern commercial LOHC systems is a significant challenge (Brückner et al., 2014; Bourane et al., 2016). Hydrogen chemically binds to LOHCs in an exothermic process during hydrogenation (at pressures of 5–7 MPa and temperatures of 100–250 °C), while hydrogen release in dehydrogenation is an endothermic process (at low pressures of about 0.1 MPa and high temperatures of 270–500 °C). As a result, hydrogenation and dehydrogenation conditions differ. Although catalysts that are active in both reactions are available, the thermodynamic characteristics of the chemical bonds must be activated, specifically, C=O/C=N/H–H and C–H/O–H/N–H bonds, leading to different reaction rates and conditions (Brückner et al., 2014; Bourane et al., 2016). The main focus of interest lies in the interaction between the catalysts utilized and the full-cycle LOHC systems rather than solely on the hydrogenation/dehydrogenation catalysts. Technology-wise, hydrogenating and dehydrogenating in a single vessel using the same catalyst or catalyst system is appealing. Furthermore, catalyst effectiveness is closely linked to the storage medium, which refers to the LOHCs employed. As a result, these catalysts are usually optimized for use with a particular set of carriers. However, the high cost and scarcity of noble metals such as platinum, palladium, and ruthenium limit their widespread use in large-scale LOHC systems. Therefore, researchers have also investigated using non-noble metal catalysts, such as iron-based or nickel-based catalysts, which are more abundant and less expensive. These alternative catalysts have shown promising results in LOHC hydrogenation/dehydrogenation reactions. However, they often suffer lower activity, selectivity, and stability than their noble metal counterparts. To address these challenges, researchers are exploring new strategies such as developing bimetallic catalysts, modifying catalyst support materials, and optimizing the reaction conditions to improve the performance of non-noble metal catalysts. New catalytic materials, such as metal-organic frameworks (MOFs), have also been investigated for potential use in LOHC systems. MOFs offer high surface area, tunable pore sizes, and excellent stability, which makes them attractive candidates for catalyst support materials. Developing effective and affordable catalytic systems for LOHCs is crucial for adopting these systems as sustainable energy carriers. Research efforts focus on improving catalyst activity, selectivity, and stability, reducing costs, and increasing scalability for commercial applications.

3.4.3 Effect of Catalyst Support

Other studies have investigated using non-oxide supports, such as carbon-based materials, for LOHC hydrogenation/dehydrogenation catalysts. Carbon nanofibers (CNFs) have been explored as support for palladium catalysts in the dehydrogenation of perhydrocarbazole, with a conversion of 95% achieved at 200 °C and 3 MPa hydrogen pressure (Yang et al., 2020). Another study used carbon nanotubes (CNTs) as a support for platinum catalysts in the dehydrogenation of perhydrodibenzyltoluene, achieving a conversion of 78.6% at 230 °C and atmospheric pressure (Wang et al., 2017). In addition to support materials, promoters have been explored to improve LOHC hydrogenation/dehydrogenation catalyst performance. For example, adding bismuth as a promoter to a platinum catalyst supported increased activity and stability of carbon in perhydrocarbazole dehydrogenation (Palkovits, Artz and Chen, 2018). Other promoters, such as tin and zinc, have also been investigated (Kalenchuk et al., 2018; Shi et al., 2020). These studies demonstrate the importance of selecting appropriate support for heterogeneous catalysts used in LOHC systems. The surface properties, pore size distribution, and acid/base nature of the support can significantly affect the catalytic activity and selectivity of the catalyst. Additionally, the interaction between the catalyst and support can affect the metal's dispersity and the catalyst's ability to absorb reagents. The development of novel catalysts, such as the dual-action Pd/Al$_2$O$_3$-YH3 catalyst and Pd/rGO catalyst, shows promise for improving the efficiency and reducing the cost of LOHC systems. For example, the Pd/Al$_2$O$_3$-YH3 catalyst achieves reversible hydrogen storage at low temperatures and short reaction times. The Pd/rGO catalyst shows significantly increased activity with a lower noble metal content than commercial catalysts (Wang et al., 2017; Palkovits, Artz and Chen, 2018; Yang et al., 2020). Overall, selecting appropriate catalyst support is crucial to optimizing LOHC systems' performance, and continued research and development in this area is imperative for advancing LOHCs as a sustainable energy storage solution. According to the study, extensively distributed Pd nanoparticles immobilized in MIL-101 (Cr) organometallic support outperformed the known commercial catalyst, 5 wt% Pd/Al$_2$O$_3$ (Shi et al., 2020), in the catalytic dehydrogenation of perhydro-N-propyl carbazole. The MIL-101 (Cr) substrate displayed a stable structure, with a high surface area specific to its composition, a sizable volume of pores, and a pore diameter ranging from 2.9 to 3.4 nm. On the other hand, when the catalytic effectiveness of several metal catalysts in the dehydrogenation process of a LOHC system employing perhydrodibenzyltoluene/dibenzyl toluene was compared, it was observed that the Pt/TiO$_2$ catalyst performed the most effectively. Hydroxyl groups and oxygen vacancies were induced on the Al$_2$O$_3$ substrate surface through conventional plasma treatment. New studies on the dehydrogenation of perhydro-N-ethyl carbazole using Pt/TiO$_2$ catalysts have revealed a significant electronic interaction between the Pt catalyst and the TiO$_2$ support. This interaction promotes the formation of active sites by causing the Pt nanoparticles deposited on

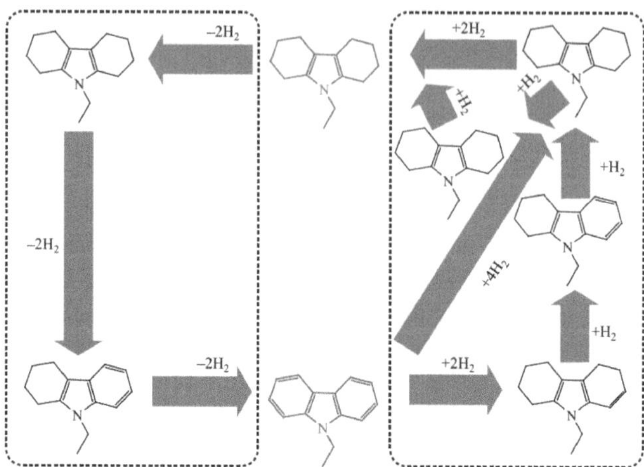

FIGURE 3.11 A diagram illustrating the reactions of hydrogenation (on the right) and dehydrogenation (on the left) in the LOHC system of N-ethyl carbazole/perhydro-N-ethyl carbazole (Zhou et al., 2020).

the TiO_2 surface to agglomerate heterogeneously. At a 1.0 wt% Pt concentration, the Pt/TiO_2 catalyst demonstrated the most effective catalytic activity in dehydrogenating perhydro-N-ethyl carbazole. When the catalyst was present, the reaction released 5.75 wt% hydrogen at 190 °C (Fei et al., 2017).

According to the hypothesized mechanism, the dehydrogenation reaction is structure-sensitive due to strong contacts between Pt and the TiO_2 substrate, which speeds up the limiting step and the overall dehydrogenation process. Figure 3.11 shows the hydrogenation and the dehydrogenation reactions in the N-ethyl carbazole/perhydro-N-ethyl carbazole LOHC system schematically (Shi et al., 2020; Zhou et al., 2020). During the N-ethyl carbazole hydrogenation study, Fei et al. discovered that Pd and Ru heterogeneous catalysts on mesoporous MoO_3 support had better activity than catalysts on standard Al_2O_3 support (Feng, Chen and Bai, 2020).

3.4.4 Alteration of the Existing Catalysts for LOHC Technologies

Efforts are underway to alter catalysts of hydrogenation and dehydrogenation. An improved process was developed to create Pt/Al_2O_3 catalysts using sulfur. The process involves sulfur absorption onto the platinum catalyst active sites, which causes undesirable side reactions. Obstructing these sites improved catalyst activity and selectivity (Modisha and Bessarabov, 2020). The efficiency of LOHC systems is determined by the balance of hydrogen storage capacity and operational stability throughout repeated charging and discharging cycles. As an illustration, Modisha and Bessarabov (Ouma, Modisha and Bessarabov, 2020) examined the complete hydrogenation of dibenzyl toluene, achieving a degree of up to 98%, utilizing a commercial nickel catalyst containing 50 wt% of Ni at a relatively low pressure of 1.5 MPa, which is lower than the commonly used range of 3–5 MPa. In contrast, perhydrodibenzyltoluene was dehydrogenated using a 1 wt% Pt/Al_2O_3 catalyst. The study aimed to determine the number of cycles the LOHC molecule could withstand and the level of by-products generated in each hydrogenation/dehydrogenation cycle. Experiments were conducted under challenging conditions to determine LOHC degradation periods. Specifically, methane production in the gaseous state exhibited an increase in the following sequence: benzene $< C_{21} <$ xylene $< C_{13}-C_{15} <$ toluene $<$ dibenzyl toluene. The experiments indicated that LOHC systems utilizing dibenzyl toluene were robust, as the level of by-products generated under challenging conditions was merely 7 mol%.

3.5 DESIGN OF HYDROGENATION AND DEHYDROGENATION REACTORS FOR LOHC

Typical reactors are suitable for hydrogenation and dehydrogenation processes within LOHC hydrogen storage systems. However, novel reactor designs may be required in certain situations to attain optimal system activity and stability and enhance fuel systems' overall energy efficiency. Research has demonstrated that storing hydrogen in a stationary LOHC system can be made more streamlined and effective by performing hydrogenation and dehydrogenation reactions within a single reactor and utilizing the same catalyst (Jorschick et al., 2017). It has been established that Pt/Al_2O_3 catalyzes the hydrogenation of dibenzyl toluene and the dehydrogenation of perhydrodibenzyltoluene in a temperature range of 290–310 °C in a reactor with an identical design to the pressure swing adsorption apparatus used for gas separation. Wunsch et al. proposed a microstructured multistage reactor with intermediate hydrogen separation during the dehydrogenation of perhydrodibenzyltoluene to intensify heat and mass transfer in the hydrogenation and dehydrogenation processes of LOHCs (Wunsch, Mohr and Pfeifer, 2018). The reactor consists of multiple stages, and each part contains a microstructured radial flow reactor that can separate hydrogen through PdAg membranes embedded in the reactor walls. Experiments were conducted on separating reaction streams to determine reaction kinetics and membrane separation capacity. $NiMo/Al_2O_3$ catalyst was used to hydrogenate phenanthrene (Zhang et al., 2016), a promising LOHC, in a 30 mL tubular microreactor. The study revealed that at 400 °C, the conversion rate could surpass 70%. Lee et al. investigated the feasibility of integrating a dehydrogenation reactor for LOHCs using dibenzyl toluene as an example of a set of low-temperature polymer-electrolyte fuel cells (Lee et al., 2020). Perhydrobenzyltoluene dehydrogenation was carried out in a reactor at 240–300 °C. To remove impurities from the hydrogen, an adsorbent (activated carbon) was used; however, even after purification, methane (100–1,400 ppm) was present in the hydrogen, and the fuel cell package capacity decreased from 39.4 to 39.0 W during dehydrogenation. When pure hydrogen was added, the stack's capacity was restored to its previous value of 39.4 W. The results indicate that the hydrogenation reactor and fuel cell package can operate together without energy loss, and using activated carbon for LOHC purification is an effective and affordable method with high efficiency. From a practical point of view, it is also interesting to carry out dehydrogenation reactions of LOHCs in membrane reactors with a hydrogen-permeable membrane. This can enhance the process efficiency while reducing the temperature levels required for the reaction (Didenko et al., 2019).

3.6 FEASIBILITY STUDIES ON THE EFFICIENCY OF LOHC STORAGE SYSTEMS

Several works are devoted to studying the technical and economic indicators of storage systems for LOHCs, some of which are analyzed in Teichmann, Arlt and Wasserscheid (2012). The report published by the U.S. Department of Energy in 2020 offers a comprehensive technical assessment of the practical usage of LOHCs in vehicles, including an examination of the expenses related to liquid organic hydrogen storage systems capable of storing up to 5.6 kg of hydrogen that can be used effectively. With the help of mathematical modeling methods, a technical and economic analysis of the possibility of practical use of a stationary energy storage system based on the LOHC system at an operating industrial plant was carried out (Eypasch et al., 2017). BMW successfully tested the proposed LOHC system in practice. Test results have shown that converting excess energy into heat is more cost-effective than storing energy using LOHCs. According to the findings, the LOHC system exhibited remarkable energy efficiency, with energy recycling rates of 69.17% during use and 88.74% when not in use. Wulf et al. conducted a study to compare the life cycles and costs of using various LOHC systems and transporting hydrogen in liquid form to evaluate the feasibility of practical applications of LOHCs as designed systems for the storage and transportation of hydrogen. It is crucial to assess the complete life cycle of LOHC operation, including environmental impact, depletion of natural resources, human health, and capital and operating costs throughout the hydrogen storage system's life cycle (Wulf and Zapp, 2018). According to the findings, none of the methods examined are ideal, and their efficiency should be evaluated case by case. Although transporting liquid hydrogen is costlier than using LOHCs, which require significant capital investments in hydrogen liquefaction equipment, environmentally, LOHCs were less favorable due to their high demand for heat during dehydrogenation and lower transport capacity than liquid hydrogen. A comparison is provided between the thermodynamic and engineering features of different approaches used for hydrogen storage on a large scale (Andersson and Grönkvist, 2019). While liquefied hydrogen storage remains a popular choice, some technologies utilizing chemical bonding, such as methanol, ammonia, and LOHCs, have shown the potential to compete with it regarding energy requirements during storage. However, these chemical bonding methods require significantly more thermal energy for hydrogen evolution. To address this, the thermal energy needed for these processes can be supplied by burning LOHCs, using third-party fuel, or part of the released hydrogen during industrial implementation.

3.7 CHALLENGES AND THE NECESSITY FOR ADDITIONAL INVESTIGATION

Although utilizing LOHCs as versatile energy carriers is an encouraging idea or framework for a future energy system that is environmentally sustainable, the creation of chemical procedures and technological apparatuses required for its execution is still in the preliminary stages. To implement the energy system as outlined, we consider the following aspects to be of the utmost importance: As mentioned earlier, N-ethyl carbazole serves as an initial example of a LOHC system that meets numerous necessary criteria and is consequently suitable as a test substance to evaluate the concept. However, for various reasons, it is not a flawless LOHC. For instance, in its pure, unloaded form, N-ethyl carbazole is solid at typical temperatures. Hence, resolving this problem would necessitate to implement technical remedies as blending or partially dehydrogenating the substance. Nevertheless, it is worth noting that some of these techniques might reduce the LOHC system's storage capacity. Any potential pair of substances considered for use in a LOHC system must thoroughly evaluate their toxicity and ecotoxicity profiles. Although current hydrocarbon-based fuels are generally considered harmful to the environment, a rigorous assessment of any LOHC system is necessary, given the considerable amount of waste generated if the concept were successfully implemented. The ideal LOHC system would be characterized by its storage density, toxicity, and cost. While large-scale aromatic hydrocarbon hydrogenation is a well-established process in various industries, developing effective onboard dehydrogenation devices is a novelty, particularly for use in the automotive sector. The automotive industry has stringent requirements, necessitating high-power-density reactors that can respond quickly to dynamic load curves. Furthermore, the number of precious metals in the catalyst used to achieve a specific hydrogen output will be a significant cost factor to consider. Furthermore, the number of precious metals in the catalyst required to achieve a specific hydrogen output will be a significant cost factor to consider. As previously mentioned, waste heat from an internal combustion engine can power the endothermic onboard dehydrogenation reaction. However, this approach may not be viable if a fuel cell is used instead since waste heat's temperature is typically insufficient. Moreover, these devices operate using high hydrogen purity to avoid damage. Fuel cells have emerged as a promising technological solution for achieving sustainable transportation in the future; it is crucial to develop solutions for integrating fuel cells with LOHC storage systems. Although it is still in the preliminary stages, the evaluation of the feasibility of utilizing the current fossil fuel infrastructure for the distribution of LOHCs appears to show potential. However, a comprehensive analysis of the necessary modifications, installations, and associated costs is still required.

3.8 CONCLUSION

This chapter discusses using LOHC as agents for the soft and more economical storage and transportation of hydrogen in fluid form. Several natural compound classes have been investigated as potential liquid hydrogen carriers, as shown in a literature review. These include cycloalkanes, polycyclic alkanes, hydrocarbons with heteroatoms, and ionic liquids. Researchers have studied different types of catalysts and reactors to achieve effective and selective dehydrogenation of these organic hydrogen carriers.

Studies have extensively examined representatives of the cyclohexane class, with methylcyclohexane considered the most promising candidate as a storage medium in LOHC systems. The suitability of these compounds has been assessed based on criteria such as their physical state (liquid in all states), toxicity, environmental safety, and ability to be reused. Heterogeneous catalysts that incorporate noble metals have demonstrated high levels of activity and selectivity and are widely used for organic compound dehydrogenation. During the dehydrogenation process of polycyclic alkanes and hydrocarbons containing heteroatoms, side reactions may occur, forming coke and partially dehydrogenated products. Various reactor types were studied to achieve high conversion rates and reduce side reactions. These included reactors with different configurations and proton exchange membranes. Utilizing LOHCs as hydrogen carriers provides the potential for high weight and volume hydrogen storage densities, low risk, and capital investment. Furthermore, existing infrastructure can be employed to transport them. Despite the technical, economic, and environmental advantages, hydrogen storage in LOHC systems has not been commercialized yet. The primary reasons are technical limitations that require significant energy to release hydrogen and dehydrogenation catalyst instability. Hydrogen storage and transportation in a liquid state offer several advantages as an alternative fuel, which could significantly contribute to energy supply. These advantages include unmatched environmental compatibility with ground transportation and aviation. Although the large-scale implementation of these applications may not occur initially, they provide an opportunity to showcase the use and safe handling of hydrogen as an energy carrier on a sufficient scale and can make a vital contribution toward developing more complex applications in the future. Although industrial LOHC technology implementation has begun, it is still in its early stages. Consequently, it is expected that this chapter will encourage scholars, engineers, and energy specialists to analyze their pilots in a way that can help verify the validity of some of the assumptions. There is still much work to be done at the molecular level to define and comprehend the chemical, technical, and technological aspects of using LOHCs as agents for hydrogen storage and transportation as an energy carrier. LOHC technology has been developed but is currently at a preliminary stage. This chapter encourages scholars, engineers, and energy specialists to conduct pilot studies validating some assumptions. Much research still needs to be carried out at the molecular level to characterize and understand the chemical, technical, and technological aspects of using LOHCs as agents for hydrogen storage and transportation as an energy carrier. Storing hydrogen as chemical compounds in the liquid phase offers many possibilities. However, to date, in practical applications, none of the storage materials that were investigated have exhibited all of the necessary performance characteristics. Therefore, developing upcoming technologies and materials is necessary to address the pressing issues of efficient hydrogen storage, which can only be accomplished through fundamental and engineering research results.

ACKNOWLEDGMENTS

This chapter has been supported by the Kazan Federal University Strategic Academic Leadership Program (PRIORITY-2030).

REFERENCES

Aakko-Saksa, P.T. et al. (2018) Liquid organic hydrogen carriers for transportation and storing of renewable energy – review and discussion, *Journal of Power Sources*, 396, pp. 803–823.

Acar, C. and Dincer, I. (2016) A review and evaluation of photoelectrode coating materials and methods for photoelectrochemical hydrogen production, *International Journal of Hydrogen Energy*, 41(19), pp. 7950–7959.

Agency, I.E. (2009) *World Energy Outlook*. OECD/IEA: Paris.

Agrafiotis, C. et al. (2014) Solar thermal reforming of methane feedstocks for hydrogen and syngas production-a review, *Renewable and Sustainable Energy Reviews*, 29, pp. 656–682.

Alhumaidan, F. et al. (2013) Hydrogen storage in liquid organic hydride: selectivity of MCH dehydrogenation over monometallic and bimetallic Pt catalysts, *International Journal of Hydrogen Energy*, 38(32), pp. 14010–14026.

Alhumaidan, F., Cresswell, D. and Garforth, A. (2011) Hydrogen storage in liquid organic hydride: producing hydrogen catalytically from methylcyclohexane, *Energy & Fuels*, 25(10), pp. 4217–4234.

Al-ShaikhAli, A.H. et al. (2015) Non-precious bimetallic catalysts for selective dehydrogenation of an organic chemical hydride system, *Chemical Communications*, 51(65), pp. 12931–12934.

Andersson, J. and Grönkvist, S. (2019) Large-scale storage of hydrogen, *International Journal of Hydrogen Energy*, 44(23), pp. 11901–11919.

Andor, M. et al. (2010) Negative strompreise und der vorrang erneuerbarer energien, *Zeitschrift für Energiewirtschaft*, 34(2), pp. 91–99.

Armaroli, N. and Balzani, V. (2011) The hydrogen issue, *Chemistry Sustainability Energy Materials*, 4, pp. 21–36.

Aslam, R. et al. (2016) Measurement of hydrogen solubility in potential liquid organic hydrogen carriers, *Journal of Chemical & Engineering Data*, 61(1), pp. 643–649.

Aziz, M., Oda, T. and Kashiwagi, T. (2019) Comparison of liquid hydrogen, methylcyclohexane and ammonia on energy efficiency and economy, *Energy Procedia*, 158, pp. 4086–4091.

Bachmann, P. et al. (2019) Dehydrogenation of the liquid organic hydrogen carrier system 2-methylindole/2-methylindoline/2-methyloctahydroindole on Pt (111), *The Journal of Chemical Physics*, 151(14), p. 144711.

Baykara, S.Z. and Bilgen, E. (1989) An overall assessment of hydrogen production by solar water thermolysis, *International Journal of Hydrogen Energy*, 14(12), pp. 881–891.

Bernskoetter, W.H. and Hazari, N. (2018) Hydrogenation and dehydrogenation reactions catalyzed by iron pincer compounds. *Pincer Compounds*. Elsevier, pp. 111–131. https://www.sciencedirect.com/science/article/abs/pii/B9780128129319000062

Bourane, A. et al. (2016) An overview of organic liquid phase hydrogen carriers, *International Journal of Hydrogen Energy*, 41(48), pp. 23075–23091.

Brayton, D.F. and Jensen, C.M. (2015) Dehydrogenation of pyrrolidine based liquid organic hydrogen carriers by an iridium pincer catalyst, an isothermal kinetic study, *International Journal of Hydrogen Energy*, 40(46), pp. 16266–16270.

Brückner, N. et al. (2014) Evaluation of industrially applied heat-transfer fluids as liquid organic hydrogen carrier systems, *ChemSusChem*, 7(1), pp. 229–235.

Bulgarin, A. et al. (2020) Purity of hydrogen released from the liquid organic hydrogen carrier compound perhydro dibenzyltoluene by catalytic dehydrogenation, *International Journal of Hydrogen Energy*, 45(1), pp. 712–720.

Campbell, P.G. et al. (2010) Hydrogen storage by Boron– nitrogen heterocycles: a simple route for spent fuel regeneration, *Journal of the American Chemical Society*, 132(10), pp. 3289–3291.

Chen, X. et al. (2020) Wet-impregnated bimetallic Pd-Ni catalysts with enhanced activity for dehydrogenation of perhydro-N-propylcarbazole, *International Journal of Hydrogen Energy*, 45(56), pp. 32168–32178.

Choi, I.Y. et al. (2016) Thermodynamic efficiencies of hydrogen storage processes using carbazole-based compounds, *International Journal of Hydrogen Energy*, 41(22), pp. 9367–9373.

Cooper, A. et al. (2010) Reversible liquid carriers for an integrated production, storage and delivery of hydrogen. In DOE Annual Merit Review Meeting, Washington, DC.

Crabtree, R.H. (2008) Hydrogen storage in liquid organic heterocycles, *Energy & Environmental Science*, 1(1), pp. 134–138.

Cui, Y. et al. (2008) The effect of substitution on the utility of piperidines and octahydroindoles for reversible hydrogen storage, *New Journal of Chemistry*, 32(6), pp. 1027–1037.

Didenko, L.P. et al. (2019) Steam reforming of methane and its mixtures with propane in a membrane reactor with industrial nickel catalyst and palladium-ruthenium foil, *Petroleum Chemistry*, 59, pp. 394–404.

Dong, Y. et al. (2015) Catalytic hydrogenation and dehydrogenation of N-ethylindole as a new heteroaromatic liquid organic hydrogen carrier, *International Journal of Hydrogen Energy*, 40(34), pp. 10918–10922.

Dong, Y. et al. (2019) Study on reversible hydrogen uptake and release of 1, 2-dimethylindole as a new liquid organic hydrogen carrier, *International Journal of Hydrogen Energy*, 44(10), pp. 4919–4929.

Durbin, D.J. and Malardier-Jugroot, C. (2013) Review of hydrogen storage techniques for on board vehicle applications, *International Journal of Hydrogen Energy*, 38, pp. 14595–14617.

Eblagon, K.M. and Tsang, S.C.E. (2014) Structure-reactivity relationship in catalytic hydrogenation of heterocyclic compounds over ruthenium black-Part A: effect of substitution of pyrrole ring and side chain in N-heterocycles, *Applied Catalysis B: Environmental*, 160, pp. 22–34.

Eblagon, K.M. et al. (2010) Hydrogenation of 9-ethylcarbazole as a prototype of a liquid hydrogen carrier, *International Journal of Hydrogen Energy*, 35(20), pp. 11609–11621.

Emelyanenko, V.N. et al. (2019) N-phenyl-carbazole as a potential liquid organic hydrogen carrier: thermochemical and computational study, *The Journal of Chemical Thermodynamics*, 132, pp. 122–128.

Energiespeicher, E.T.G.T.F. (2009) *Energiespeicher in Stromversorgungssystemen mit hohem Anteil erneuerbarer Energieträger-Bedeutung*. Stand der Technik, Handlungsbedarf: Bedeutung.

Eypasch, M. et al. (2017) Model-based techno-economic evaluation of an electricity storage system based on liquid organic hydrogen carriers, *Applied Energy*, 185, pp. 320–330.

Fei, S. et al. (2017) A study on the catalytic hydrogenation of N-ethylcarbazole on the mesoporous Pd/MoO$_3$ catalyst, *International Journal of Hydrogen Energy*, 42(41), pp. 25942–25950.

Feng, Z., Chen, X. and Bai, X. (2020) Catalytic dehydrogenation of liquid organic hydrogen carrier dodecahydro-N-ethylcarbazole over palladium catalysts supported on different supports, *Environmental Science and Pollution Research*, 27, pp. 36172–36185.

Geburtig, D. et al. (2016) Chemical utilization of hydrogen from fluctuating energy sources-catalytic transfer hydrogenation from charged liquid organic hydrogen carrier systems, *International Journal of Hydrogen Energy*, 41(2), pp. 1010–1017.

Gianotti, E. et al. (2018) High-purity hydrogen generation via dehydrogenation of organic carriers: a review on the catalytic process, *ACS Catalysis*, 8(5), pp. 4660–4680.

Goebel, E.D. et al. (1984) Geology, composition, isotopes of naturally occurring rich gas from wells near Junction City, Kans, *Oil & Gas Journal*, 82(19), pp. 215–222.

Hamayun, M.H. et al. (2020) Simulation study to investigate the effects of operational conditions on methylcyclohexane dehydrogenation for hydrogen production, *Energies*, 13(1), p. 206.

Haupt, A. and Müller, K. (2017) Integration of a LOHC storage into a heat-controlled CHP system, *Energy*, 118, pp. 1123–1130.

He, T., Pei, Q. and Chen, P. (2015) Liquid organic hydrogen carriers, *Journal of Energy Chemistry*, 24(5), pp. 587–594.

Hernandez-Santoyo, J. and Sanchez-Cifuentes, A. (2010) DENA-Netzstudie, Integration erneuerbarer Energien in die deutsche Stromversorgung im Zeitraum 2015–2020 mit Ausblick 2025, *Deutsche Energie-Agentur GmbH*, 35, pp. 139–153.

Hoisl, B. (2014) *Integration and Test of MOF/UML-based Domain-specific Modeling Languages*. WU Vienna University of Economics and Business: Wien, Austria.

Itoh, N. et al. (2000) Electrochemical coupling of benzene hydrogenation and water electrolysis, *Catalysis Today*, 56(1–3), pp. 307–314.

Jiang, N. et al. (2012) Effect of hydrogen spillover in decalin dehydrogenation over supported Pt catalysts, *Applied Catalysis A: General*, 425, pp. 62–67.

Jorschick, H. et al. (2017) Hydrogen storage using a hot pressure swing reactor, *Energy & Environmental Science*, 10(7), pp. 1652–1659.

Jorschick, H. et al. (2019) Hydrogenation of liquid organic hydrogen carrier systems using multicomponent gas mixtures, *International Journal of Hydrogen Energy*, 44(59), pp. 31172–31182.

Jorschick, H. et al. (2020) Benzyltoluene/dibenzyltoluene-based mixtures as suitable liquid organic hydrogen carrier systems for low temperature applications, *International Journal of Hydrogen Energy*, 45(29), pp. 14897–14906.

Kalenchuk, A.N. et al. (2018) Dehydrogenation of polycyclic naphthenes on a Pt/C catalyst for hydrogen storage in liquid organic hydrogen carriers, *Fuel Processing Technology*, 169, pp. 94–100.

Kariya, N., Fukuoka, A. and Ichikawa, M. (2002) Efficient evolution of hydrogen from liquid cycloalkanes over Pt-containing catalysts supported on active carbons under "wet-dry multiphase conditions", *Applied Catalysis A: General*, 233(1–2), pp. 91–102.

Kikuchi, Y. et al. (2019) Battery-assisted low-cost hydrogen production from solar energy: rational target setting for future technology systems, *International Journal of Hydrogen Energy*, 44(3), pp. 1451–1465.

Klvana, D. et al. (1988) Catalytic storage of hydrogen: Hydrogenation of toluene over a nickel/silica aerogel catalyst in integral flow conditions, *Applied Catalysis*, 42(1), pp. 121–130.

Kothandaraman, J. et al. (2017) Efficient reversible hydrogen carrier system based on amine reforming of methanol, *Journal of the American Chemical Society*, 139(7), pp. 2549–2552.

Kustov, L.M., Tarasov, A.L. and Kirichenko, O.A. (2017) Microwave-activated dehydrogenation of perhydro-N-ethylcarbazol over bimetallic Pd-M/TiO$_2$ catalysts as the second stage of hydrogen storage in liquid substrates, *International Journal of Hydrogen Energy*, 42(43), pp. 26723–26729.

Latapí, M., Davíðsdóttir, B. and Jóhannsdóttir, L. (2023) Drivers and barriers for the large-scale adoption of hydrogen fuel cells by Nordic shipping companies, *International Journal of Hydrogen Energy*, 48(15), pp. 6099–6119.

Lee, S. et al. (2020) Connected evaluation of polymer electrolyte membrane fuel cell with dehydrogenation reactor of liquid organic hydrogen carrier, *International Journal of Hydrogen Energy*, 45(24), pp. 13398–13405.

Leinweber, A. and Müller, K. (2018) Hydrogenation of the liquid organic hydrogen carrier compound monobenzyl toluene: Reaction pathway and kinetic effects, *Energy Technology*, 6(3), pp. 513–520.

Li, X. et al. (2014) Effects of carbon support on microwave-assisted catalytic dehydrogenation of decalin, *Carbon*, 67, pp. 775–783.

Li, L. et al. (2016) Hydrogen storage and release from a new promising liquid organic hydrogen storage carrier (LOHC): 2-methylindole, *International Journal of Hydrogen Energy*, 41(36), pp. 16129–16134.

Luo, W., Zakharov, L.N. and Liu, S.-Y. (2011) 1, 2-BN cyclohexane: synthesis, structure, dynamics, and reactivity, *Journal of the American Chemical Society*, 133(33), pp. 13006–13009.

Makepeace, J.W. et al. (2019) Reversible ammonia-based and liquid organic hydrogen carriers for high-density hydrogen storage: recent progress, *International Journal of Hydrogen Energy*, 44(15), pp. 7746–7767.

Markiewicz, M. et al. (2015) Environmental and health impact assessment of liquid organic hydrogen carrier (LOHC) systems-challenges and preliminary results, *Energy & Environmental Science*, 8(3), pp. 1035–1045.

Mazloomi, K. and Gomes, C. (2012) Hydrogen as an energy carrier: prospects and challenges, *Renewable and Sustainable Energy Reviews*, 16(5), pp. 3024–3033.

McQueen, S. et al. (2020) *Department of Energy Hydrogen Program Plan*. US Department of Energy (USDOE): Washington, DC.

Mehranfar, A., Izadyar, M. and Esmaeili, A.A. (2015) Hydrogen storage by N-ethylcarbazol as a new liquid organic hydrogen carrier: a DFT study on the mechanism, *International Journal of Hydrogen Energy*, 40(17), pp. 5797–5806.

Mizuno, Y. et al. (2016) Economic analysis on international hydrogen energy carrier supply chains, *Japan Society of Energy and Resources*, 38(3), pp. 11–17.

Modisha, P. and Bessarabov, D. (2020) Stress tolerance assessment of dibenzyltoluene-based liquid organic hydrogen carriers, *Sustainable Energy & Fuels*, 4(9), pp. 4662–4670.

Modisha, P. et al. (2019) Evaluation of catalyst activity for release of hydrogen from liquid organic hydrogen carriers, *International Journal of Hydrogen Energy*, 44(39), pp. 21926–21935.

Modisha, P.M. et al. (2019) The prospect of hydrogen storage using liquid organic hydrogen carriers, *Energy & Fuels*, 33(4), pp. 2778–2796.

Müller, K. (2019) Technologies for the storage of hydrogen part 1: hydrogen storage in the narrower sense, *Chemical and Biomolecular Engineering Reviews*, 6(3), pp. 72–80.

Müller, K. et al. (2012) Amine borane based hydrogen carriers: an evaluation, *Energy & Fuels*, 26(6), pp. 3691–3696.

Müller, K. et al. (2015) Liquid organic hydrogen carriers: thermophysical and thermochemical studies of benzyl-and dibenzyl-toluene derivatives, *Industrial & Engineering Chemistry Research*, 54(32), pp. 7967–7976.

Müller, K. et al. (2016) Experimental assessment of the degree of hydrogen loading for the dibenzyl toluene based LOHC system, *International Journal of Hydrogen Energy*, 41(47), pp. 22097–22103.

Müller, K., Stark, K., Emelyanenko, V.N., Varfolomeev, M.A., Zaitsau, D., Shoifet, E., Schick, C., Verevkin, S. and Arlt, W. (2015) Liquid organic hydrogen carriers: thermophysical and thermochemical studies of benzyl- and dibenzyl-toluene derivatives, *Industrial & Engineering Chemistry Research*, 54, pp. 7967–7976.

Müller, K., Völkl, J. and Arlt, W. (2013) Thermodynamic evaluation of potential organic hydrogen carriers, *Energy Technology*, 1(1), pp. 20–24.

Nafchi, F.M. et al. (2018) Performance assessment of a solar hydrogen and electricity production plant using high temperature PEM electrolyzer and energy storage, *International Journal of Hydrogen Energy*, 43(11), pp. 5820–5831.

Neupert, U. et al. (no date) Fraunhofer-Institut für Naturwissenschaftlich-Technische Fachaufsätze, *Energiespeicher* https://www.researchgate.net/profile/Ulrik-Neupert-2/publication/26920440_Energiespeicher_Technische_Grundlagen_und_energiewirtschaftliches_Potenzial/links/65773cd5ea5f7f02055f97d1/Energiespeicher-Technische-Grundlagen-und-energiewirtschaftliches-Potenzial.pdf

Niermann, M. et al. (2019) Liquid organic hydrogen carrier (LOHC)-assessment based on chemical and economic properties, *International Journal of Hydrogen Energy*, 44(13), pp. 6631–6654.

Niermann, M. et al. (2019) Liquid organic hydrogen carriers (LOHCs)-techno-economic analysis of LOHCs in a defined process chain, *Energy & Environmental Science*, 12(1), pp. 290–307.

Ninomiya, W. et al. (2006) Dehydrogenation of tetralin on Pd/C and Te-Pd/C catalysts in the liquid-film state under distillation conditions, *Catalysis Letters*, 110, pp. 191–194.

Oh, J. et al. (2018) 2-(N-Methylbenzyl) pyridine: a potential liquid organic hydrogen carrier with Fast H2 release and stable activity in consecutive cycles, *ChemSusChem*, 11(4), pp. 661–665.

Okada, Y. et al. (2006) Development of dehydrogenation catalyst for hydrogen generation in organic chemical hydride method, *International Journal of Hydrogen Energy*, 31(10), pp. 1348–1356.

Okada, Y., Mikuriya, T. and Yasui, T.M. (2015) Large scale hydrogen energy storage transportation technology, SPERA system. *KEM Enjiniyaringu*, 60, pp. 187–193.

Ouma, C.N.M., Modisha, P.M. and Bessarabov, D. (2019) Catalytic dehydrogenation of the liquid organic hydrogen carrier octahydroindole on Pt (1 1 1) surface: Ab initio insights from density functional theory calculations, *Applied Surface Science*, 471, pp. 1034–1040.

Ouma, C.N.M., Modisha, P.M. and Bessarabov, D. (2020) Catalytic dehydrogenation onset of liquid organic hydrogen carrier, perhydro-dibenzyltoluene: the effect of Pd and Pt subsurface configurations, *Computational Materials Science*, 172, p. 109332.

Palkovits, R., Artz, J. and Chen, X. (2018) A study on the dehydrogenation of loaded liquid organic hydrogen carriers (LOHC) with heterogeneous catalysts, *Chemie Ingenieur Technik*, 90(9), p. 1171.

Peng, L. et al. (2019) Chimney effect of the interface in metal oxide/metal composite catalysts on the hydrogen evolution reaction, *Applied Catalysis B: Environmental*, 245, pp. 122–129.

Pez, G.P. et al. (2008) *Hydrogen Storage by Reversible Hydrogenation of PI-Conjugated Substrates*. Air Products and Chemicals, Inc.: Allentown, PA (United States).

Pieper, C. and Rubel, H. (2011) Electricity storage: Making large-scale adoption of wind and solar energies a reality, In: G. Mennillo (ed). *Balanced Growth: Finding Strategies for Sustainable Development*. Springer: Berlin. pp. 163–181. https://link.springer.com/chapter/10.1007/978-3-642-24653-1_11

Preuster, P., Papp, C. and Wasserscheid, P. (2017) Liquid organic hydrogen carriers (LOHCs): toward a hydrogen-free hydrogen economy, *Accounts of Chemical Research*, 50(1), pp. 74–85.

Rao, P.C. and Yoon, M. (2020) Potential liquid-organic hydrogen carrier (LOHC) systems: a review on recent progress, *Energies*, 13(22), p. 6040.

Reuß, M. et al. (2017) Seasonal storage and alternative carriers: a flexible hydrogen supply chain model, *Applied Energy*, 200, pp. 290–302.

Rivard, E., Trudeau, M. and Zaghib, K. (2019) Hydrogen storage for mobility: a review, *Materials*, 12(12), p. 1973.

Sartbaeva, A. et al. (2008) Hydrogen nexus in a sustainable energy future, *Energy and Environmental Science*, 1, pp. 79–85.

Scherer, G.W.H. and Newson, E. (1998) Analysis of the seasonal energy storage of hydrogen in liquid organic hydrides, *International Journal of Hydrogen Energy*, 23(1), pp. 19–25.

Schneider, M.J. (2015) Hydrogen storage and distribution via liquid organic carriers, In Bridging Renewable Electricity with Transportation Fuels Workshop; Brown Palace Hotel: Denver, CO, USA.

Shi, L. et al. (2019) Integration of hydrogenation and dehydrogenation based on dibenzyltoluene as liquid organic hydrogen energy carrier, *International Journal of Hydrogen Energy*, 44(11), pp. 5345–5354.

Shi, L. et al. (2020) Pt catalysts supported on H_2 and O_2 plasma-treated Al_2O_3 for hydrogenation and dehydrogenation of the liquid organic hydrogen carrier pair dibenzyltoluene and perhydrodibenzyltoluene, *ACS Catalysis*, 10(18), pp. 10661–10671.

Shukla, A., Karmakar, S. and Biniwale, R.B. (2012) Hydrogen delivery through liquid organic hydrides: considerations for a potential technology, *International Journal of Hydrogen Energy*, 37(4), pp. 3719–3726.

Sinfelt, J.H. (1977) Heterogeneous catalysis: some recent developments: some properties of metal catalysts and recent advances in alloys and bimetallic systems are discussed, *Science*, 195(4279), pp. 641–646.

Sinfelt, J.H. (2000) The turnover frequency of methylcyclohexane dehydrogenation to toluene on a Pt reforming catalyst, *Journal of Molecular Catalysis A: Chemical*, 163(1–2), pp. 123–128.

Sobota, M. et al. (2011) Dehydrogenation of dodecahydro-N-ethylcarbazole on Pd/Al2O3 model catalysts, *Chemistry-A European Journal*, 17(41), pp. 11542–11552.

Sotoodeh, F. and Smith, K.J. (2010) Kinetics of hydrogen uptake and release from heteroaromatic compounds for hydrogen storage, *Industrial & Engineering Chemistry Research*, 49(3), pp. 1018–1026.

Sotoodeh, F., Huber, B.J.M. and Smith, K.J. (2012) Dehydrogenation kinetics and catalysis of organic heteroaromatics for hydrogen storage, *International Journal of Hydrogen Energy*, 37(3), pp. 2715–2722.

Staffell, I. et al. (2019) The role of hydrogen and fuel cells in the global energy system, *Energy & Environmental Science*, 12(2), pp. 463–491.

Stark, K. et al. (2015) Liquid organic hydrogen carriers: thermophysical and thermochemical studies of carbazole partly and fully hydrogenated derivatives, *Industrial & Engineering Chemistry Research*, 54(32), pp. 7953–7966.

Stark, K. et al. (2016) Melting points of potential liquid organic hydrogen carrier systems consisting of N-alkylcarbazoles, *Journal of Chemical & Engineering Data*, 61(4), pp. 1441–1448.

Suttisawat, Y. et al. (2012) Microwave effect in the dehydrogenation of tetralin and decalin with a fixed-bed reactor, *International Journal of Hydrogen Energy*, 37(4), pp. 3242–3250.

Tarasov, B.P. et al. (2021) Metal hydride hydrogen storage and compression systems for energy storage technologies, *International Journal of Hydrogen Energy*, 46(25), pp. 13647–13657.

Taube, M. et al. (1983) A system of hydrogen-powered vehicles with liquid organic hydrides, *International Journal of Hydrogen Energy*, 8(3), pp. 213–225.

Teichmann, D. et al. (2011) A future energy supply based on liquid organic hydrogen carriers (LOHC), *Energy & Environmental Science*, 4(8), pp. 2767–2773.

Teichmann, D. et al. (2012) Energy storage in residential and commercial buildings via liquid organic hydrogen carriers (LOHC), *Energy & Environmental Science*, 5(10), pp. 9044–9054.

Teichmann, D., Arlt, W. and Wasserscheid, P. (2012) Liquid organic hydrogen carriers as an efficient vector for the transport and storage of renewable energy, *International Journal of Hydrogen Energy*, 37(23), pp. 18118–18132.

Usman, M.R. (2011) The catalytic dehydrogenation of methylcyclohexane over monometallic catalysts for on-board hydrogen storage, production, and utilization, *Energy Sources, Part A: Recovery, Utilization, and Environmental Effects*, 33(24), pp. 2231–2238.

Verevkin, S.P., Pimerzin, A.A. and Sun, L.-X. (2020) Liquid organic hydrogen carriers: hydrogen storage by di-phenyl ether derivatives: An experimental and theoretical study, *The Journal of Chemical Thermodynamics*, 144, p. 106057.

Verevkin, S.P., Siewert, R. and Pimerzin, A.A. (2020) Furfuryl alcohol as a potential liquid organic hydrogen carrier (LOHC): thermochemical and computational study, *Fuel*, 266, p. 117067.

Von Wild, J. et al. (2010) Liquid organic hydrogen carriers (LOHC): An auspicious alternative to conventional hydrogen storage technologies, In: D. Stolten and T. Grube (eds). 18th World Hydrogen Energy Conference. Forschungszentrum Jlich GmbH: Essen, Germany.

von Wild, J., Freymann, R. and Zenner, M. (2008) Potentiale von alternativen Wasserstoff-Speicherungstechnologien, *VDI-Berichte*. VDI Verlag GmbH: German.

Wang, H., Zhou, X. and Ouyang, M. (2016) Efficiency analysis of novel liquid organic hydrogen carrier technology and comparison with high pressure storage pathway, *International Journal of Hydrogen Energy*, 41(40), pp. 18062–18071.

Wang, B. et al. (2017) Palladium supported on reduced graphene oxide as a high-performance catalyst for the dehydrogenation of dodecahydro-N-ethylcarbazole, *Carbon*, 122, pp. 9–18.

Wang, B. et al. (2019) Component controlled synthesis of bimetallic PdCu nanoparticles supported on reduced graphene oxide for dehydrogenation of dodecahydro-N-ethylcarbazole, *Applied Catalysis B: Environmental*, 251, pp. 261–272.

Wijayanta, A.T. et al. (2019) Liquid hydrogen, methylcyclohexane, and ammonia as potential hydrogen storage: comparison review, *International Journal of Hydrogen Energy*, 44(29), pp. 15026–15044.

Wolf, E. (2015) Large-scale hydrogen energy storage, In: P.T. Moseley and J. Garche (eds). *Electrochemical Energy Storage For Renewable Sources And Grid Balancing*. Elsevier: Amsterdam, The Netherlands, pp. 129–142.

Wu, W. and Wang, Z.L. (2016) Piezotronics and piezo-phototronics for adaptive electronics and optoelectronics, *Nature Reviews Materials*, 1(7), pp. 1–17.

Wulf, C. and Zapp, P. (2018) Assessment of system variations for hydrogen transport by liquid organic hydrogen carriers, *International Journal of Hydrogen Energy*, 43(26), pp. 11884–11895.

Wunsch, A., Mohr, M. and Pfeifer, P. (2018) Intensified LOHC-dehydrogenation using multi-stage microstructures and Pd-based membranes, *Membranes*, 8(4), p. 112.

Yang, X. et al. (2020) A YH3 promoted palladium catalyst for reversible hydrogen storage of N-ethylcarbazole, *International Journal of Hydrogen Energy*, 45(58), pp. 33657–33662.

Yanxing, Z. et al. (2019) Thermodynamics analysis of hydrogen storage based on compressed gaseous hydrogen, liquid hydrogen and cryo-compressed hydrogen, *International Journal of Hydrogen Energy*, 44(31), pp. 16833–16840.

Yu, H. et al. (2020) Bimetallic Ru-Ni/TiO$_2$ catalysts for hydrogenation of N-ethylcarbazole: Role of TiO$_2$ crystal structure, *Journal of Energy Chemistry*, 40, pp. 188–195.

Yu, H. et al. (2021) LaNi5. 5 particles for reversible hydrogen storage in N-ethylcarbazole, *Nano Energy*, 80, p. 105476.

Zhang, D. et al. (2016) Catalytic hydrogenation of phenanthrene over NiMo/Al$_2$O$_3$ catalysts as hydrogen storage intermediate, *International Journal of Hydrogen Energy*, 41(27), pp. 11675–11681.

Zhao, H.Y., Oyama, S.T. and Naeemi, E.D. (2010) Hydrogen storage using heterocyclic compounds: the hydrogenation of 2-methylthiophene, *Catalysis Today*, 149(1–2), pp. 172–184.

Zhou, L. et al. (2020) Recent developments of effective catalysts for hydrogen storage technology using N-Ethylcarbazole, *Catalysts*, 10(6), p. 648.

Zhu, Q.-L. and Xu, Q. (2015) Liquid organic and inorganic chemical hydrides for high-capacity hydrogen storage, *Energy & Environmental Science*, 8(2), pp. 478–512.

Zhu, T. et al. (2019) A highly active bifunctional Ru-Pd catalyst for hydrogenation and dehydrogenation of liquid organic hydrogen carriers, *Journal of Catalysis*, 378, pp. 382–391.

4 Pressurized Gaseous Hydrogen Storage

Maneesh Kumar and Sachidananda Sen

4.1 INTRODUCTION

With an increase in energy demand and while countering the adverse impacts of CO_2 emissions, the utilization of H_2 is rapidly getting attention. At the same time, efforts are being exerted to reduce its cost. It is evident that the reduction in CO_2 emission can reduce global warming (Kumar et al., 2020, 2023), and hence the increase in global energy demand can be met by producing H_2 from solar radiation. It can also be produced from the splitting of H_2O molecules. There are various other processes through which the production of H_2 is done, viz. thermochemical process, photo-electrochemical process, photo-biological process, etc. These processes are followed by molecular H_2 synthesis. Table 4.1 shows the production of H_2 from various types of sources. Similarly, various other processes are used to store the H_2 in different forms, such as gaseous storage, cryogenic storage, solid storage system, electrochemical storage of hydrogen, etc. Many organizations support the research and developments in this area to produce H_2 more economically and through some low carbon emission pathways. As mentioned earlier, the most challenging task for H_2 production is its cost. Many countries are focusing on developing new technologies to produce H_2 in an economically viable way. In the U.S., the target is set by the Department of Energy (DOE) to reduce the H_2 production cost to \$2/kg by 2025 and \$1/kg by 2030 using low carbon emission approaches.

The sun and water can produce hydrogen through various processes (Shi and Yu, 2016). The oxidation of water requires input energy as this process is energetically unfavorable. However, efficient oxidation occurs with an oxygenic photosynthetic

TABLE 4.1
Production of Hydrogen from Various Sources

S. No	Fossil Resources	Biomass/Bio-Waste	H_2O Splitting
1.	Low-cost, large-scale H_2 production with carbon capture unit system (CCUS)	This includes biogas reforming, and fermentation of waste streams	Electrolyzer can be grid-tied or directly coupled with the renewables
2.	New options include by-product production, such as solid carbon	By-product benefits include clean water, electricity, and chemicals	New direct-water splitting technologies offer long-term options

DOI: 10.1201/9781003382553-6

process (Zhang et al., 2017a–c). The production of H_2 and O_2 can be achieved by splitting H_2O in the photosystem because of the photosynthetic capability of the microalgae (Yang et al., 2014). A reduced content of ferredoxin, nicotinamide adenine dinucleotide phosphate (NADPH), and adenosine triphosphate (ATP) is utilized by the green microalgae and the cyanobacteria to produce H_2 under anaerobic conditions (Fakhimi and Tavakoli, 2019). In this process, the electrons are realized from the H_2O through light energy, and the reduction of H^+ to H_2 occurs. Many pieces of literature also show the production of H_2 through direct photolysis. The addition of magnesium hydroxide $Mg(OH)_2$ can maintain the feasibility of algae. The presence of $Mg(OH)_2$ also maintains the anaerobic reaction environment of algae (Chen et al., 2020). Various other methods are available to produce H_2, such as the use of biomass, water stream, and through nonsterile substances (Chi et al., 2011). Apart from this, various waste feedstock, such as wastewater from food industries, biomass, organic waste, etc., can be used for the production of H_2 (Turhal et al., 2019). Various streams of waste materials used for the production of H_2 cause efficient and effective dark fermentation (Sharma et al., 2020a), and also nonsterile conditions are used for H_2 synthesis.

4.2 HYDROGEN DELIVERY

A feasible overall hydrogen infrastructure should be capable of delivering H_2 from the production site to the end-user for its utilization. These end-user facilities may include industries, electricity generation, or even fueling stations. The overall infrastructure of H_2 incorporates transportation pipelines, plants for the liquefaction process, trucks, storage facilities, etc. The technology used for H_2 delivery is currently in a mature stage. Several companies in the U.S. commercially deliver the bulk amount of hydrogen. However, the increase in H_2 demand globally needs a local expansion of the existing H_2 infrastructure. It also requires the inclusion of new and advanced technologies. These technologies may include the chemical carriers to transport H_2 at very-high-density fueling technologies. It is evident that to transport the bulk quantity, the H_2 should either be pressurized and delivered in the form of compressed gas or liquid. The site where the H_2 is being produced has an impact on the cost. For instance, if hydrogen is produced on a larger scale at a centrally located facility, the cost of production shall be less, but the cost of transportation may be high. Similarly, for the distributed small-scale H_2 production facilities, the cost of production shall be high, but the cost of delivery of H_2 is low.

4.2.1 Various Strategies Involved in Hydrogen Delivery

- Transportation of H_2 from the production site to the end-user through pipelines and roads.
- H_2 is transported in cryogenic liquid tankers or gaseous tube trailers.
- For the high-demand areas, pipelines are established.
- For the areas where the demand for H_2 is comparatively low, liquefaction plants, liquid tankers, and tube trailers are used.
- At the user end, the infrastructures for compression, storage, and contaminant detection technologies need to be established.

- For example, the site which is being utilized to dispense H_2 to medium as well as heavy-duty fuel cell vehicles uses compressed H_2 at 350–750 bar of pressure and is dispensed at 10 kg/min.

4.2.1.1 Delivery of Gaseous Hydrogen

Trucks and pipelines are the two most common ways to deliver gaseous hydrogen. The gaseous H_2 is produced at low pressure, which is approximately 20–30 bar, and hence should be compressed before transportation. Trucks that are used for the transportation of gaseous hydrogen are called tube trailers. A pressure of 180 bar or more is used to compress the gaseous hydrogen and then fill the long cylinders. These cylinders are stacked on tube trailers for transportation to longer distances. Pipelines are also used for the transportation of gaseous hydrogen. It is generally used for long-distance and large-volume transport. Most of the current hydrogen pipelines are installed at oil refiners. The primary reason behind this is the utilization of H_2 for petroleum upgrades.

4.2.1.1.1 Compression of Gaseous Hydrogen

Typically, the H_2 is produced at low pressures (~20 to 30 bar). It is then compressed for transportation purposes. Usually, centrifugal and positive displacement compressors (PDC) are used for gaseous H_2 compression. PDCs are either reciprocating or rotary type.

4.2.1.1.2 Reciprocating Compressors (RCs)

The compressors often use a linear drive motor to provide a back-and-forth motion. The compression of H_2 is done by reducing the volume of hydrogen. Some of the RCs are commonly used to get a very high compression ratio.

4.2.1.1.3 Rotary Compressors

In these compressors, the motion is provided through the rotation of gears, screws, or rollers. The compression of H_2 is a difficult task for PDCs because of the requirement of tolerance levels to prevent leakage.

4.2.1.1.4 Centrifugal Compressors

These types of compressors are usually incorporated for pipeline applications. It is due to their moderate compression ratio and high throughput. Hydrogen is compressed by a turbine which rotates at very high speeds. The tip speed of these compressors is very high (~3 times) as compared to natural gas compressors. There are certain alternatives to mechanical compression. These are electrochemical reactions (ECRs), ionic liquids, metal hydrates, etc. The ECR approach utilize a proton exchange membrane which is surrounded by the electrodes. An external supply is used to drive the dissociation process in H_2 at the anode and then recombine it at the cathode at high pressure shown in Figure 4.1. An electrochemical compressor is used for this purpose. The humid H_2 at a slightly higher pressure than the inlet pressure is moved toward the adsorption-desorption compressor, which further compresses the H_2 at 70 MPa for storage purposes.

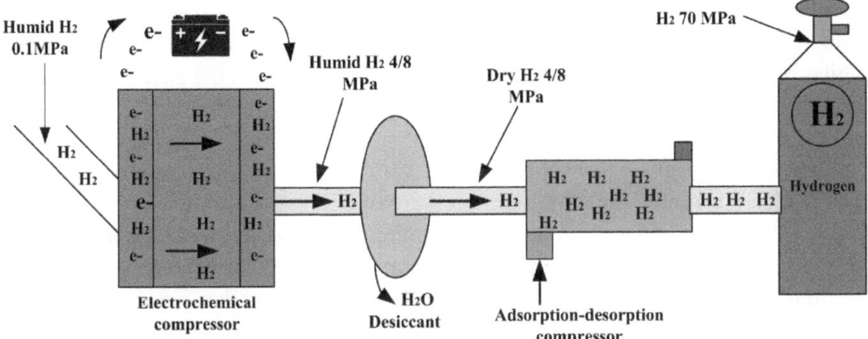

FIGURE 4.1 Hydrogen production through centrifugal compressors (Giuseppe et al., 2020).

In metal hydrate-based compressors, the metals are used, which form hydrides through exothermic reactions. This process later releases H_2 when heat is applied at high pressures.

4.2.1.1.5 Hydrogen Pipelines

The transportation of H_2 is also done through pipelines in a similar way as natural gas. As an estimation, approximately 1,600 miles of H_2 pipelines are in operation in the U.S. These pipelines are located within the large H_2 user areas such as oil refineries and chemical plants, more specifically near to Gulf Coast regions. The utilization of existing pipelines for the transportation of H_2 is a low-cost solution for delivering a large amount of H_2. The construction of a new pipeline requires a high capital investment and is, therefore, a big barrier to expanding the hydrogen-delivering pipelines.

The recent research mainly focuses on mitigating the technical issues associated with pipeline transmission. These include:

- The use of H_2 for steel embrittlement and welds for pipeline fabrications.
- Hydrogen suffusion and leakages.
- Exploring ways of cost reduction, reliability, and more appropriate and feasible H_2 compression.

The use of fiber-reinforced polymer pipelines is considered a more appropriate solution for the distribution of H_2 to usable sites. Moreover, the cost of installation of fiber-reinforced pipelines is less (~20%) as compared to steel-based pipelines as the welding requirements are less.

The rapid expansion of the H_2 delivery infrastructure is possible through the use of a certain part of the natural gas delivery infrastructure for H_2 accommodation. This can be made possible by a slight change in the pipeline of the natural gas for carrying a mixture of hydrogen and natural gas (approximately 15% of H_2 (up to)). However, change in the natural gas existing pipeline infrastructure for pure H_2 extraction may be a difficult task. The current research area covers both the aforementioned approaches.

4.2.1.1.6   Hydrogen Tube Trailer

Tube trailers are trucks that carry gaseous H_2 for longer distances. A high pressure, which is approximately 180 bar or higher, is used to compress the gaseous hydrogen. Thereafter, the compressed H_2 is kept in long cylinders which are installed on the trailers hauled by the truck.

This assembly gives the appearance of very long tubes and is therefore named tube trailer. Currently, the maximum pressure of 250 bar is allowed for tube trailers in the U.S. by the transport department regulations. However, some exemptions are provided for further higher pressures, which are up to 500 bar in some cases or may be slightly higher. The most commonly used type of tube trailer is made up of steel. These trailers can carry up to a 380 kg load on board, and their load-carrying capacity is limited by the weight of steel. The recent research on tube trailers includes the manufacturing of composite storage vessels that can carry 550–900 kg of H_2 per trailer. These kinds of trailers are being used for compressed natural gas transportation in some countries.

4.3   HYDROGEN STORAGE

Hydrogen storage is critical as it is considered a substitute for fossil fuels. The H_2 is also equally important to reduce carbon footprints and achieving sustainable development. Currently, H_2 is stored in various ways. It can be stored physically either in gaseous or liquid form. In gaseous form, the storage of H_2 requires approximately 350–700 bar pressure tanks. Whereas the liquid storage of H_2 needs cryogenic temperatures. H_2 can also be stored on solid surfaces either by adsorption or absorption processes. Hydrogen is the key contributor in various H_2 and fuel cell-related applications. These applications include portable power, stationary power, and also in transportation. H_2 has the highest energy per mass capacity among various types of fuels. The low ambient temperature density results in a small energy per unit volume. Therefore, the wide development of advanced storage technologies is undoubtedly an emerging area. Figure 4.2 shows a schematic of hydrogen storage. It can be seen from the figure that hydrogen can be stored either in physical-based storage or in material-based storage. The physical-based storage is divided into three types: compressed gas, cold or cryogenic compressed manner, or liquid form.

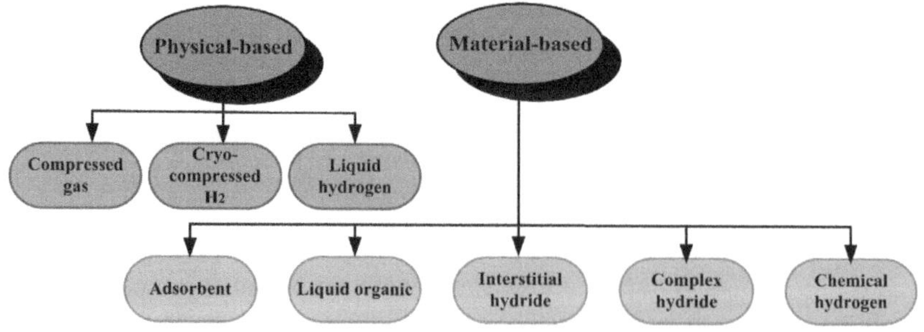

FIGURE 4.2   Typical hydrogen storage methods (Gómez-Gualdrón et al., 2016).

The material-based storage is categorized in five ways: adsorbent, liquid organic, interstitial hydride, complex hydride, and chemical hydrogen. The material-based research is currently focusing on metal hydrides, chemical hydrogen storage, and sorbent material. In this chapter, we will mainly focus on pressurized gaseous hydrogen storage.

4.3.1 Gaseous Storage

The gaseous H_2 storage is done in the form of a pressured gas. For this purpose, high pressure is required, which is about 35–79 MPa at low-temperature conditions. This technique has a drawback of 11%–12% of H_2 wastage from the total hydrogen content. Generally, steel or aluminum containers are used to store commercial purpose H_2. However, sometimes the leakage can be seen in the vessel on account of the unsubstantial weight of H_2 (Feng et al., 2020). The carbon fiber-reinforced vessels are more suitable for the gaseous H_2 as compared to steel or aluminum vessel. Geological areas such as large caves or caverns of salts are suitable for a large amount of H_2 storage for a longer time period (Nam et al., 2013). This process of storage has the benefit of preventing undesired H_2 leakage even under very high pressures.

4.3.2 Cryogenic Storage

Storing of H_2 in the liquid state is done by cooling it to 20 K. In cryogenic storage, a lesser H_2 volume is required. In this state, the energy density of H_2 is raised from 2.50 MJ/L to 8.0 MJ/L (Gómez-Gualdrón et al., 2016). The energy density of H_2 at a boiling point of 20 K is about 1.80 times larger than the energy density that is obtained by compression of 70.0 MPa, although this method has some disadvantages, such as the requirement of very high energy for storing H_2 (Hou et al., 2016).

4.3.3 H_2 Storage in Solid Form

4.3.3.1 Storage in Metal Hydrides

This type of storage is not suitable for large-scale H_2 storage because of various reasons. It is quite expensive compared to liquid or gaseous storage. Moreover, this technique requires a large volume and a substantial storage system as well. Other than these, the safety aspects are of much concern with this type of storage. However, it can be efficient where the characteristics of a reversible release and absorb process are required that possess solid-state materials (Gholami et al., 2016). H_2 has the property to be absorbed physically by the material such as fullerenes, carbon nano-tubes, metal-organic materials, and also activated carbon. These types of materials possess high reversibility, but at the same time, they are considered to be very marginally efficient since the capacity to store H_2 is somehow low at ambient conditions. Therefore, these materials require low temperatures for the storage of H_2 (Zhang et al., 2019).

The storage by metal hydrides is considered to be one of the promising and advanced technology while considering its safety features. This technique also possesses the need for low-pressure equipment, and also high reversibility during the process of hydrogenation and dehydrogenation (Sulaiman et al., 2016). The metal hydrides are being utilized for various applications like storage, sensors, making switch mirrors, and also for battery making.

However, a fast absorption takes place on the metal hydride surface; at the same time, the low aggregation may lead to surface instability. The synthesized composite exhibits comprehensive thermal as well as mechanical stability with the help of the prevention of aggregation of nanoparticles. These nanoparticles are considered to be coated with MgO while dealing with the H_2 absorption or desorption process (Zhang et al., 2017a–c). An excellent reversible chemical storage medium of H_2 characteristics is shown by intermetallic hydrides of AB_5 compositions, where "A" is rare earth metals, "B" is Ni or Co. Moreover, the absorption of H_2 in metal hydrides is considered to be an exothermic process. This process is required for an efficient and effective cooling of complete storage tanks. Although with few drawbacks, the H_2 storage in metal hydrides is still advantageous from the viewpoint of safety (Schlapbach et al., 2001). Also, the metal hydride concept benefits both higher storage capacity as well as improved thermal management.

4.3.3.2 Storage in Microporous Materials

H_2 has the ability to interact with the surface of so many microporous materials with the help of weak Van der Waals forces that form a monolayer. The energy required for these intersections is so low (of the order of 0.01–0.1 eV) (Züttel, 2003) that there are no chemical bonds developed between H_2 molecules. The content of the absorbed H_2 on a microporous material's surface is approximately proportional to the specific area of the microporous surface. It also depends on the distribution size and the average micropore width (Rzepka et al., 1998; Farha et al., 2010).

4.3.4 High-Pressure Gaseous H_2 Storage Vessel

It is observed that the density of H_2 increases with an increase in pressure exerted for the storage. The storage of high-pressure gaseous hydrogen (HPGH$_2$) is done with the help of an increase in the pressure to get a high density of storage. The pressure of 35.0–71.0 MPa (Von Helmolt et al., 2007) is considered to be ideal for an onboard H_2 system under the consideration of driving range, energy consumption by compression, infrastructural investment, and some other critical factors. For fast and direct refueling through the difference in pressure, H_2 at the refueling station is generally used at 40–75 MPa of pressure. The high-pressure H_2 storage vessel is considered to be the most critical equipment of the overall HPGH$_2$ system. With respect to various utilization scenarios, the HPGH$_2$ shall be bifurcated into three main types. These are stationary vessel, vehicular vessels, and bulk transportation vessels.

4.3.4.1 Stationary HPGH$_2$ Storage Vessel System

This type of storage vessel system generally carries H_2 in refueling establishments when cheaper and large-scale H_2 storage is needed. Currently, two types of stationary HPGH$_2$ vessels are in use. These are seamless H_2 storage vessels and multifunctional layered HPGH$_2$ stationary vessels. The seamless H_2 storage vessels are manufactured with the help of high-strength seamless tubes under various specifications provided by the boiler and pressure vessels code of the American Society of Mechanical Engineers (ASME). In order to store a large volume of H_2, a multi-vessel system with a number of valves and an interconnected pipe system are used. However, with a maximum pressure of 65 MPa, the volume reaches 0.41 m³ (Zheng et al., 2008). It means a large number of vessels are required to assemble for the large H_2 refueling station. There is a constraint on the volume of seamless H_2 storage vessels on account of the limitations of the maximum diameter size of walled tubes of the seamless tubes. To overcome this issue, steel can be used to expand the maximum limit of the working pressure. However, the effect of H_2 increases as the strength of steel or the H_2 pressure increases. And hence, the seamless H_2 storage vessels are quite more prone to H_2 embrittlement when the operating pressure increases (Xu et al., 2009; Zheng et al., 1991). To get rid of this, Zheng et al. proposed a stationary H_2 storage vessel consisting of a flat steel ribbon wound cylinder (FSRWC). This vessel is a multifunctional layered type of H_2 storage vessel and contains two double-layered hemispherical heads (Zheng et al., 1993; Yang et al., 2010). An FSRWC is manufactured in a multiple-shell system. These are an inner shell, a layered shell, and also a protective shell over the vessel. These types of vessels are free from size limitations either on the length of the cylinder or on the thickness of the shell. With this, it is feasible to make an H_2 storage vessel with higher and higher pressure. There are various materials that can be used to resolve the H_2 embrittlement issue. Several experimental validations are available which show that the major reason of failure of FSRWC is always leakage rather than any brittle fracture (Xu et al., 2009).

4.3.4.2 Vehicular HPSH$_2$ Storage Vessel

The requirement of a vehicular HPSH$_2$ storage vessel is obtained because of the onboard H_2 supply need. The U.S. DOE in 2003 suggested that the gravimetric and the volumetric densities of an onboard H_2 storage system must not be less than 6.0% weight of H_2 and 60.0 kgH$_2$/m³, respectively, to complete a driving range of 500 km in one-time fill. Thereafter a number of H_2-powered vehicles are tested and validated.

With consideration of the current H_2-powered vehicle bottom-line performance, variations in the basic architecture of the H_2 storage system were revised in 2009 (Yang et al., 2010). The amount of H_2 to be dosed by its weight (gravimetric density) must not be less than 5.50 wt% of hydrogen. The volumetric density should not be less than 40.0 kgH$_2$/m³. For achieving this, Type IV and Type V vessels are used. The Type IV vessels are fully wrapped vessels with a provision of the metallic liner, and Type V vessels are fully wrapped vessels with a non-metallic liner. The liner of these vessels is made up of high-density polyethylene, which is wrapped with carbon fibers. A new Type IV lightweight high-pressure gaseous H_2 storage vessel is manufactured using QUANTUM technologies. The developed vessel was named "TriShield" with a maximum operating pressure of 35.0 MPa in 2000. Similarly, in 2002, a Type IV H_2 storage vessel was developed named "Tuff-shell" with 70 MPa of working pressure (Zheng et al., 2004). In recent years, the use of QUANTUM has increased rapidly and now it is possible to create a 70 MPa Type IV H_2 storage vessel having a volume of 129 L that is capable to store about 5 kg of H_2. The Type IV vessel is of high composite material in it and also has high gravimetric storage density, although the compatibility and the robustness of polymeric liner for an overall life span of the $HPGH_2$ storage facility should be validated under various system conditions. Various researchers from the Commission for Atomic Energy, France (CEA) and Toyota Motors are in the way of developing innovative and suitable liners with the help of polyamide and polyurethane as primary components (Guidebook, 2009).

4.3.4.3 $HPGH_2$ Bulk Transportation Storage Vessel

This type of vessel is generally used to transport the H_2 from its origin to the end-user or to the H_2 refueling centers. The bulk transportation of highly pressurized H_2 consists of a large number of seamless high-pressure gaseous vessels in which the pressure is maintained between 16.0 and 20.0 MPa. In a single trip, the weight of H_2 is not more than 280 kg (Zheng et al., 2006). The transportation of cryogenic H_2 via rails is more economical as compared to tube trailers via roads. However, an almost negligible amount of H_2 is transported via rails because of the lack of timely scheduling and transport of H_2. Other reasons are lack of rail car capacity to handle the cryogenic H_2.

4.4 CHALLENGES IN H_2 STORAGE

The high-density H_2 is a challenge for portable as well as stationary applications. This is also a challenge for transportation applications. At present, most of the typical storage systems require a large-volume system that can store the H_2 in its gaseous form. The fuel cell-powered vehicles have an appropriate quantity of H_2 to power a vehicle for a driving range of more than 300 miles. Hydrogen storage is a significant challenge in the development of hydrogen-based energy systems. Hydrogen is a very light and volatile gas, which makes it difficult to store and transport. There are several methods of hydrogen storage, including compressed gas storage, liquid hydrogen storage, and solid-state storage, but each has its own set of challenges.

For example, compressed gas storage requires high-pressure tanks, which can be heavy and expensive, while liquid hydrogen storage requires extremely low temperatures, which can be difficult to maintain. Solid-state storage, on the other hand, is still in the early stages of development and has not yet been commercialized.

Despite these challenges, many researchers and companies are working on developing new and innovative methods of hydrogen storage, which could help to make hydrogen a more practical and accessible energy source in the future. Some of these methods include using advanced materials to store hydrogen at lower pressures, developing new catalysts to enable more efficient hydrogen production and storage, and using hydrogen carriers to transport hydrogen more easily.

4.5 CONCLUSION

Some of the key takeaways about pressurized hydrogen storage are that it is a promising technology for storing and transporting hydrogen, but it also presents specific challenges. One of the main advantages of pressurized hydrogen storage is that it is relatively simple and straightforward, and it can be used in a wide range of applications. However, it also requires high-pressure vessels that can be expensive and heavy, and there are concerns about the safety of pressurized hydrogen storage. The costs associated with the various processes, such as H_2 storage, transportation, and also H_2 distribution, increase the overall technology cost. The reduction in the H_2 storage cost very critically impacts the development and implementation of H_2-powered automotive H_2 applications such as fuel cells. It is well proven that high-pressure H_2 storage is the most suitable type of storage under decentralized facilities. These facilities can be H_2 refueling stations. In decentralized facilities, mechanical compressors, such as metal hydride, adsorption-desorption compressors, and electrochemical compressors, can be suitable as a replacement for mechanical compressors. The non-mechanical compressors have various other advantages, such as the absence of moving parts reduces the installation and maintenance costs as well as loss due to frictions. The metal hydride-based compressors provide storage safety along with high H_2 compression. These types of compressors are also known as thermally driven compressors as they need high heat exchange. Adsorption-desorption type of compressors depend on the ability of H_2 to combine marginally to the solid porous surface such as carbon materials. These compressors are also thermally driven. Overall, pressurized hydrogen storage has a lot of potential and will continue to be an important area of research and development in the future.

REFERENCES

Chen, J., Li, J., Li, Q., Wang, S., Wang, L., Liu, H., Fan, C., 2020. Engineering a chemoenzymatic cascade for sustainable photobiological hydrogen production with green algae. *Energy Environ. Sci.* 13, 2064–2068. https://doi.org/10.1039/d0ee00993h.

Chi, Z., Zheng, Y., Ma, J., Chen, S., 2011. Oleaginous yeast Cryptococcus curvatus culture with dark fermentation hydrogen production effluent as feedstock for microbial lipid production. *Int. J. Hydrogen Energy.* 36, 9542–9550. https://doi.org/10.1016/j.ijhydene.2011.04.124.

Fakhimi, N., Tavakoli, O., 2019. Improving hydrogen production using co-cultivation of bacteria with Chlamydomonas reinhardtii microalga. *Mater. Sci. Energy Technol.* 2, 1–7. https://doi.org/10.1016/j.mset.2018.09.003.

Farha, O.K., Özgür Yazaydın, A., Eryazici, I., Malliakas, C.D., Hauser, B.G., Kanatzidis, M.G., Nguyen, S.T., Snurr, R.Q., Hupp, J.T., 2010. De novo synthesis of a metal-organic framework material featuring ultrahigh surface area and gas storage capacities. *Nat. Chem.* 2, 944–948.

Feng, S., Shang, Y., Wang, Z., Kang, Z., Wang, R., Jiang, J., Fan, L., Fan, W., Liu, Z., Kong, G., Feng, Y., Hu, S., Guo, H., Sun, D., 2020. Fabrication of a hydrogen-bonded organic framework membrane through solution processing for pressure-regulated gas separation. *Angew. Chem. Int. Ed.* 59, 3840–3845. https://doi.org/10.1002/anie.201914548.

Gómez-Gualdrón, D.A., Colón, Y.J., Zhang, X., Wang, T.C., Chen, Y.S., Hupp, J.T., Yildirim, T., Farha, O.K., Zhang, J., Snurr, R.Q., 2016. Evaluating topologically diverse metal-organic frameworks for cryo-adsorbed hydrogen storage. *Energy Environ. Sci.* 9, 3279–3289. https://doi.org/10.1039/c6ee02104b.

Gholami, T., Salavati-Niasari, M., 2016. Effects of copper: aluminum ratio in CuO/Al_2O_3 nanocomposite: electrochemical hydrogen storage capacity, band gap and morphology. *Int. J. Hydrogen Energy.* 41, 15141–15148. https://doi.org/10.1016/j.ijhydene.2016.06.191.

Giuseppe, S., Gaël, M., et al., 2020. Towards non-mechanical hybrid hydrogen compression for decentralized hydrogen facilities. *Energies.* 13, 3145. https://doi.org/10.3390/en13123145.

Hou, X.X., Sulic, M., Ortmann, J.P., Cai, M., Chakraborty, A., 2016. Experimental and numerical investigation of the cryogenic hydrogen storage processes over MOF-5. *Int. J. Hydrogen Energy*. 41, 4026–4038. https://doi.org/10.1016/j. ijhydene.2015.12.187.

Kumar, M., Diwania, S., Sen, S. et al., 2023. Emission-averse techno-economical study for an isolated microgrid system with solar energy and battery storage. *Electr. Eng.* 105, 1883–1896. https://doi.org/10.1007/s00202-023-01785-8.

Kumar, M., Tyagi, B., 2020. Multi-variable constrained non-linear optimal planning and operation problem for isolated microgrids with stochasticity in wind, solar, and load demand data. *IET Gener. Transm. Distrib.* 14(11), 2181–2190. https://doi.org/10.1049/iet-gtd.2019.0643.

Kumar, M., Tyagi, B., 2020. An optimal multivariable constrained nonlinear (MVCNL) stochastic microgrid planning and operation problem with renewable penetration. *IEEE Syst. J.* 14(3), 4143–4154. https://doi.org/10.1109/JSYST.2019.2963729.

Nam, T.H., Kim, J.G., Choi, Y.S., 2013. Electrochemical hydrogen discharge of high-strength low alloy steel for high-pressure gaseous hydrogen storage tank: effect of discharging temperature. *Int. J. Hydrogen Energy*. 38, 999–1003. https://doi.org/10.1016/j. ijhydene.2012.10.100.

Rzepka, M., Lamp, P., de la Casa-Lillo, M.A., 1998. Physisorption of hydrogen on microporous carbon and carbon nanotubes. *J. Phys. Chem. B*. 102, 10894–10898.

Schlapbach, L., Züttel, A., 2001. Hydrogen-storage materials for mobile applications. *Nature*. 414, 353.

Sharma, S., Basu, S., 2020. Highly reusable visible light active hierarchical porous WO_3/SiO_2 monolith in centimeter length scale for enhanced photocatalytic degradation of toxic pollutants. *Separ. Purif. Technol.* 231, 115916. https://doi.org/10.1016/j. seppur.2019.115916.

Sharma, S., Basu, S., Shetti, N.P., Aminabhavi, T.M., 2020a. Waste-to-energy nexus for circular economy and environmental protection: recent trends in hydrogen energy. *Sci. Total Environ.* 713, 136633. https://doi.org/10.1016/j.scitotenv.2020.136633.

Shi, X.Y., Yu, H.Q., 2016. Simultaneous metabolism of benzoate and photo-biological hydrogen production by Lyngbya sp. *Renew. Energy*. 95, 474–477. https://doi.org/10.1016/j. renene.2016.04.051.

Sulaiman, N.N., Ismail, M., 2016. Enhanced hydrogen storage properties of MgH2 co-catalyzed with K2NiF6 and CNTs. *Dalton Trans.* 45, 19380–19388. https://doi.org/10.1039/c6dt03646e.

Turhal, S., Turanbaev, M., Argun, H., 2019. Hydrogen production from melon and watermelon mixture by dark fermentation. *Int. J. Hydrogen Energy*. 44, 18811–18817. https://doi. org/10.1016/j.ijhydene.2018.10.011.

Von Helmolt, R., Eberle, U., 2007. Fuel cell vehicles: status 2007. *J. Power Sources*. 165(2), 833e43.

Xu, P., Zheng, J., Liu, P., Chen, R., Kai, F., Li, L., 2009. Risk identification and control of stationary high-pressure hydrogen storage vessels. *J. Loss Prevent Proc.* 22(6), 950e3.

Yang, J., Sudik, A., Wolverton, C., Siegel. D.J., 2010. High capacity hydrogen storage materials: attributes for automotive applications and techniques for materials discovery. *Chem. Soc. Rev.* 39(2), 656e75.

Yang, D., Zhang, Y., Barupal, D.K., Fan, X., Gustafson, R., Guo, R., Fiehn, O., 2014. Metabolomics of photobiological hydrogen production induced by CCCP in Chlamydomonas reinhardtii. *Int. J. Hydrogen Energy*. 39, 150–158. https://doi. org/10.1016/j.ijhydene.2013.09.116.

Zhang, H., Ding, Q., He, D., Liu, H., Liu, W., Li, Z., Yang, B., Zhang, X., Lei, L., Jin, S., 2016. A p-Si/NiCoSe:X core/shell nanopillar array photocathode for enhanced photoelectrochemical hydrogen production. *Energy Environ. Sci.* 9, 3113–3119. https://doi. org/10.1039/c6ee02215d.

Zhang, J., Qi, H., He, Z., Yu, X., Ruan, L., 2017a. Investigation of light transfer procedure and photobiological hydrogen production of microalgae in photobioreactors at different locations of China. *Int. J. Hydrogen Energy.* 42, 19709–19722. https://doi.org/10.1016/j.ijhydene.2017.06.079.

Zhang, J., Yu, Z., Gao, Z., Ge, H., Zhao, S., Chen, C., Chen, S., Tong, X., Wang, M., Zheng, Z., Qin, Y., 2017b. Porous TiO_2 Nanotubes with spatially separated platinum and CoO_xCo catalysts produced by atomic layer deposition for photocatalytic hydrogen production. *Angew. Chem. Int. Ed.* 56, 816–820. https://doi.org/10.1002/anie.201611137.

Zhang, J., Zhu, Y., Lin, H., Liu, Y., Zhang, Y., Li, S., Ma, Z., Li, L., 2017c. Metal hydride nanoparticles with ultrahigh structural stability and hydrogen storage activity derived from microencapsulated nano confinement. *Adv. Mater.* 29, 1–6. https://doi. org/10.1002/adma.201700760.

Zhang, J., Zhu, Y., Yao, L., Xu, C., Liu, Y., Li, L., 2019. State of the art multi-strategy improvement of Mg-based hydrides for hydrogen storage. *J. Alloys Compd.* 782, 796–823. https://doi.org/10.1016/j.jallcom.2018.12.217.

Zheng, J., Zhu, G., Huang, Z., 1991. The optimum design of ribbon wound pressure vessel by using adaptive random search method. *J. Zhejiang Univ. (Engineering Science).* 25(6), 673e80.

Zheng, J., Zhu, G., Wang, L., Huang, Z., 1993. Design criteria for unique ribbon wound pressure vessel. *J. Zhejiang Univ. (Engineering Science).* 27(3), 324e33.

Zheng, J., Fu, Q., Kai, F., Chen, C., 2004. The state of art of light-weight high pressure hydrogen storage tank. *Acta Energiae Solaris Sinica.* 25(5), 576e81.

Zheng, J., Kai, F., Liu, Z., Chen, R., Chen, C., 2006. Risk assessment and control of high pressure hydrogen equipment. *Acta Energiae Solaris Sinica.* 27(11), 1168e74.

Zheng, J., Li, L., Chen, R., Xu, P., Kai, F., 2008. High pressure steel storage vessels used in hydrogen refueling station. *J. Pressure Vessel Technol.* 130(1), 14503.

Züttel, A., 2003. Materials for hydrogen storage. *Mater. Today.* 6, 24–33.

5 Low-Temperature Liquefaction Hydrogen Storage

John Owolabi, Abdurrazzaq Ahmad, and Chinenye Azie

5.1 INTRODUCTION

Hydrogen holds immense potential as a key energy carrier in paving the way toward a sustainable future. With its abundant availability and clean combustion properties, hydrogen offers a promising solution to address the challenges of climate change and energy transition (Yuranov et al., 2018). Liquefied hydrogen (LH_2) is a highly versatile and sought-after fuel source that has been gaining traction in recent years due to its numerous advantages over conventional fuels. LH_2 is a clear, colorless liquid that is formed when hydrogen gas is cooled to extremely low temperatures ($-253°C$ or $-423.67°F$) at standard atmospheric pressure (Aziz, 2021). This process of cooling hydrogen gas to its liquid state is known as liquefaction and is achieved by compressing the gas to a high pressure and then cooling it down to a very low temperature. The resulting liquid hydrogen has a density of approximately 70.8 kg/m³, which is about one-eighth the density of water, making it a highly lightweight fuel source (Stock, 2021). LH_2 has a very low boiling point and can vaporize rapidly when exposed to ambient temperatures and pressure. Therefore, special storage facilities are required to store and transport LH_2, which must be maintained at extremely low temperatures to keep the fuel in its liquid state. Some of these facilities like the cryogenic storage ones involve storing LH_2 in specially designed insulated (typically consisting of several layers of materials, such as perlite, foam glass, or vacuum insulation panels plus stainless steel or aluminum) tanks designed to minimize heat transfer and able to maintain the extremely low temperature required to keep it in a liquid state (Timmerhaus, Flynn, Timmerhaus, & Flynn, 1989).

Low-temperature liquefaction hydrogen storage is a method of storing hydrogen as a liquid at very low temperatures, typically below $-253°C$, which is the boiling point of hydrogen (Tan, Kim, Yasuda, & Nogita, 2023). This method is a promising option for storing large amounts of hydrogen for use in various applications, such as fuel cell vehicles, power generation, and chemical production. The low-temperature liquefaction process involves compressing hydrogen gas and cooling it down to below its boiling point using cryogenic cooling systems. The liquefied hydrogen is then stored in insulated tanks at very low temperatures, which can be maintained using advanced insulation materials and active cooling systems (Ghorbani, Zendehboudi,

DOI: 10.1201/9781003382553-7

Saady, & Dusseault, 2023). One of the primary challenges associated with this process is the high energy required to cool hydrogen gas down to its boiling point. This energy requirement is significant and often requires significant refrigeration resources. One of the primary advantages of low-temperature liquefaction hydrogen storage is its high energy density (Ghorbani et al., 2023). Hydrogen in liquid form has a much higher energy density than in its gaseous state. This allows for the storage of large amounts of hydrogen in a relatively small volume, making it a preferred option for fuel cell vehicles and other applications that require compact storage systems. However, this high energy density comes with the drawback of significant boil-off losses if the storage system is not properly insulated or if the cooling system fails. To address the challenges associated with low-temperature liquefaction hydrogen storage, ongoing research is focused on improving the efficiency of cryogenic cooling systems, as well as developing advanced insulation materials and tank designs that can minimize heat transfer and improve safety (Zhang et al., 2023). There is also research focused on developing new materials that can store hydrogen at lower temperatures and higher pressures, which could lead to more compact and efficient storage systems. One area of research is focused on the use of advanced adsorbent materials for low-temperature hydrogen storage. This involves using materials such as metal-organic frameworks (MOFs) and zeolites to trap hydrogen molecules at low temperatures and high pressures. Another area of research is focused on the use of cryo-compressed hydrogen storage (Bosu & Rajamohan, 2023), which involves compressing hydrogen gas at low temperatures before liquefying it, which can improve the efficiency of the liquefaction process and reduce energy requirements. There is also research focused on the use of hybrid hydrogen storage systems, which combine low-temperature liquefaction with other storage methods such as high-pressure storage or solid-state hydrogen storage. These hybrid systems have the potential to provide improved storage efficiency and safety compared to single storage methods (Chong, Wong, Rajkumar, Rajkumar, & Isa, 2016). Another area of research is focused on the development of new cryogenic cooling systems, such as cryocoolers and pulse tube refrigerators, which can improve the efficiency and reliability of low-temperature liquefaction hydrogen storage. There is also research focused on improving the performance of insulation materials, such as vacuum insulation panels and aerogels, which can reduce heat transfer and minimize boil-off losses (Zhang et al., 2023). Low-temperature liquefaction hydrogen storage is a promising option for storing large amounts of hydrogen for use in a variety of applications. However, the high energy required to cool hydrogen gas down to its boiling point, as well as the significant boil-off losses associated with this storage method, must be addressed to make it a practical and cost-effective option. Ongoing research and development efforts are focused on improving the efficiency and safety of this storage method, which could lead to significant advances in hydrogen storage technology in the future.

5.2 HYDROGEN PRODUCTION

In recent years, there has been increasing interest in hydrogen production as a potential solution to global energy and environmental challenges. Liang Yin and Yonglin Ju (2020) presented a novel cryogenic hydrogen liquefaction process configuration,

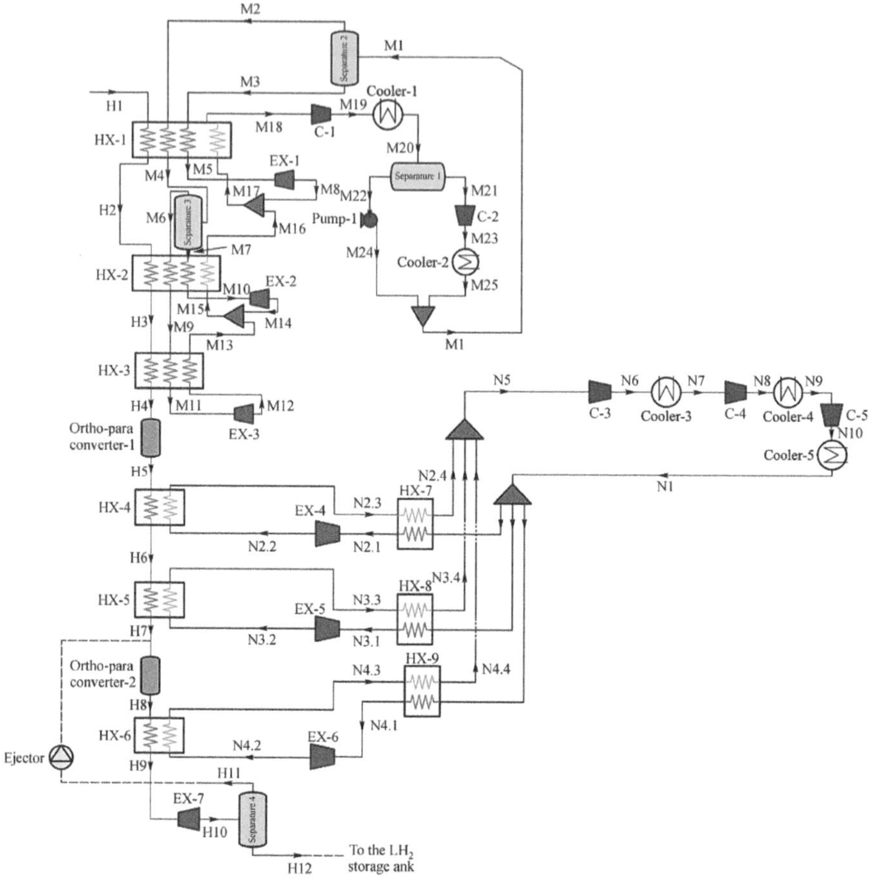

FIGURE 5.1 Conceptual diagram for a liquefaction process of hydrogen (Yin & Ju, 2020).

depicted in Figure 5.1. Their study described a multi-component mixed refrigerant (MR) system that was simulated using Aspen HYSYS software, with nine components for the precooling cycle. The MR system with three components (10% neon, 6.5% hydrogen, and 83.5% helium) was employed to produce 300 TPD of LH_2 using the cryogenic J-B cycle (Yin & Ju, 2020).

The process and the operating conditions of the equipment were optimized to achieve the best performance. Table 5.1 presents an analysis of the distribution of exergy loss (I_i) in each process, providing valuable insights into the locations of significant exergy losses. The exergy performance of individual components can be comprehensively assessed, allowing for a detailed evaluation (Wang et al., 2022). This approach enhances the understanding of the system's efficiency by identifying areas where exergy losses occur and enabling further optimization efforts to improve overall performance.

Similarly, Table 5.1 presents equations for analysis of the distribution of exergy loss (I_i) in each process, providing valuable insights into the locations of significant exergy losses. The exergy performance of individual component can be comprehensively

TABLE 5.1

A Novel Method for Calculating Exergy Losses Associated with Each Component in CcH$_2$ (Wang et al., 2022)

Component	Exergy Loss (I_i)
Compressor	$W - E_{\text{out}}; -\Sigma_{\text{in}}$
After cooler	$E_{\text{in}} - E_{\text{out}}$
Mixer	$E_{\text{out}}; -\Sigma_{E_{\text{in}}}$
Heat exchanger	$\Sigma_{(Ec,\,\text{in}-Ec,\,\text{out})} - \Sigma_{(EW,\,\text{out}-EW,\,\text{in})}$
JT element	$E_{\text{in}} - E_{\text{out}}$
Expander	$E_{\text{in}} - E_{\text{out}} - 0.8 W_{\text{EP}}$

assessed (Wang et al., 2022). This approach enhances the understanding of the system's efficiency by identifying areas where exergy losses occur and enabling further optimization efforts to improve overall performance.

5.2.1 Steam Methane Reforming

Steam methane reforming (SMR) is presently the predominant approach for hydrogen production, responsible for more than 95% of the world's hydrogen production (Chen, Qi, Zhang, Su, & Somorjai, 2020). SMR involves reacting natural gas (primarily methane) with steam at high temperatures and pressures over a catalyst to produce hydrogen and carbon monoxide. The carbon monoxide is then reacted with steam in a secondary reaction to produce additional hydrogen and carbon dioxide. The resulting hydrogen is typically purified and compressed for use in various applications. The main technical challenges associated with SMR include the management of the high-temperature and high-pressure conditions required for the reaction, as well as the capture and storage of the carbon dioxide produced as a by-product (Yue et al., 2021).

5.2.2 Electrolysis

Electrolysis is the process of using an electric current to split water into hydrogen and oxygen. Electrolysis can be performed using either a membrane or a solid oxide electrolyzer (Ni, Leung, & Leung, 2008). In a membrane electrolyzer, an electric current is passed through a solution of water and an electrolyte, causing hydrogen ions to migrate to the cathode and oxygen ions to migrate to the anode. The hydrogen and oxygen are then collected and separated. In a solid oxide electrolyzer, an electric current is passed through a solid electrolyte, causing water vapor to be split into hydrogen and oxygen at the anode and cathode, respectively. The main technical challenges associated with electrolysis include the high energy requirements for the process and the need for a reliable source of electricity (Grigoriev, Fateev, Bessarabov, & Millet, 2020).

5.2.3 Biomass Gasification

Biomass gasification involves converting organic matter (such as wood, agricultural waste, or municipal solid waste) into a gas that can be used for hydrogen production. Gasification is achieved by heating organic matter in the presence of a gasifying agent (such as air or oxygen) to produce a syngas (a mixture of hydrogen, carbon monoxide, and carbon dioxide) (Monteiro & Brito, 2023). The syngas is then purified to produce hydrogen. The main technical challenges associated with biomass gasification include the variability of the feedstock, which can affect the quality and composition of the syngas, as well as the need to manage the high-temperature and high-pressure conditions required for the gasification process (Hwang, Maharjan, & Cho, 2023).

5.2.4 Photoelectrochemical (PEC) Water Splitting

PEC water splitting represents a highly promising technique for the production of hydrogen by utilizing solar energy to separate water molecules into hydrogen and oxygen components (Hwang et al., 2023). The process entails harnessing the capabilities of a semiconductor material to capture solar energy and produce electron-hole pairs. These generated pairs are subsequently utilized to facilitate the process of water splitting. The main technical challenges associated with PEC water splitting include the need for efficient and stable semiconductor materials as well as the need to optimize the design and configuration of the PEC cell to maximize the efficiency of the process (Hwang et al., 2023). Hydrogen production is a complex process that requires careful management of a variety of technical challenges. Each of the methods discussed above has its own unique advantages and disadvantages, and the choice of method will depend on a variety of factors, including the availability of feedstock, the desired purity of the hydrogen, and the cost and efficiency of the production process. With continued research and development, hydrogen production has the potential to become an increasingly important source of clean, renewable energy in the future (Kalamaras & Efstathiou, 2013).

5.3 STRATEGIES FOR HYDROGEN LIQUEFACTION

Liquefaction of hydrogen is a process that involves cooling hydrogen gas to extremely low temperatures until it condenses into a liquid state (Al Ghafri et al., 2022). This is achieved through a combination of cooling and compression techniques. The strategies for liquefaction of hydrogen include:

5.3.1 The Linde-Hampson Cycle

This cycle is a multistage process that uses a combination of expansion and compression to achieve cooling and liquefaction of a gas. In this cycle, the compressed gas is first passed through a Joule-Thomson valve, which cools the gas due to expansion. The cooled gas is then passed through a heat exchanger, where it is further cooled by a refrigerant such as nitrogen or helium (Fast, 2012). The gas is then compressed to a higher pressure and passed through another heat exchanger where it is further cooled by the refrigerant (Kanoglu, Yilmaz, & Abusoglu, 2016). This cycle is repeated several times, with each stage producing a progressively cooler gas, until the gas is cooled to its liquefaction temperature. The Linde-Hampson cycle is more complex and energy intensive.

5.3.2 Brayton Cycle

The Brayton cycle is another refrigeration cycle that can be used for hydrogen liquefaction. This cycle uses a combination of compressors, heat exchangers, turbines, and a regenerator to compress and expand the hydrogen gas, resulting in liquefaction. The cycle starts by compressing hydrogen gas using a compressor to a pressure of around 15–25 bar (Yartys et al., 2021). The compressed gas is then passed through a heat exchanger where it is cooled by a coolant such as helium or nitrogen. After the gas is cooled, it is passed through a first turbine, known as the expansion turbine, where it is expanded to a lower pressure and temperature. The turbine generates mechanical energy that can be used to drive the compressor. The expanded gas is then passed through a regenerator, which is a heat exchanger that uses a matrix of materials with high thermal conductivity, such as stainless steel wire mesh or ceramic, to recover and transfer heat between the gas streams (Popov et al., 2019). The regenerator pre-cools the incoming high-pressure gas and reheats the outgoing low-pressure gas before it is expanded further in the second turbine. The second turbine, known as the cold turbine, further expands the gas to a very low pressure and temperature, which causes the gas to condense into a liquid. The liquid hydrogen is then collected and stored in a cryogenic container (Popov et al., 2019). The final stage of the cycle involves passing the cold exhaust gas from the second turbine through the regenerator, where it is used to cool the incoming high-pressure gas. The cold exhaust gas is then released to the atmosphere, while the incoming high-pressure gas is compressed again by the compressor, and the cycle starts over. The Brayton cycle has the advantage of being a simple and efficient process that can achieve high liquefaction rates for hydrogen. However, it is limited by the low density of hydrogen gas, which requires large equipment and high operating pressures, as well as the need for a regenerator to recover heat and improve efficiency (Chang, Kim, & Choi, 2020).

5.3.3 Claude Cycle

This cycle is similar to the Brayton cycle but uses multiple stages of compression and cooling to liquefy the hydrogen. The Claude cycle is a modified version of the classic refrigeration cycle that is used for liquefaction of hydrogen gas (Yilmaz, 2020). The cycle consists of four main components: a compressor, a heat exchanger, a turbine, and a second heat exchanger. The cycle operates by compressing hydrogen gas, cooling it, expanding it, and then cooling it again, resulting in liquefaction of the gas. The first stage of the cycle involves compressing the hydrogen gas using a multistage reciprocating compressor to a pressure of around 200–250 bar

(Dornheim et al., 2022). The compressed gas is then passed through the first heat exchanger, also known as the pre-cooler, where it is cooled by a coolant such as helium or nitrogen. The cooled gas is then further cooled by passing it through the second heat exchanger, also known as the main heat exchanger. In this heat exchanger, the hydrogen gas is cooled to a temperature of around 20–30 K (Xu, Xu, Zheng, Chen, & Wang, 2020) by exchanging heat with a cryogenic fluid such as liquid nitrogen. After the gas is cooled, it is expanded through a turbine to reduce its pressure and temperature. The turbine generates mechanical energy that can be used to drive the compressor. The expanded gas is then passed through the second heat exchanger again to further cool it. The cooled gas is then passed through another turbine, known as the turbo-expander, which further reduces its pressure and temperature. The turbo-expander also generates additional mechanical energy that can be used to drive other equipment, such as an electric generator (Rusanov, Solovey, & Lototskyy, 2020). The final stage of the cycle involves passing the expanded and cooled hydrogen gas through the main heat exchanger again to reduce its temperature to around 18–20 K. At this temperature and pressure, the hydrogen gas condenses into a liquid, which is collected and stored in a cryogenic container (Kanoglu et al., 2016).

5.3.4 Magnetic Refrigeration Cycle

This refrigeration cycle is a promising technology that uses the magnetocaloric effect to produce cooling and achieve liquefaction of hydrogen. The magnetic refrigeration cycle works by using a magnetic material, such as gadolinium, that undergoes a reversible temperature change in the presence of a magnetic field (Numazawa, Kamiya, Utaki, & Matsumoto, 2014). This material is typically in the form of a solid block or a powder that is contained within a refrigerant fluid that circulates through a series of heat exchangers. The heat exchangers are used to transfer heat between the refrigerant fluid and the hydrogen gas being liquefied. The hydrogen gas is first compressed and then cooled through a heat exchanger to remove some of the heat. The cooled gas is then passed through a magnetic field and meets the magnetic material (Balli, Jandl, Fournier, & Kedous-Lebouc, 2017). The magnetic material absorbs heat from the hydrogen gas, causing the gas to further cool down. The now-cold hydrogen gas is then passed through another heat exchanger where it is further cooled by the refrigerant fluid, which absorbs the heat released by the magnetic material. This cycle is repeated several times, with each cycle producing a progressively colder hydrogen gas until the gas reaches its liquefaction temperature of −253°C (−423°F) (Ehrhart, Klebanoff, Mohmand, & Markt, 2021). At this temperature, the hydrogen gas condenses into a liquid, which can be stored in a cryogenic container. The advantage of using the magnetic refrigeration cycle for hydrogen liquefaction is that it does not require any harmful chemicals, such as chlorofluorocarbons (CFCs), and is highly efficient (Zhang et al., 2019). However, this technology is still in the early stages of development and there are several challenges that need to be overcome before it can become a practical alternative to traditional liquefaction methods, including the high cost of the magnetic materials and the need for large and expensive magnetic field generators.

5.3.5 HYBRID REFRIGERATION

Hybrid refrigeration is a method of hydrogen liquefaction that combines the use of different refrigeration cycles to achieve high efficiency and low operating costs (Zhang et al., 2023). In this process, a combination of a Joule-Thomson cycle, a Brayton cycle, and a Claude cycle is used to cool and compress the hydrogen gas. In hybrid refrigeration, the different refrigeration cycles are combined to optimize their respective advantages and minimize their disadvantages. For example, the Joule-Thomson cycle can provide significant cooling at low cost but is limited by its low efficiency and inability to achieve low temperatures. The Brayton cycle can achieve higher efficiencies and lower temperatures but requires more energy and equipment (Lee et al., 2021). The Claude cycle can achieve the lowest temperatures and highest efficiencies but is the most expensive and complex. Hence, the hybrid refrigeration cycle. Some of the hybrid cycles for hydrogen liquefaction include:

i. Brayton-Claude cycle: This cycle combines elements of the Brayton cycle and Claude cycle. In this cycle, the compressed hydrogen gas is first cooled in a heat exchanger (as in the Claude cycle) (Zhang et al., 2023) and then further cooled through expansion in a turbine (as in the Brayton cycle). This process can be repeated to achieve further cooling and liquefaction of the hydrogen.

ii. Linde-Hampson-Brayton cycle: This cycle combines elements of the Linde-Hampson cycle and the Brayton cycle (Ellison, 1994). In this cycle, the compressed hydrogen gas is first expanded through a Joule-Thomson valve (as in the Linde-Hampson cycle) to achieve cooling. The cooled gas is then compressed and cooled further in a Brayton cycle compressor and heat exchanger.

iii. Brayton-magnetic refrigeration cycle: This cycle combines elements of the Brayton cycle and magnetic refrigeration cycle. In this cycle, the compressed hydrogen gas is first cooled through a heat exchanger (as in the Brayton cycle) (Zhang et al., 2023) and then further cooled through magnetic refrigeration. The cooled gas can then be expanded and further cooled through a Brayton cycle turbine and heat exchanger.

These hybrid cycles can offer advantages such as improved efficiency, reduced energy consumption, and increased reliability compared to traditional cycles (Dincer & Acar, 2015). However, they can also be more complex and require more advanced equipment and control systems. The specific hybrid cycle used for hydrogen liquefaction will depend on the specific requirements and constraints of the application.

5.4 STRATEGIES FOR HYDROGEN STORAGE

Hydrogen has the potential to be a major source of clean energy due to its high energy density and ability to produce zero emission power. However, storing and transporting hydrogen gas is challenging due to its low density and high flammability

(Borowski & Karlikowska, 2023). To overcome these challenges, various strategies have been developed for hydrogen storage. The following are the most common strategies for hydrogen storage.

5.4.1 Compressed Gas Storage

Compressed gas storage is the most common method for hydrogen storage. It involves compressing hydrogen gas to a high pressure, typically between 350 and 700 bar (Gangloff Jr, Kast, Morrison, & Marcinkoski, 2017), to increase its energy density and reduce its volume. The compressed hydrogen gas is stored in high-pressure cylinders or tanks made of high-strength materials, such as carbon fiber-reinforced polymers or metal alloys. Compressed gas storage is a mature and well-established technology, but it has some disadvantages, including the requirement for heavy and bulky high-pressure storage tanks, which can limit the range and payload of vehicles (Rivard, Trudeau, & Zaghib, 2019). Additionally, the compression process can generate heat, which can reduce the efficiency of the storage system.

5.4.2 Cryogenic Storage

Cryogenic storage involves reducing the temperature of hydrogen gas below its boiling point of −252.8°C (Nikolaidis & Poullikkas, 2017) to produce a liquid form, which allows for higher energy density storage. The liquid hydrogen is then stored in cryogenic tanks, typically made of stainless steel, aluminum, or composites. Cryogenic storage offers a higher energy density than compressed gas storage and is well-suited for applications requiring large amounts of hydrogen, such as space launch vehicles. Wang et al. (2022) presented simulation results of density plots, as shown in Figure 5.2, to illustrate the calculated hydrogen density and cooling power requirement for cryo-compressed hydrogen (CcH$_2$) under various storage conditions using REFPROP simulation software. The green region represents favorable

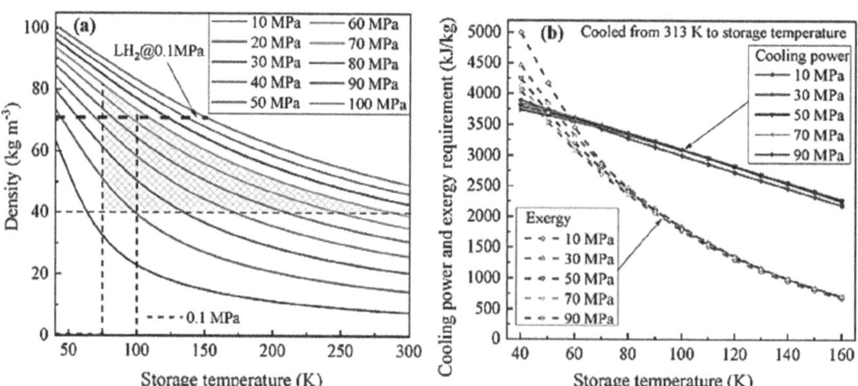

FIGURE 5.2 Density variations of CcH$_2$ (carbon compound H$_2$) under different conditions. (a) Providing insights into the changes in the substance's mass per unit volume. (b) Showcases the cooling power and exergy requirement of CcH$_2$ (Wang et al., 2022).

hydrogen density which is approximately 70% of the density of liquid hydrogen (LH_2) (Aziz, 2021). The density plot provides insights into the changes in CcH_2 density across the investigated range of conditions. Meanwhile, the cooling power and exergy requirement plot demonstrates the relationship between these two parameters for CcH_2, allowing for a better understanding of its behavior and performance under different scenarios (Klopčič, Grimmer, Winkler, Sartory, & Trattner, 2023).

However, it has some disadvantages, including the requirement for expensive cryogenic equipment, which can limit its widespread adoption.

5.4.3 Chemical Storage

Chemical storage involves storing hydrogen in a chemical form that can be easily released when needed. The two most common methods for chemical storage are metal hydride and chemical hydride storage (Eberle, Felderhoff, & Schueth, 2009). Metal hydride storage is a fascinating technique that entails the absorption of hydrogen gas into a solid metal structure, resulting in the formation of a metal hydride. This metal hydride can subsequently release the stored hydrogen gas through the application of heat or exposure to a catalyst. This innovative approach offers great potential for efficient and controllable hydrogen storage and release, contributing to advancements in clean energy technologies. Chemical hydride storage involves using a chemical reaction to release hydrogen gas from a chemical compound (Muhammed et al., 2023). The chemical reaction can be reversed to store hydrogen gas in the chemical compound. Chemical storage offers the advantages of high energy density and easy transportability, but it also has some disadvantages, including the requirement for complex and expensive storage systems and the need for additional energy input to release the stored hydrogen (Muhammed et al., 2023).

5.4.4 Underground Storage

Underground storage involves storing hydrogen gas in underground caverns or depleted oil and gas reservoirs. The hydrogen gas is compressed and injected into the underground storage facilities, where it is stored until needed (Tarkowski, 2019). Underground storage offers the advantage of large-scale storage capacity and low environmental impact, but it has some disadvantages, including the requirement for expensive and complex infrastructure and the potential for leakage and safety concerns. Hydrogen storage is a critical technology for the widespread adoption of hydrogen as a clean energy source. Compressed gas storage is the most common method, but cryogenic storage, chemical storage, and underground storage offer alternative strategies with their own unique advantages and disadvantages (Assareh & Ghafouri, 2023; Wan, Zhang, & Xu, 2020). The choice of hydrogen storage method depends on the application and specific requirements, including energy density, cost, safety, and environmental impact.

Furthermore, in Figure 5.3, Wang et al. demonstrated the relationship between temperature (T) and enthalpy (h) for hydrogen, indicating that the specific heat capacity (c_p) of hydrogen remains relatively constant, exhibiting a linear trend (Wang et al., 2022). The temperature changes concerning enthalpy and specific heat capacity for

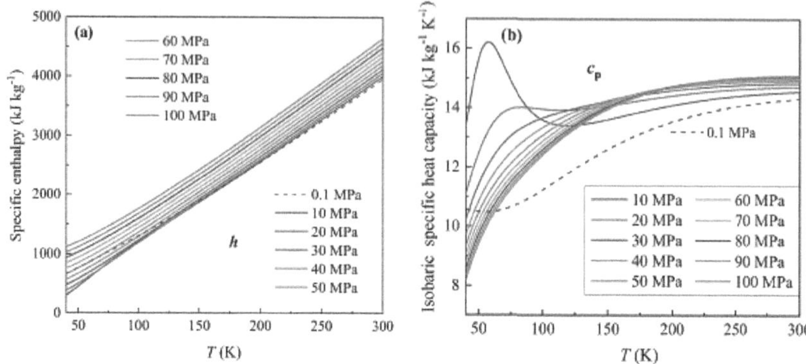

FIGURE 5.3 The *T-h* (a) and *T-c$_p$* (b) diagrams of relationships of hydrogen at different pressure levels (Wang et al., 2022).

hydrogen under varying pressures. The highlights cover the thermodynamic behavior of hydrogen and allow for a better understanding of its properties at different pressure conditions. This behavior is particularly prominent at temperatures above 120 K and at higher hydrogen pressures (pH$_2$) (Hirscher et al., 2020). Consequently, for a hydrogen stream with a fixed flow rate, the overall heat capacity (m·c_p) experiences minimal variation due to the nearly constant c_p. As a result, there is no need to modify the cooling process for different hydrogen pressures, as the cooling characteristics of the hydrogen stream (CcH$_2$) remain consistent (Wang et al., 2022).

5.5 CONDITIONS FOR LOW-TEMPERATURE LIQUEFACTION HYDROGEN STORAGE

Hydrogen has been recognized as a potential energy carrier due to its high energy content and clean combustion products. However, the low volumetric energy density of hydrogen at ambient conditions has been a major impediment to its widespread adoption (Zhang et al., 2023). Therefore, efficient and safe storage of hydrogen is of paramount importance for its use as an energy carrier. Low-temperature hydrogen storage has been proposed as a viable solution to address this challenge.

5.5.1 CONDITIONS

The conditions for low-temperature hydrogen storage depend on several factors, including the physical and chemical properties of the storage materials, the operating temperature, and the pressure (Hassan, Ramadan, Saleh, & Hissel, 2021). One of the most promising low-temperature storage methods is cryogenic hydrogen storage, which involves storing hydrogen at temperatures below its boiling point (−252.8°C) in a cryogenic liquid state. This method offers high-density storage but requires complex and expensive cryogenic equipment. Another low-temperature hydrogen storage method is adsorption-based storage, which involves the use of porous materials to trap and store hydrogen through physisorption and/or chemisorption mechanisms

(Muthukumar et al., 2023). The efficiency of adsorption-based storage depends on the properties of the storage materials, such as pore size, surface area, and chemical composition. The operating temperature also plays a crucial role in adsorption-based storage, as higher temperatures can lead to desorption of the stored hydrogen. Metal hydrides are another class of materials that can be used for low-temperature hydrogen storage (Le, Kim, Park, & Tran, 2024). These materials can store hydrogen through chemical reactions in which hydrogen is absorbed into the metal lattice to form a metal hydride. The amount of hydrogen stored in metal hydrides depends on factors such as the metal used, the particle size, and the temperature and pressure of the storage system (Desai, Uddin, Rahman, & Asmatulu, 2023). The main advantage of metal hydrides is their ability to store large amounts of hydrogen in a small volume, but they can be expensive and may require high temperatures to release the stored hydrogen.

Finally, liquid organic hydrogen carriers (LOHCs) have emerged as a promising low-temperature hydrogen storage method. LOHCs involve the use of hydrogen-rich organic compounds, such as cyclohexane or toluene, which can store hydrogen through reversible hydrogenation and dehydrogenation reactions (Chu, Wu, Luo, Cao, & Zhang, 2023). LOHCs offer high energy density and are compatible with existing liquid fuel infrastructure, but they require catalysts and may be costly to produce. Therefore, the conditions for low-temperature hydrogen storage depend on various factors, such as the storage materials, operating temperature, and pressure (Gianotti, Taillades-Jacquin, Rozière, & Jones, 2018). Cryogenic hydrogen storage, adsorption-based storage, metal hydrides, and LOHCs are all promising methods for low-temperature hydrogen storage, each with their own advantages and challenges. The development of efficient and cost-effective low-temperature hydrogen storage systems is critical for the widespread adoption of hydrogen as a clean and sustainable energy carrier (Muhammed et al., 2023).

5.5.2 Challenges

The storage and transportation of hydrogen remain a significant challenge. One of the most promising methods for hydrogen storage is low-temperature hydrogen storage, which involves storing hydrogen at temperatures below −253°C (Mohan, Sharma, Kumar, & Gayathri, 2019). While low-temperature hydrogen storage has significant advantages over other hydrogen storage methods, it also faces several challenges that must be addressed to make it a viable option for widespread use. The following are challenges associated with low-temperature hydrogen storage systems:

5.5.2.1 Cryogenic Storage

The most significant challenge associated with low-temperature hydrogen storage systems is the need for cryogenic storage. Hydrogen must be stored at extremely low temperatures, which requires expensive cryogenic equipment (Yang, Hunger, Berrettoni, Sprecher, & Wang, 2023). Cryogenic storage tanks are bulky, heavy, and costly, making it challenging to transport hydrogen over long distances. Moreover, the thermal insulation required for cryogenic storage tanks is susceptible to wear and tear, leading to the possibility of leaks and the release of hydrogen into the environment.

5.5.2.2 Boil-Off Losses

Boil-off losses are another significant challenge associated with low-temperature hydrogen storage systems. When hydrogen is stored at such low temperatures, it tends to boil off, leading to significant losses over time (Morales-Ospino, Celzard, & Fierro, 2023). These losses can be reduced by improving the thermal insulation of storage tanks and implementing effective recovery systems to capture the boil-off hydrogen.

5.5.2.3 Material Compatibility

The low-temperature storage of hydrogen can cause materials to become brittle, leading to cracks and leaks in storage tanks. This issue can be addressed by using materials that are compatible with low-temperature hydrogen storage conditions (Hassan et al., 2021). However, finding suitable materials can be challenging, and the cost of these materials can be prohibitive.

5.5.2.4 Safety Concerns

Low-temperature hydrogen storage systems pose several safety concerns that must be addressed. Cryogenic storage tanks can be prone to leaks, and the release of hydrogen into the environment can lead to fire or explosion (Abohamzeh, Salehi, Sheikholeslami, Abbassi, & Khan, 2021). Moreover, the handling of cryogenic hydrogen can be hazardous, and personnel must be adequately trained to handle it safely.

5.5.2.5 Infrastructure

Another significant challenge associated with low-temperature hydrogen storage systems is the lack of infrastructure to support their widespread use. Cryogenic storage tanks require specialized equipment for handling and transportation, and the cost of building this infrastructure can be prohibitive. Low-temperature hydrogen storage has the potential to be an effective method for storing hydrogen, but it also faces several challenges (Ratnakar et al., 2021). Cryogenic storage, boil-off losses, material compatibility, safety concerns, and infrastructure are all significant hurdles that must be addressed. However, with continued research and development, these challenges can be overcome, making low-temperature hydrogen storage a viable option for widespread use (Ratnakar et al., 2021).

5.5.3 Comparisons of Low Liquefaction with Other Storage Systems

Hydrogen is a promising fuel for transportation and power generation due to its high energy content and zero emissions when used in fuel cells. However, hydrogen storage remains a significant challenge due to its low energy density per unit volume compared to traditional fuels (Hassan et al., 2021). Various methods have been proposed for hydrogen storage, including high-pressure gas storage, cryogenic liquid storage, solid-state storage, and chemical storage. In recent years, low-temperature liquefaction (LTL) has emerged as a promising method for hydrogen storage when comparing LTL with other methods for hydrogen storage systems.

5.5.3.1 High-Pressure Gas Storage

High-pressure gas storage is the most common method of hydrogen storage today. Hydrogen is compressed and stored in high-pressure tanks, typically at pressures of 350–700 bar (Ahluwalia, Hua, & Peng, 2012). While this method is relatively simple and inexpensive, it has a low energy density per unit volume, which limits the driving range of hydrogen vehicles. Additionally, high-pressure tanks are heavy and require careful handling due to the risk of explosion.

5.5.3.2 Cryogenic Liquid Storage

Cryogenic liquid storage involves cooling hydrogen to very low temperatures (−253°C) to liquefy it. The resulting liquid has a much higher energy density per unit volume than compressed gas, but it requires complex and expensive cryogenic equipment to maintain the low temperature (Zohuri & Zohuri, 2019). Additionally, cryogenic storage tanks have a high heat transfer rate, which leads to hydrogen losses due to evaporation. Cryogenic storage is also not suitable for portable applications due to the large size and weight of the storage tanks. Cryogenic storage offers high storage capacity and energy density but requires significant energy consumption for liquefaction and refrigeration (Zohuri & Zohuri, 2019). The infrastructure required for this method is also expensive and complex, limiting its applicability in some contexts.

5.5.3.3 Solid-State Storage

Solid-state storage involves storing hydrogen in a solid material such as metal hydrides, carbon nanotubes, or MOFs (Zacharia & Rather, 2015). While this method has a high theoretical energy density, it suffers from low hydrogen storage capacity, slow kinetics, and a high cost. Additionally, solid-state storage materials often require high temperatures or pressures to release the stored hydrogen, which limits their practicality.

5.5.3.4 Chemical Storage

Chemical storage involves storing hydrogen in a chemical compound such as ammonia or methanol, which involves the reversible chemical reaction of hydrogen with a solid-state storage material (Gianotti, Taillades-Jacquin, Rozière, & Jones, 2018). While this method has a high energy density per unit volume, it requires complex and expensive equipment for hydrogen production and separation. Additionally, chemical storage suffers from low efficiency due to the energy required for hydrogen production and separation. This method has the potential to offer high storage capacity and energy density, but the practicality of this method is currently limited due to slow kinetics and the need for high-temperature operation (Lin, Salari, Arava, Ajayan, & Grinstaff, 2016).

5.5.3.5 Low-Temperature Liquefaction

Low-temperature liquefaction (LTL) involves cooling hydrogen to very low temperatures (−253°C) to liquefy it. The resulting liquid has a much higher energy density per unit volume than compressed gas, and it can be stored in insulated containers at atmospheric pressure (Gianotti, Taillades-Jacquin, Rozière, & Jones, 2018). LTL has several advantages over other storage methods, including high energy density, low weight and volume, and low cost. LTL also has a high storage capacity and a low heat transfer rate, which minimizes hydrogen losses due to evaporation (Gianotti, Taillades-Jacquin, Rozière, & Jones, 2018). LTL is also suitable for portable applications due to its compact and lightweight storage containers.

Low-temperature liquefaction is a promising method for hydrogen storage due to its high energy density, low weight and volume, and low cost (Gianotti, Taillades-Jacquin, Rozière, & Jones, 2018). While other methods of hydrogen storage have their advantages, they suffer from limitations such as low energy density, high cost, or low storage capacity. LTL has the potential to overcome these limitations and enable the widespread use of hydrogen as a clean and efficient fuel.

5.6 PERSPECTIVES ON HYDROGEN STORAGE

Hydrogen storage is one of the most critical challenges in the deployment of hydrogen as a clean energy carrier. Hydrogen is a gas that has a low volumetric density and a low energy density at atmospheric pressure and room temperature (Hassan et al., 2023). To store hydrogen efficiently, it needs to be compressed or liquefied, which requires energy and can result in safety concerns. There are several perspectives on hydrogen storage that need to be considered, including safety, cost, efficiency, and scalability.

5.6.1 SAFETY

Safety is a critical concern in the development and deployment of hydrogen storage systems. Hydrogen is a highly flammable gas, and its storage and handling require careful consideration of safety risks. Storage technologies should be designed to prevent hydrogen leaks and ensure safe operation under various conditions (Abohamzeh et al., 2021). Additionally, safety regulations and codes need to be developed and implemented to ensure the safe use of hydrogen in various applications.

5.6.2 COST

The cost of hydrogen storage is a critical factor in the commercial viability of hydrogen as a clean energy carrier. The cost of storage technologies can vary significantly depending on the technology used, the scale of the system, and the level of integration with other energy systems (Rivard et al., 2019). To achieve cost competitiveness, hydrogen storage technologies need to be developed and optimized to reduce the capital and operating costs of the systems.

5.6.3 EFFICIENCY

Efficiency is a critical factor in the deployment of hydrogen storage systems. The energy required to store hydrogen needs to be minimized to improve the overall efficiency of the energy system. Additionally, the energy efficiency of the storage technology needs to be optimized to ensure that the stored hydrogen can be converted back into electricity or heat with high efficiency (Yang et al., 2023). Technologies that enable the conversion of excess renewable energy into hydrogen need to be developed and optimized to ensure that the process is energy efficient.

5.6.4 SCALABILITY

Scalability is an essential consideration in the development of hydrogen storage systems (Yang et al., 2023). Hydrogen storage systems need to be scalable to meet the growing demand for hydrogen in various applications. Technologies that can be deployed at different scales, from small-scale systems for residential and commercial applications to large-scale systems for industrial and utility-scale applications, need to be developed to ensure that hydrogen storage can be deployed widely and rapidly. To achieve widespread adoption of hydrogen, significant advancements in hydrogen storage technologies are needed to improve safety, reduce costs, increase efficiency, and enable scalability (Hassan et al., 2023). The development and deployment of hydrogen storage technologies require collaboration between industry, academia, and government to address the technical and economic challenges and accelerate the transition to a hydrogen-based energy system.

5.7 PROS AND CONS OF HYDROGEN STORAGE

Hydrogen storage is a critical challenge in the deployment of hydrogen as a clean energy carrier. Hydrogen has several advantages over conventional fossil fuels, including its abundance, high energy density, and clean burning properties (Hassan et al., 2023). However, hydrogen storage also has some pros and cons that need to be considered when evaluating the feasibility of hydrogen as a clean energy carrier.

5.7.1 PROS OF HYDROGEN STORAGE

5.7.1.1 High Energy Density

Hydrogen has a high energy density, which means that a small volume of hydrogen can store a large amount of energy (McCay & Shafiee, 2020). This makes hydrogen an attractive energy carrier for applications that require high energy density, such as transportation and grid-scale energy storage.

5.7.1.2 Renewable Energy Storage

Hydrogen can be produced from renewable energy sources such as solar, wind, and hydropower (McCay & Shafiee, 2020). This makes hydrogen an attractive option for storing excess renewable energy and using it to meet energy demand when renewable energy production is low.

5.7.1.3 Clean Burning Properties

Hydrogen has clean burning properties, producing only water as a by-product when used in fuel cells or combustion engines (Teoh et al., 2023). This makes hydrogen an attractive alternative to conventional fossil fuels, which produce harmful emissions that contribute to air pollution and climate change.

5.7.1.4 Scalability

Hydrogen storage systems can be designed to scale from small-scale systems for residential and commercial applications to large-scale systems for industrial and utility-scale applications (Kharel & Shabani, 2018). This makes hydrogen an attractive option for meeting a wide range of energy needs and applications.

5.7.2 CONS OF HYDROGEN STORAGE

5.7.2.1 Safety Concerns

Hydrogen is a highly flammable gas and requires careful consideration of safety risks. Storage technologies should be designed to prevent hydrogen leaks and ensure safe operation under various conditions (Abohamzeh et al., 2021). Additionally, safety regulations and codes need to be developed and implemented to ensure the safe use of hydrogen in various applications.

5.7.2.2 Energy Intensive

The storage of hydrogen requires energy, which can result in a loss of efficiency in the overall energy system (Preuster, Alekseev, & Wasserscheid, 2017). Additionally, the production of hydrogen itself can be energy intensive, requiring electricity or fossil fuels to produce hydrogen through processes such as electrolysis or SMF.

5.7.2.3 Infrastructure Development

The widespread deployment of hydrogen as a clean energy carrier requires significant investment in infrastructure development. This includes the development of hydrogen production facilities, storage and transportation systems, and refueling stations (Yang et al., 2023). The cost of infrastructure development can be a significant barrier to the widespread adoption of hydrogen.

5.7.2.4 Material Compatibility

Hydrogen is a highly reactive gas and can cause embrittlement and degradation in some materials. Therefore, hydrogen storage systems need to be designed with materials that are compatible with hydrogen to ensure long-term reliability (Hosseini & Wahid, 2020). Hydrogen storage has both pros and cons that need to be considered when evaluating the feasibility of hydrogen as a clean energy carrier. While hydrogen has several advantages over conventional fossil fuels, significant advancements in hydrogen storage technologies are needed to improve safety, reduce costs, increase efficiency, and enable scalability (Hosseini & Wahid, 2020). The development and deployment of hydrogen storage technologies require collaboration between industry, academia, and government to address the technical and economic challenges and accelerate the transition to a hydrogen-based energy system.

5.8 COMMERCIAL AND PILOT PLANTS WORLDWIDE

Hydrogen storage technologies are crucial for the commercial deployment of hydrogen as a clean energy carrier. Several commercial and pilot plants have been developed worldwide to demonstrate the feasibility of various hydrogen storage technologies for different applications (Hassan et al., 2023). Hydrogen storage is a critical component of a hydrogen-based energy system, and several commercial and pilot plants have been developed worldwide to demonstrate the feasibility of hydrogen storage technologies. These plants vary in size, technology, and application and provide valuable insights into the commercialization potential of hydrogen storage. The following are some of the commercial and pilot plants for hydrogen storage technologies worldwide.

5.8.1 UNDERGROUND HYDROGEN STORAGE, GERMANY

The underground hydrogen storage plant in Germany is one of the largest hydrogen storage facilities in the world. The plant stores hydrogen in underground salt caverns, which can store up to 1.5 million cubic meters of hydrogen (Sambo et al., 2022). The stored hydrogen is used for grid-scale energy storage and is supplied to customers through a pipeline network.

5.8.2 ITM POWER ELECTROLYSIS AND REFUELLING STATION, UK

The ITM Power Electrolysis and Refuelling Station in the UK is a pilot plant that demonstrates the use of hydrogen as a fuel for transportation (Ling-Chin et al., 2023). The plant uses renewable energy sources such as solar and wind power to produce hydrogen through electrolysis. The produced hydrogen is then used to refuel hydrogen fuel cell vehicles.

5.8.3 HYBALANCE PILOT PLANT, DENMARK

The HyBalance Pilot Plant in Denmark is a pilot plant that demonstrates the use of hydrogen as a fuel for power generation (Liao, Liu, & Qing, 2020). The plant uses renewable energy sources, such as wind power, to produce hydrogen through electrolysis. The produced hydrogen is then used to generate electricity through a fuel cell system, which can provide backup power to the grid.

5.8.4 H_2V INDUSTRY PILOT PLANT, FRANCE

The H_2V Industry Pilot Plant in France is a pilot plant that demonstrates the use of hydrogen as a feedstock for industrial processes (Delaval et al., 2022). The plant uses natural gas to produce hydrogen through SMF. The produced hydrogen is then used in various industrial processes, such as the production of ammonia and methanol.

5.8.5 HYDROGENICS CORPORATION, CANADA

Hydrogenics Corporation is a Canadian company that specializes in the design, development, and manufacture of hydrogen storage and fuel cell systems (Razi & Dincer, 2022). The company has installed several hydrogen storage systems worldwide, including the Energiepark Mainz in Germany, which is the world's largest hydrogen energy storage facility. The facility uses an alkaline electrolyzer to produce hydrogen from renewable energy sources and stores the hydrogen in underground salt caverns.

5.8.6   McPhy Energy, France

McPhy Energy is a French company that develops and manufactures hydrogen storage and production systems. The company has installed several hydrogen storage systems in France and Europe, including a hydrogen refueling station in Grenoble, France, and a hydrogen storage system for a fuel cell electric vehicle fleet (Ling-Chin et al., 2023) in Hamburg, Germany.

5.8.7   Air Liquide, France

Air Liquide is a French multinational company that produces industrial gases and develops hydrogen storage and fuel cell systems (Léon, 2008). The company has installed several hydrogen storage and distribution systems worldwide, including a hydrogen storage and distribution system for fuel cell electric buses in Rotterdam, the Netherlands.

5.8.8   Linde, Germany

Linde is a German multinational company that produces industrial gases and develops hydrogen storage and distribution systems (Preuster et al., 2017). The company has installed several hydrogen storage and distribution systems worldwide, including a hydrogen storage and distribution system for fuel cell electric buses in London, UK, and a hydrogen refueling station for fuel cell electric vehicles in California, USA.

5.8.9   Hydrogenious Technologies, Germany

Hydrogenious Technologies is a German company that specializes in the development of LOHCs for hydrogen storage and transportation (Preuster, Papp, & Wasserscheid, 2017). The company has installed several pilot plants in Germany, including a hydrogen storage and transportation system for a fuel cell electric vehicle fleet in Munich.

5.8.10   Toshiba, Japan

Toshiba is a Japanese multinational company that develops and manufactures hydrogen storage and fuel cell systems (Gielen, Taibi, & Miranda, 2019). The company has installed several hydrogen storage and distribution systems in Japan, including a hydrogen refueling station for fuel cell electric vehicles in Tokyo.

5.8.11   ITM Power, UK

ITM Power is a UK-based company that specializes in the development of hydrogen storage and production systems. The company has installed several hydrogen refueling stations in the UK, including one in London, and is developing a 5 MW electrolyzer system for the Gigastack project in the UK (Wan, Zhang, & Xu 2020). Commercial and pilot plants for hydrogen storage have been developed worldwide, demonstrating the feasibility and potential of hydrogen storage technologies. These plants vary in size, technology, and application and provide valuable insights into the commercialization potential of hydrogen storage. The development and deployment of hydrogen storage technologies require collaboration between industry, academia, and government to address the technical and economic challenges and accelerate the transition to a hydrogen-based energy system.

5.9　CURRENT STATUS

Low-temperature liquefaction is one of the most promising hydrogen storage technologies, as it provides a high storage density and enables the use of hydrogen as a liquid fuel. In low-temperature liquefaction hydrogen storage, hydrogen gas is cooled down to a temperature below its boiling point (−252.87°C) (Nikolaidis & Poullikkas, 2017) to form a liquid that can be stored and transported in cryogenic tanks. The current status of low-temperature liquefaction hydrogen storage worldwide includes, but are not limited to:

5.9.1　JAPAN

Japan is one of the leading countries in low-temperature liquefaction hydrogen storage. The country has developed several technologies for hydrogen liquefaction, including the magnetic refrigeration system, which uses a magnetic field to cool hydrogen gas, and the Joule-Thomson cooling system, which uses a combination of expansion and cooling to liquefy hydrogen gas. In 2020, Japan launched the world's first liquefied hydrogen carrier, named Suiso Frontier, which can transport up to 1,250 cubic meters of liquid hydrogen at −253°C (Xu, Zhao, Hillmansen, & Roberts, 2022).

5.9.2　EUROPE

Europe has also made significant progress in low-temperature liquefaction hydrogen storage. The European Union is funding several research projects focused on developing low-temperature liquefaction technologies, including the HYDRAITE project, which aims to develop a high-performance hydrogen liquefaction system using liquid nitrogen as a cooling medium. In addition, several European countries, including Germany, France, and Norway, are investing in the development of hydrogen liquefaction facilities to support the growing demand for hydrogen as a fuel.

5.9.3　UNITED STATES

The United States is also actively developing low-temperature liquefaction hydrogen storage technologies. The Department of Energy (DOE) is funding several research projects focused on developing cost-effective hydrogen liquefaction systems, including the Liquefied Hydrogen Infrastructure (LH_2) project, which aims to demonstrate the feasibility of using liquid hydrogen as a fuel for transportation (Gielen et al., 2019). In addition, several private companies, including Air Products and Linde, are investing in the development of hydrogen liquefaction facilities to support the growing demand for hydrogen as a fuel.

5.9.4 CHINA

China is also investing heavily in low-temperature liquefaction hydrogen storage. The country is developing several technologies for hydrogen liquefaction, including the magnetic refrigeration system and the Joule-Thomson cooling system. In addition, several Chinese companies, including Sinohydro and CNPC, are investing in the development of hydrogen liquefaction facilities to support the growing demand for hydrogen as a fuel. Low-temperature liquefaction hydrogen storage is a promising technology that enables the use of hydrogen as a liquid fuel (Hwang et al., 2023). Japan, Europe, the United States, and China are actively developing and investing in low-temperature liquefaction hydrogen storage technologies, which indicates the increasing demand for hydrogen as a fuel and the potential for widespread adoption of hydrogen as a clean energy source. However, the development and deployment of low-temperature liquefaction hydrogen storage systems still face technical and economic challenges, including high capital and operating costs, safety concerns, and the need for a well-established hydrogen infrastructure.

5.10　POTENTIALS AND PREDICTABLE FUTURE

Low-temperature liquefaction hydrogen storage has the potential to play a significant role in the transition toward a low-carbon economy. The ability to store hydrogen in liquid form at a high density makes it a viable alternative to conventional fuels for a wide range of applications, including transportation, power generation, and industrial processes (Zhang et al., 2023). These are the potentials and predictable future for low-temperature liquefaction hydrogen storage.

5.10.1 TRANSPORTATION

One of the most promising applications for low-temperature liquefaction hydrogen storage is in the transportation sector. Hydrogen has long been touted as a clean alternative to gasoline and diesel, but its low energy density in gaseous form has limited its viability as a fuel for transportation (McCay & Shafiee, 2020). By liquefying hydrogen, it can be stored in a compact and portable form, making it a practical fuel for a range of vehicles, including cars, buses, and trucks. As the demand for zero emission vehicles continues to grow, low-temperature liquefaction hydrogen storage is expected to play an increasingly important role in the transportation sector.

5.10.2 POWER GENERATION

Low-temperature liquefaction hydrogen storage can also be used to store hydrogen for power generation. In this application, hydrogen is used to fuel a turbine or a fuel cell to generate electricity (Aziz, 2021). By storing hydrogen in liquid form, it is possible to maintain a steady supply of fuel for the power generation system, which can improve its efficiency and reduce its environmental impact. This application is particularly promising for remote locations where grid-connected power is not available.

5.10.3 INDUSTRIAL PROCESSES

Hydrogen is widely used in industrial processes such as refining, chemical production, and metallurgy. By liquefying hydrogen, it can be stored and transported more efficiently, reducing the cost and environmental impact of these processes (Aziz, 2021). In addition, the ability to store hydrogen in liquid form makes it possible to maintain a steady supply of fuel for these processes, which can improve their efficiency and reduce their environmental impact. The future of low-temperature liquefaction hydrogen storage is bright, with increasing investment and development in this technology. As the demand for clean energy sources continues to grow, hydrogen is expected to play an increasingly important role in the transition toward a low-carbon economy. In addition, advances in low-temperature liquefaction technologies, such as the development of more efficient cooling systems and the use of renewable energy sources for hydrogen production, are expected to make this technology more cost-effective and environmentally sustainable (Muhammed et al., 2023). However, the widespread adoption of low-temperature liquefaction hydrogen storage still faces several challenges. These include the high cost of infrastructure development and the need for a well-established hydrogen supply chain. In addition, safety concerns associated with the handling and storage of hydrogen in its liquid form must be addressed. Nonetheless, with increasing investment and development in this technology, low-temperature liquefaction hydrogen storage is poised to play a significant role in the future of clean energy.

5.11 CONCLUSIONS

In this chapter, we have extensively discussed the low-temperature liquefaction hydrogen storage technology, which has emerged as a promising solution to overcome the storage and transportation challenges of hydrogen fuel. The importance of low-temperature liquefaction hydrogen storage technology is evident as it provides a high-density storage solution for hydrogen, making transportation and distribution feasible. This technology is a viable option to store and transport large volumes of hydrogen for industrial, commercial, and residential applications. It offers an efficient and cost-effective solution that can significantly contribute to the development of the hydrogen economy. However, there are still some challenges that need to be addressed to ensure the widespread adoption of this technology. One of the major challenges is the high energy consumption during the liquefaction process, which affects the overall efficiency and economic viability of the technology. The development of advanced materials and processes for low-temperature liquefaction could help address this challenge. Furthermore, there is a need for research and development to explore the potential of low-temperature liquefaction hydrogen storage technology in combination with other technologies such as fuel cells and renewable energy sources. This could help in developing an integrated energy system that is sustainable and environmentally friendly.

ABBREVIATIONS

CcH2　Cryo-compressed Hydrogen
CFCs　Chlorofluorocarbons
LH2　Liquefied Hydrogen
LTL　Low-Temperature Liquefaction
MOFs　Metal-Organic Frameworks
PEC　Photoelectrochemical
SMR　Steam Methane Reforming

REFERENCES

Abohamzeh, E., Salehi, F., Sheikholeslami, M., Abbassi, R., & Khan, F. (2021). Review of hydrogen safety during storage, transmission, and applications processes. *Journal of Loss Prevention in the Process Industries, 72*, 104569.

Ahluwalia, R. K., Hua, T., & Peng, J. (2012). On-board and off-board performance of hydrogen storage options for light-duty vehicles. *International Journal of Hydrogen Energy, 37*(3), 2891–2910.

Al Ghafri, S. Z., Munro, S., Cardella, U., Funke, T., Notardonato, W., Trusler, J. M., & Pearce, G. (2022). Hydrogen liquefaction: a review of the fundamental physics, engineering practice and future opportunities. *Energy & Environmental Science, 15*(7), 2690–2731.

Assareh, E., & Ghafouri, A. (2023). An innovative compressed air energy storage (CAES) using hydrogen energy integrated with geothermal and solar energy technologies: a comprehensive techno-economic analysis-different climate areas-using artificial intelligent (AI). *International Journal of Hydrogen Energy, 48*(34), 12600–12621.

Aziz, M. (2021). Liquid hydrogen: a review on liquefaction, storage, transportation, and safety. *Energies, 14*(18), 5917.

Balli, M., Jandl, S., Fournier, P., & Kedous-Lebouc, A. (2017). Advanced materials for magnetic cooling: fundamentals and practical aspects. *Applied Physics Reviews, 4*(2), 021305.

Borowski, P. F., & Karlikowska, B. (2023). Clean hydrogen is a challenge for enterprises in the era of low-emission and zero-emission economy. *Energies, 16*(3), 1171.

Bosu, S., & Rajamohan, N. (2023). Recent advancements in hydrogen storage-comparative review on methods, operating conditions and challenges. *International Journal of Hydrogen Energy, 52*(2), 352–370.

Chang, H.-M., Kim, B. H., & Choi, B. (2020). Hydrogen liquefaction process with Brayton refrigeration cycle to utilize the cold energy of LNG. *Cryogenics, 108*, 103093.

Chen, L., Qi, Z., Zhang, S., Su, J., & Somorjai, G. A. (2020). Catalytic hydrogen production from methane: a review on recent progress and prospect. *Catalysts, 10*(8), 858.

Chong, L. W., Wong, Y. W., Rajkumar, R. K., Rajkumar, R. K., & Isa, D. (2016). Hybrid energy storage systems and control strategies for stand-alone renewable energy power systems. *Renewable and Sustainable Energy Reviews, 66*, 174–189.

Chu, C., Wu, K., Luo, B., Cao, Q., & Zhang, H. (2023). Hydrogen storage by liquid organic hydrogen carriers: catalyst, renewable carrier, and technology-a review. *Carbon Resources Conversion, 6*(4), 334–351.

Delaval, B., Rapson, T., Sharma, R., Hugh-Jones, W., McClure, E., Temminghoff, M., & Srinivasan, V. (2022). *Hydrogen RD&D Collaboration Opportunities*. United Kingdom: CSIRO.

Desai, F. J., Uddin, M. N., Rahman, M. M., & Asmatulu, R. (2023). A critical review on improving hydrogen storage properties of metal hydride via nanostructuring and integrating carbonaceous materials. *International Journal of Hydrogen Energy, 48*(75), 29256–29294.

Dincer, I., & Acar, C. (2015). Review and evaluation of hydrogen production methods for better sustainability. *International Journal of Hydrogen Energy*, *40*(34), 11094–11111.

Dornheim, M., Baetcke, L., Akiba, E., Ares, J.-R., Autrey, T., Barale, J., & Charbonnier, V. (2022). Research and development of hydrogen carrier based solutions for hydrogen compression and storage. *Progress in Energy*, *4*(4), 042005.

Eberle, U., Felderhoff, M., & Schueth, F. (2009). Chemical and physical solutions for hydrogen storage. *Angewandte Chemie International Edition*, *48*(36), 6608–6630.

Ehrhart, B. D., Klebanoff, L. E., Mohmand, J. A., & Markt, C. (2021). *Study of Hydrogen Fuel Cell Technology for Rail Propulsion and Review of Relevant Industry Standards*. Washington, DC: U.S. Department of Transportation.

Ellison, W. (1994). *Advanced Regenerators for Very Low Temperature Cryocoolers*. Scottsdale, AZ: General Pneumatics Corporation Western Research Center.

Fast, R. W. (2012). *Advances in Cryogenic Engineering* (Vol. 37). Berlin, Germany: Springer Science & Business Media.

Gangloff Jr, J. J., Kast, J., Morrison, G., & Marcinkoski, J. (2017). Design space assessment of hydrogen storage onboard medium and heavy duty fuel cell electric trucks. *Journal of Electrochemical Energy Conversion and Storage*, *14*(2), 021001.

Ghorbani, B., Zendehboudi, S., Saady, N. M. C., & Dusseault, M. B. (2023). Hydrogen storage in North America: status, prospects, and challenges. *Journal of Environmental Chemical Engineering*, *11*(14), 109957.

Gianotti, E., Taillades-Jacquin, M. l., Rozière, J., & Jones, D. J. (2018). High-purity hydrogen generation via dehydrogenation of organic carriers: a review on the catalytic process. *ACS Catalysis*, *8*(5), 4660–4680.

Gielen, D., Taibi, E., & Miranda, R. (2019). *Hydrogen: A Reviewable Energy Perspective*. Report prepared for the 2nd Hydrogen Energy Ministerial Meeting in Tokyo, Japan.

Grigoriev, S., Fateev, V., Bessarabov, D., & Millet, P. (2020). Current status, research trends, and challenges in water electrolysis science and technology. *International Journal of Hydrogen Energy*, *45*(49), 26036–26058.

Hassan, I., Ramadan, H. S., Saleh, M. A., & Hissel, D. (2021). Hydrogen storage technologies for stationary and mobile applications: review, analysis and perspectives. *Renewable and Sustainable Energy Reviews*, *149*, 111311.

Hassan, Q., Algburi, S., Sameen, A. Z., Jaszczur, M., & Salman, H. M. (2023). Hydrogen as an energy carrier: properties, storage methods, challenges, and future implications. *Environment Systems and Decisions*, *2023*, 1–24.

Hassan, Q., Sameen, A. Z., Salman, H. M., Jaszczur, M., & Al-Jiboory, A. K. (2023). Hydrogen energy future: advancements in storage technologies and implications for sustainability. *Journal of Energy Storage*, *72*, 108404.

Hirscher, M., Yartys, V. A., Baricco, M., von Colbe, J. B., Blanchard, D., Bowman Jr, R. C., & Chen, P. (2020). Materials for hydrogen-based energy storage-past, recent progress and future outlook. *Journal of Alloys and Compounds*, *827*, 153548.

Hosseini, S. E., & Wahid, M. A. (2020). Hydrogen from solar energy, a clean energy carrier from a sustainable source of energy. *International Journal of Energy Research*, *44*(6), 4110–4131.

Hwang, J., Maharjan, K., & Cho, H. (2023). A review of hydrogen utilization in power generation and transportation sectors: achievements and future challenges. *International Journal of Hydrogen Energy*, *48*(74), 28629–28648.

Kalamaras, C. M., & Efstathiou, A. M. (2013). Hydrogen production technologies: current state and future developments. *Paper Presented at the Conference papers in Science*, *2013*(Special Issue), 690627.

Kanoglu, M., Yilmaz, C., & Abusoglu, A. (2016). Geothermal energy use in absorption pre-cooling for Claude hydrogen liquefaction cycle. *International Journal of Hydrogen Energy, 41*(26), 11185–11200.

Kharel, S., & Shabani, B. (2018). Hydrogen as a long-term large-scale energy storage solution to support renewables. *Energies, 11*(10), 2825.

Klopčič, N., Grimmer, I., Winkler, F., Sartory, M., & Trattner, A. (2023). A review on metal hydride materials for hydrogen storage. *Journal of Energy Storage, 72*, 108456.

Le, T. H., Kim, M. P., Park, C. H., & Tran, Q. N. (2024). Recent developments in materials for physical hydrogen storage: a review. *Materials, 17*(3), 666.

Lee, K.-S., Yoon, J.-I., Son, C.-H., Lee, J.-H., Moon, C.-G., Yoo, W.-J., & Lee, B.-C. (2021). Performance characteristics of a Joule-Thomson refrigeration system with mixed refrigerant composition. *Heat Transfer Engineering, 42*(13–14), 1087–1096.

Léon, A. (2008). *Hydrogen Technology: Mobile and Portable Applications*. Berlin, Germany: Springer Science & Business Media.

Liao, M., Liu, C., & Qing, Z. (2020). *A Recent Overview of Power-to-Gas Projects*. Paper presented at the 2020 IEEE 4th Conference on Energy Internet and Energy System Integration (EI2), Wuhan, China.

Lin, X., Salari, M., Arava, L. M. R., Ajayan, P. M., & Grinstaff, M. W. (2016). High temperature electrical energy storage: advances, challenges, and frontiers. *Chemical Society Reviews, 45*(21), 5848–5887.

Ling-Chin, J., Giampieri, A., Wilks, M., Lau, S. W., Bacon, E., Sheppard, I., & Roskilly, A. P. (2023). Technology roadmap for hydrogen-fuelled transportation in the UK. *International Journal of Hydrogen Energy, 52*(part B), 705–733.

McCay, M. H., & Shafiee, S. (2020). Hydrogen: an energy carrier. In Trevor, M. L. (ed) *Future Energy* (pp. 475–493). Amsterdam, Netherlands: Elsevier.

Mohan, M., Sharma, V. K., Kumar, E. A., & Gayathri, V. (2019). Hydrogen storage in carbon materials-a review. *Energy Storage, 1*(2), e35.

Monteiro, E., & Brito, P. S. (2023). Hydrogen supply chain: Current status and prospects. *Energy Storage, 5*(7), e466.

Morales-Ospino, R., Celzard, A., & Fierro, V. (2023). Strategies to recover and minimize boil-off losses during liquid hydrogen storage. *Renewable and Sustainable Energy Reviews, 182*, 113360.

N. S., Gbadamosi, A. O., Epelle, E. I., Abdulrasheed, A. A., Haq, B., Patil, S., & Kamal, M. S. (2023). Hydrogen production, transportation, utilization, and storage: recent advances towards sustainable energy. *Journal of Energy Storage, 73*, 109207.

Muthukumar, P., Kumar, A., Afzal, M., Bhogilla, S., Sharma, P., Parida, A., & Jain, I. (2023). Review on large-scale hydrogen storage systems for better sustainability. *International Journal of Hydrogen Energy, 48*(85), 33223–33259.

Ni, M., Leung, M. K., & Leung, D. Y. (2008). Technological development of hydrogen production by solid oxide electrolyzer cell (SOEC). *International Journal of Hydrogen Energy, 33*(9), 2337–2354.

Nikolaidis, P., & Poullikkas, A. (2017). A comparative overview of hydrogen production processes. *Renewable and Sustainable Energy Reviews, 67*, 597–611.

Numazawa, T., Kamiya, K., Utaki, T., & Matsumoto, K. (2014). Magnetic refrigerator for hydrogen liquefaction. *Cryogenics, 62*, 185–192.

Popov, D., Fikiin, K., Stankov, B., Alvarez, G., Youbi-Idrissi, M., Damas, A., & Brown, T. (2019). Cryogenic heat exchangers for process cooling and renewable energy storage: a review. *Applied Thermal Engineering, 153*, 275–290.

Preuster, P., Alekseev, A., & Wasserscheid, P. (2017). Hydrogen storage technologies for future energy systems. *Annual Review of Chemical and Biomolecular Engineering, 8*, 445–471.

Preuster, P., Papp, C., & Wasserscheid, P. (2017). Liquid organic hydrogen carriers (LOHCs): toward a hydrogen-free hydrogen economy. *Accounts of Chemical Research, 50*(1), 74–85.

Ratnakar, R. R., Gupta, N., Zhang, K., van Doorne, C., Fesmire, J., Dindoruk, B., & Balakotaiah, V. (2021). Hydrogen supply chain and challenges in large-scale LH2 storage and transportation. *International Journal of Hydrogen Energy, 46*(47), 24149–24168.

Razi, F., & Dincer, I. (2022). Challenges, opportunities and future directions in hydrogen sector development in Canada. *International Journal of Hydrogen Energy, 47*(15), 9083–9102.

Rivard, E., Trudeau, M., & Zaghib, K. (2019). Hydrogen storage for mobility: a review. *Materials, 12*(12), 1973.

Rusanov, A. V., Solovey, V. V., & Lototskyy, M. V. (2020). Thermodynamic features of metal hydride thermal sorption compressors and perspectives of their application in hydrogen liquefaction systems. *Journal of Physics Energy, 2*, 021007.

Sambo, C., Dudun, A., Samuel, S. A., Esenenjor, P., Muhammed, N. S., & Haq, B. (2022). A review on worldwide underground hydrogen storage operating and potential fields. *International Journal of Hydrogen Energy, 47*(54), 22840–22880.

Stock, S. (2021). *The Potential of Biomass-Derived Activated Carbons for Hydrogen Storage.* Leoben, Austria: University of Leoben,

Tan, X. F., Kim, M., Yasuda, K., & Nogita, K. (2023). Strategies to enhance hydrogen storage performances in bulk Mg-based hydrides. *Journal of Materials Science & Technology, 153*(1), 139–158.

Tarkowski, R. (2019). Underground hydrogen storage: characteristics and prospects. *Renewable and Sustainable Energy Reviews, 105*, 86–94.

Teoh, Y. H., How, H. G., Le, T. D., Nguyen, H. T., Loo, D. L., Rashid, T., & Sher, F. (2023). A review on production and implementation of hydrogen as a green fuel in internal combustion engines. *Fuel, 333*, 126525.

Timmerhaus, K. D., Flynn, T. M., Timmerhaus, K. D., & Flynn, T. M. (1989). Storage and transfer systems. *Cryogenic Process Engineering, 1*, 377–476.

Wan, L., Zhang, W., & Xu, Z. (2020). *Overview of Key Technologies and Applications of Hydrogen Energy Storage in Integrated Energy Systems.* Paper presented at the 2020 12th IEEE PES Asia-Pacific Power and Energy Engineering Conference (APPEEC), Nanjing, China.

Wang, H., Zhao, Y., Dong, X., Yang, J., Guo, H., & Gong, M. (2022). Thermodynamic analysis of low-temperature and high-pressure (cryo-compressed) hydrogen storage processes cooled by mixed-refrigerants. *International Journal of Hydrogen Energy, 47*(67), 28932–28944.

Xu, X., Xu, H., Zheng, J., Chen, L., & Wang, J. (2020). A high-efficiency liquid hydrogen storage system cooled by a fuel-cell-driven refrigerator for hydrogen combustion heat recovery. *Energy Conversion and Management, 226*, 113496.

Xu, Z., Zhao, N., Hillmansen, S., & Roberts, C. (2022). Techno-economic analysis of hydrogen storage technologies for railway engineering. *Energies, 15*(17), 6467

Yang, M., Hunger, R., Berrettoni, S., Sprecher, B., & Wang, B. (2023). A review of hydrogen storage and transport technologies. *Clean Energy, 7*(1), 190–216.

Yang, Y., Tong, L., Liu, Y., Guo, W., Wang, L., Qiu, Y., & Ding, Y. (2023). A novel integrated system of hydrogen liquefaction process and liquid air energy storage (LAES): energy, exergy, and economic analysis. *Energy Conversion and Management, 280*, 116799.

Yartys, V. A., Lototskyy, M. V., Linkov, V., Pasupathi, S., Davids, M. W., Tolj, I., & Taube, K. (2021). HYDRIDE4MOBILITY: an EU HORIZON 2020 project on hydrogen powered fuel cell utility vehicles using metal hydrides in hydrogen storage and refuelling systems. *International Journal of Hydrogen Energy, 46*(72), 35896–35909.

Yilmaz, C. (2020). Life cycle cost assessment of a geothermal power assisted hydrogen energy system. *Geothermics, 83*, 101737.

Yin, L., & Ju, Y. (2020). Review on the design and optimization of hydrogen liquefaction processes. *Frontiers in Energy, 14*, 530–544.

Yue, M., Lambert, H., Pahon, E., Roche, R., Jemei, S., & Hissel, D. (2021). Hydrogen energy systems: a critical review of technologies, applications, trends and challenges. *Renewable and Sustainable Energy Reviews, 146*, 111180.

Yuranov, I., Autissier, N., Sordakis, K., Dalebrook, A. F., Grasemann, M., Orava, V., & Laurenczy, G. (2018). Heterogeneous catalytic reactor for hydrogen production from formic acid and its use in polymer electrolyte fuel cells. *ACS Sustainable Chemistry & Engineering, 6*(5), 6635–6643.

Zacharia, R., & Rather, S. U. (2015). Review of solid state hydrogen storage methods adopting different kinds of novel materials. *Journal of Nanomaterials, 2015*, 1–18.

Zhang, H., Gimaev, R., Kovalev, B., Kamilov, K., Zverev, V., & Tishin, A. (2019). Review on the materials and devices for magnetic refrigeration in the temperature range of nitrogen and hydrogen liquefaction. *Physica B: Condensed Matter, 558*, 65–73.

Zhang, T., Uratani, J., Huang, Y., Xu, L., Griffiths, S., & Ding, Y. (2023). Hydrogen liquefaction and storage: recent progress and perspectives. *Renewable and Sustainable Energy Reviews, 176*, 113204.

Zohuri, B., & Zohuri, B. (2019). Cryogenics and liquid hydrogen storage. *Hydrogen Energy: Challenges and Solutions for a Cleaner Future, 4*, 121–139.

6 Carbonaceous Materials for Hydrogen Storage

Beatriz Oliveira Nascimento,
Bianca Ferreira dos Santos, Moises Bastos-Neto,
and Diana Cristina Silva de Azevedo

6.1 INTRODUCTION

In recent years, the decarbonization of the global energy matrix has become a major concern and hydrogen has been suggested as the best alternative to replace fossil fuels owing to its high gravimetric energy density (120–142 MJ/kg) and clean combustion/oxidation (zero emissions of greenhouse gases). However, its application on large scale still faces critical technological and economic challenges, particularly regarding storage and transportation due to its low volumetric energy density (0.01 MJ/L at 0°C and 1 bar) (Prasek et al. 2011; Epelle et al. 2022; Ramirez-Vidal et al. 2021).

There are several methods to store hydrogen, including as compressed gas, cryogenic liquid, cryo-compressed and solid-state storage (absorption or adsorption) (Llewellyn 2019). Hydrogen compression is a well-established technology and the most applied; however, it requires high pressures to raise hydrogen density, which leads to high operational costs and security concerns. In liquefaction, although higher hydrogen storage capacity can be achieved, high costs and hydrogen boil-off are important drawbacks. Cryo-compressed is basically a combination of the previously mentioned techniques, combining low temperatures (~−253°C) with high pressures (~300 bar), which results in higher safety and storage densities, although infrastructure improvements are required for large-scale applications (Zheng et al. 2021; Mohan et al. 2019; Epelle et al. 2022). Lastly, hydrogen storage in the solid phase is considered a safer and more practical method, owing to the milder temperatures and pressures commonly applied, in addition to allowing for the storage of significant amounts of hydrogen in smaller volumes (Zhang et al. 2022; Epelle et al. 2022).

Furthermore, critical aspects related to the use of materials as hydrogen storage media are the gravimetric and volumetric capacities. Researchers often use the targets outlined by the U.S. Department of Energy (DOE) for onboard hydrogen storage as a reference in their studies. The latest targets for the complete system (tank, material, valves, regulators, etc.) are 6.5 wt% and 50 g/L for the gravimetric and volumetric capacities, respectively, which take into account the storage of 5.6 kg of usable hydrogen (U.S. Department of Energy 2017).

Regarding solid phase storage options, physisorption in nanoporous materials has promising advantages such as faster adsorption/desorption kinetics and complete reversibility. Therefore, several materials have been investigated for hydrogen storage, including zeolites, carbons, metal-organic frameworks (MOFs), covalent organic frameworks (COFs), porous aromatic frameworks (PAFs) and nanoporous organic polymers (Zheng et al. 2021; Xia, Yang, and Zhu 2013; Broom et al. 2019).

Among them, carbons-based materials stand out due to their easy and cheap preparation, high porosity and low density, besides their excellent thermal and chemical stability. In addition, its surface chemistry and textural properties (specific surface area, pore volume, pore size, etc.) can be easily tailored through synthesis strategies, enabling the optimization of their hydrogen adsorption capacity (Blankenship and Mokaya 2022; Xia, Yang, and Zhu 2013).

Furthermore, it is important to mention that nanoporous materials generally show significant performance only at cryogenic temperatures (~−196°C), hence significant research is underway to work on solutions/strategies that allow for their application at mild conditions of temperature and pressure (Xia, Yang, and Zhu 2013). Thus, this chapter aims to provide a background and some perspectives regarding hydrogen storage in the solid phase using carbon-based materials.

6.2 CARBON

Carbon is an essential element to human life, being able to combine with itself and other elements. In addition, it can form sp^3, sp^2 and sp hybrid bonds and, therefore, it exists in different allotropic forms (diamond, graphite, fullerene, carbynes), which are the basis of a variety of carbon materials with different structures and properties (Krueger 2010; Lozano-Castelló et al. 2013).

Generally, carbon materials can be synthesized from several precursors and offer diversity in structure, composition, pore size and specific surface area, among others. Moreover, their properties can be easily adjusted by modification of the methods of synthesis or post-synthesis procedures. These aspects make them interesting materials for several applications, including hydrogen storage (Llewellyn 2019; Vergara-Rubio et al. 2022; Zheng et al. 2021).

In these senses, different types of carbon-based materials have been investigated for hydrogen storage, including activated carbon, graphene, fullerene, carbon nanotubes and carbon nanofibers, as shown in Figure 6.1 (Lee et al. 2022; Mohan et al. 2019; Xia, Yang, and Zhu 2013). In the following, their major characteristics and methods of synthesis will be briefly presented.

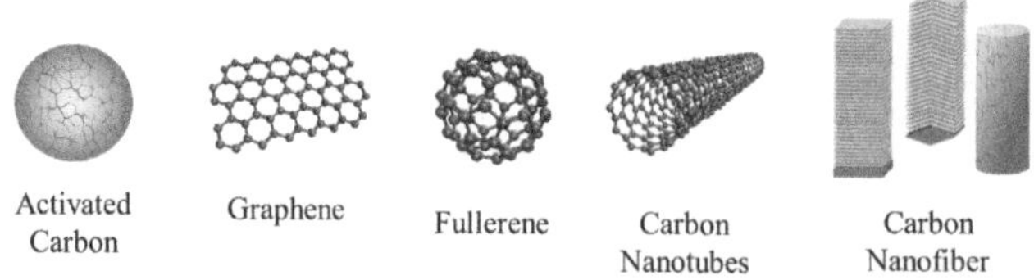

FIGURE 6.1 Structures of different carbon-based materials.

6.2.1 ACTIVATED CARBON

Activated carbon is a term used to assign porous carbon materials that exhibit a well-developed porosity and high surface area (Rodriguez-Reinoso, 2002). It is amorphous and has high chemical and thermal stability; besides, its surface chemistry can be easily modified (Marsh and Rodríguez-Reinoso 2006; Rosas et al. 2014) to target desired adsorption/catalysis properties. Industrial production of activated carbon started at the beginning of the 20th century in powder form and currently it can be found as grains, fibers, pellets, monoliths, cloths or felts (Sevilla and Mokaya 2014; Kostoglou et al. 2017; Vergara-Rubio et al. 2022).

6.2.1.1 Precursors

Activated carbon precursors are ordinarily raw materials rich in the element carbon in its composition (lignocellulosic feedstocks, biomass wastes, peat, anthracite, among others). They should be widely available, inexpensive and nonhazardous to the environment. In addition, the (atomic) O/C ratio of the precursors is considered relevant to pore development, because precursors with higher O/C ratios normally generate materials with higher specific surface areas. However, this parameter should be carefully evaluated since pore size distribution (PSD) widening and pore collapse are concurrent effects, hindering storage/capture of small molecules (e.g., hydrogen) (González-García 2018; Blankenship and Mokaya 2022).

Basically, activated carbon can be synthesized from natural (wood, coal, coconut shell, fruit seeds, agricultural by-products, biomass, etc.) and synthetic (polymers) carbon-rich sources (Blankenship and Mokaya 2022; Bedia et al. 2020; Llewellyn 2019; Xia, Yang, and Zhu 2013).

In recent years, the use of lignocellulosic biomass waste (rice husk, corncob, palm shell, etc.) as carbon precursors has been greatly encouraged, since it is renewable, inexpensive and contributes to waste disposal. The main structural polymers present in these materials are lignin, cellulose and hemicellulose; however, lignin is considered the most important for activated carbon synthesis due to its high content of carbon (Kopac, 2021; González-García, 2018).

6.2.1.2 Methods of Synthesis

Activated carbon is usually produced by chemical or physical activation, comprising one or two different thermal steps. In general terms, the chemical activation is a one-step process, where a mixture of the carbon precursor and a chemical agent is heated, whereas the physical activation is a two-step process that involves a previous pyrolysis of the precursor under inert atmosphere (or carbonization), followed by a partial gasification with an oxidizing agent (Rodriguez-Reinoso 2002; Sevilla and Mokaya 2014).

Despite this classification, the combination of these processes is also a possibility, which has been frequently reported in different arrangements. In some cases, chemical agents (e.g., NaOH and KOH) may dissolve the organic matter of soft precursors (e.g., biomass waste), leading to the impossibility of the activation and, therefore, initial carbonization is required (Bedia et al., 2020). Another situation is the application of two activation steps to improve porosity development and pore size homogeneity. In this case, a weak chemical activation (low proportion of the chemical agent) is applied to generate an early microporosity, which is followed by physical activation to properly develop the previous pore structure (Prauchner and Rodríguez-Reinoso 2008).

6.2.1.2.1   Chemical Activation

The chemical activation involves precursor impregnation with a chemical agent, pyrolysis at temperatures around 400°C–900°C, washing to remove the remaining activating agents and by-products that could be filling pores and drying (Bedia et al., 2020). It normally produces materials with particularly high surface areas (over 2,000 m²/g) and well-developed microporosity, which are very important features for gas storage (H_2, CH_4 or CO_2) (Wang and Kaskel 2012; Ströbel et al. 2006; Sevilla and Mokaya 2014).

In addition, it could be performed using acidic (H_3PO_4, H_2SO_4, etc.), alkaline (NaOH, KOH, etc.), neutral ($FeCl_3$, $MgCl_2$, $ZnCl_2$, etc.) and self-activating (potassium citrate, sodium tartrate, etc.) compounds. However, the most widely used activating agents are $ZnCl_2$, H_3PO_4 and alkali hydroxides (e.g., KOH and NaOH) (Gao et al. 2020; Ania and Raymundo-Piñero 2019; Bedia et al. 2020). Among them, KOH is mostly used to synthesize activated carbons for the adsorption of small molecules, since it generates exceptional carbons with high specific surface area and pore volume, great microporosity and tunable pore size distribution (Blankenship and Mokaya 2022).

Overall, the activating agent affects carbon yield (often referred to as "burn-off degree"), specific surface area, pore size distribution, morphology, chemical composition and distribution of elements on the carbon surface (Gao et al. 2020). Nevertheless, porosity development is affected not only by the activating agent, but also by activating agent/precursor ratio, pyrolysis temperature, heating rate, activation time and mixture method (Sevilla and Mokaya 2014).

6.2.1.2.2   Physical Activation

In physical activation, the previous pyrolysis of the carbon precursor aims to eliminate bulk volatile compounds, whereas the gasification removes pyrolysis subproducts trapped within the pores, and lastly, parts of the carbon skeleton are burned leading to the development of micropores (Sevilla and Mokaya 2014; Ania and Raymundo-Piñero 2019).

In addition, the gasification is performed at high temperatures (~750°C–1,100°C) using CO_2, steam, oxygen or their mixtures, which leads to the formation of surface oxygen complexes (Ania and Raymundo-Piñero 2019; Wang and Kaskel 2012). In fact, the gasification temperature depends on the activating agent used, ranging from 300°C to 450°C for steam and CO_2, and 700°C–900°C for pure O_2 due to its higher reactivity (Bedia et al. 2020).

Moreover, the extension of the activation process is represented by the percentage weight loss of the carbon precursor after carbonization, which is known as burn-off number, and depends on the oxidizing agent and the activation temperature (Sevilla and Mokaya 2014; Sing 2013).

Finally, the chemical and physical properties of the resulting activated carbon are closely related to the nature of the precursor, oxidizing agent, activation temperature and activation time. In general, as the temperature/time activation ratio increases, porosity development improves; however, higher porosity development usually leads to a wider pore size distribution. Regarding the oxidizing agents, steam and CO_2 are usually preferred over air (or O_2) due to the highly exothermic nature of the reaction of carbon with oxygen, which leads to excessive burning and, thus, lower activated carbon yields. Besides, steam is commonly considered more reactive than CO_2 and hence requires lower activating temperatures. However, there is still conflicting information with respect to the pattern of porosity that is generated using these agents (Sevilla and Mokaya 2014; Blankenship and Mokaya 2022).

6.2.1.3 Use as Hydrogen Storage Media

The use of activated carbons for hydrogen storage has been widely reported over the years, due to their high specific surface areas (~3,000 m²/g) and narrow pore sizes (abundance of pores <~1 nm); this value is close to the optimal pore size (~0.6 nm) recommended for hydrogen storage (Xia, Yang, and Zhu 2013).

The uptake of hydrogen on activated carbons at cryogenic temperatures has been often addressed in publications, being strongly influenced by specific surface area and micropore volume (Bicil and Doğan 2021; Fierro et al. 2010; Heo and Park 2015; Sethia and Sayari 2016; Wang, Gao, and Hu 2009). In this case, the uptakes range from 1 to 3 wt% and 2 to 7 wt% at atmospheric pressure (1 bar) and moderate pressures (10–100 bar), respectively. At room temperature (~25°C), the uptake of hydrogen is usually very low (<1 wt%), although significant improvements might be achieved by making use of high pressure (>100 bar) and carbon surface modifications (Sevilla and Mokaya 2014; Zheng et al. 2021).

6.2.2 GRAPHENE

Essentially, graphene is a two-dimensional nanomaterial composed of sp_2 hybridized carbon atoms arranged in hexagons, which possess many attractive features, such as high mechanical strength (Young's modulus of ~1 TPa) and thermal stability (~3,000–5,000 W/m/K), besides good optical transmittance (97.7%) and huge theoretical specific surface area (~2,630 m²/g) (Zhu, Li, and Wang 2013; Ferrari et al. 2015; Gutiérrez-Cruz et al. 2022).

Graphene was originally isolated in 2004 by researchers from the University of Manchester (UK) and, since then, it has been investigated for a diversity of applications owing to their promising characteristics (Novoselov et al. 2004; Abbas et al. 2022).

6.2.2.1 Methods of Synthesis

A simple method to obtain graphene is the micromechanical cleavage (or exfoliation) of graphite, which was applied in 2004 to isolate it for the first time (Novoselov et al. 2004). Graphite is basically composed of multiple graphene layers stacked up through van der Waals interactions (Krueger 2010; Sing 2013). Thus, the mechanical exfoliation consists of applying an external force that overcomes

TABLE 6.1

Major Aspects of Different Methods Used to Synthesize Graphene

Method	Advantages	Drawbacks	References
Mechanical exfoliation	High-quality monolayer, not scalable, large size	Low yield, time consuming	Gutiérrez-Cruz et al. (2022); Kumar et al. (2021); Novoselov et al. (2004)
Chemical exfoliation	Low cost, high yield, scalable	Large number of defects, low quality	Abbas et al. (2022); Ferrari et al. (2015)
Electrochemical exfoliation	Low cost, scalable, environmentally friendly	Low quality, slight oxidation	Gutiérrez-Cruz et al. (2022); Prekodravac et al. (2021)
Epitaxial growth	Low defects, high quality	High cost, low yield	Ferrari et al. (2015); Gutiérrez-Cruz et al. (2022)
Chemical vapor deposition	High quality, diversity of carbon precursors	Expensive, complex procedures, small surface	Jastrzębski and Kula (2021); Kumar et al. (2021)
Chemical synthesis (sonication)	Controlled size, simplicity, cost-effective	Time-consuming, poor reproducibility	Amiri et al. (2018); Ferrari et al. (2015)
Pyrolysis	Simplicity, scalable	Low yield, impurities in the final product	Shams et al. (2015)

these interactions to peel off a layer and, consequently, obtain graphene layers (Gutiérrez-Cruz et al. 2022). Moreover, this method is inadequate for large-scale applications, although it remains the preferred method for fundamental studies in lab scale (Ferrari et al. 2015).

Therefore, many methods have been explored to synthesize graphene, including distinct graphite exfoliation techniques (mechanical, chemical or electrochemical), chemical synthesis (sonication or reduction of graphene oxide), chemical vapor deposition, epitaxial growth and pyrolysis (Ghany, Elsherif, and Handal 2017; Prekodravac et al. 2021). Chemical vapor deposition has been the most applied method to produce high-quality graphene over the years, even though its application faces many drawbacks such as elevated cost, which makes its application on a large scale difficult, and generation of hazardous by-products (Prekodravac et al. 2021). By contrast, the chemical reduction of graphene oxide is cost-effective and considered one of the best-established methods to produce graphene on a large scale; however, impurities (oxy moieties) that affect the electronic properties of graphene are generated during this process (Titirici et al. 2015). Table 6.1 summarizes some aspects of different methods used to synthesize graphene.

6.2.2.2   Use as Hydrogen Storage Media

Pristine graphene exhibits weak affinity toward hydrogen molecules; therefore, its hydrogen uptake is low and needs improvement. Hence, besides the use of graphene as the basic material for hydrogen sorption/desorption processes, it has been investigated as (i) support for metal catalysts that will assist hydrogen sorption/desorption on graphene itself and (ii) additive to modify other storage materials (e.g., composites based on graphene and metal hydride) (Li, Zhao, and Chen 2015; Jastrzębski and Kula 2021).

6.2.3 Fullerene

Fullerenes were first described in 1985 by Kroto and coworkers (1985). Thereafter, these materials were studied for several applications as delivery systems, nanosensors and chemical catalysts. The fullerenes are composed of five- and six-membered carbon rings, which are arranged in a truncated icosahedron form. For this purpose, 12 pentagonal rings and 20 hexagonal rings are necessary to form the fullerene type C60 and the sphere core could be assumed as a pore (Marsh and Rodríguez-Reinoso 2006; Pan et al. 2020; Mojica, Alonso, and Méndez 2013).

Fullerene spherical structure implies modifications on sp^2 hybridized graphene plane orbital, thus producing an orbital intermediate between the sp^2 and sp^3, which are the orbitals attributed to the graphene and diamond bonds, respectively (Pan et al. 2020; Mojica, Alonso, and Méndez 2013). According to theoretical calculations, the C60 fullerenes own two kinds of C–C bonds. Single bonds on the five-membered rings and double bonds on the shared edge between two six-membered rings. The angle between the two σ bonds and the σ and π bonds is 106° and 101.64°, respectively. There are differences in the average bond length, for the single bonds is 0.145 nm and for the double bonds is 0.141 nm. The C60 molecules present an anisotropic state and, in solid form, are thermodynamically disordered (Pan et al. 2020).

Wang and co-workers synthesized the smallest fullerene type by an argon ionized irradiation method (Wang et al. 2001). The C20 fullerene is formed by 12 five-membered rings and 20 carbons atoms. In addition, the fullerene group includes C70, C76, C84, C240 and C540 in spherical form. Furthermore, the fullerene structure was the stepping stone in the development of a new family of materials based on the carbonaceous pyrolysis gases under different conditions. These materials take part in the nanotubes in several formats as nano-horns, necklaces and others (Marsh and Rodríguez-Reinoso 2006).

6.2.3.1 Carbon Nanotubes

In 1991, Iijima described the carbon nanotube synthesis by the arc discharge method (Iijima 1991). Carbon nanotubes (CNT) have hybridized orbital sp^2 and stemming from this atomic bonding configuration makes it possible to obtain different structures and properties. A single nanotube is composed of a one-atom-thick sheet of graphite, which is denominated graphene, rolled into a tube. Thus, the CNT presents many graphene properties as the arrangement in a honeycomb (hexagonal) pattern (Gupta, Gupta, and Sharma 2019; Prasek et al. 2011).

CNT could have its structure classified as long or short regarding its length (it presents the diameter in the order of nanometers while the length is in the order of micrometers). CNT could be single-walled, double-walled or multi-walled (related to the number of concentric tubes present in the structure), in open or close types

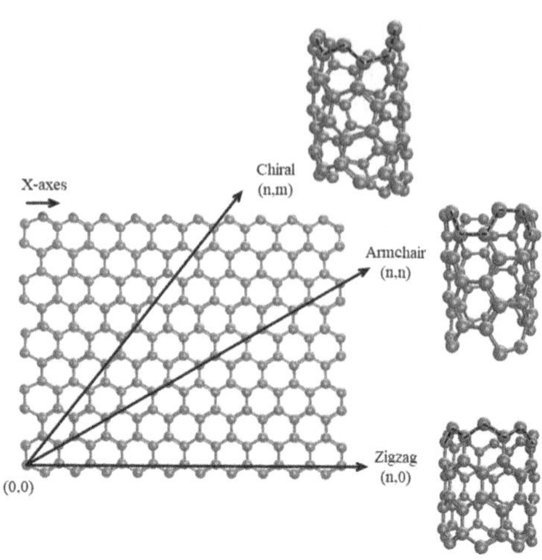

FIGURE 6.2 Schematic drawn for CNT construction from a graphene sheet in different configurations.

(it depends on whether it has open or closed ends by a half fullerene), and, at last, in a spiral format (Gupta, Gupta, and Sharma 2019). The crystallographic configuration is normally used to classify the nanotubes as zigzag, armchair and chiral types. The manner in which the graphene sheet is rolled up is represented by the indices (n, m), as shown in Figure 6.2. These numbers, n and m, are the unit vector quantity along the graphene crystalline lattice and provide the extent of nanotube twisting. In a plane where the x-axis is the aligned hexagons, when $m = 0$ is the zigzag tube, in a symmetric case $n = m$ is the armchair, and in other cases is a chiral nanotube (Gupta, Gupta, and Sharma 2019; Prasek et al. 2011; Mohan et al. 2019). A single-walled nanotube (SWNT or SWCNT) in accordance with the chirality could have either a metal or semiconductor character, for the same tube diameter could present a band gap between 0.4 and 2 eV (Gupta, Gupta, and Sharma 2019).

Multi-walled nanotube (MWNT or MWCNT) is formed by an ensemble of concentric tubes with different diameters. Each tube is separated from the other by 0.344 nm at least; due to the space restriction the arrangement of the tubes into each other is not always possible. MWNT could be defined as two types: the Russian doll model (concentric SWNT) and the parchment model (spiral around itself format). MWNT presents a better tensile strength than the SWNT due to the multilayer character, which provides also a major chemical resistance (Gupta, Gupta, and Sharma 2019).

CNT structure could be adjustable to provide electrical, mechanical and thermal unique properties. Because of that, the CNT presents several applications in conductive films, solar cells, fuel cells, supercapacitors, separation membranes and filters, purification systems, among others (Gupta, Gupta, and Sharma 2019; Prasek et al. 2011).

6.2.3.2 Methods of Synthesis

The first fullerene was synthesized by carbon vaporization, using focused pulsed laser irradiation over a rotating graphite disk under a high helium flow (Kroto et al. 1985). In 1990, Krätschmer and coworkers developed a fullerene purification method, which made it possible to obtain a reasonable quantity of C60 to measure properties for several applications. In this case, fullerene production also was made by graphite evaporation in a helium atmosphere, whereby it required drastic conditions of temperature and pressure (Krätschmer et al. 1990; Mojica, Alonso, and Méndez 2013).

Thereafter, the studies were focused on the development of new synthesis methods, which could be better controlled, with low cost and a high yield (Mojica, Alonso, and Méndez 2013). The commercially available and most employed techniques for fullerene synthesis are pyrolysis and arc plasma. In the arc plasma process, high-purity graphite rods are required, as well as high power and a large amount of helium. This method provides better control over the fullerene species generated. On the other hand, the pyrolysis process involves the combustion of a hydrocarbon in a controlled oxygen atmosphere under low pressure. In the latter method, a higher fullerene quantity is obtained per hour depending on the precursor material. Toluene was the most common hydrocarbon used initially, but the 1,2,3,4-tetrahydronaphthalene (tetralin) presented a higher processing rate. Table 6.2 presents the typical synthesis parameter values for the main fullerene production (Anctil et al. 2011).

Fullerene yield based on feedstock material is limited, about 1 and 2 wt%. It requires a large amount of material to obtain the desired product due to the formation of more stable compounds like graphite or amorphous carbons in a solid state, or even in gas form as carbon dioxide, methane, among others (Anctil et al. 2011).

The relationship between the properties and the CNT structure provides many application possibilities but introduces a chemistry challenge in controlling the nanotube diameter and chirality to obtain tunable properties (Dai 2002). Carbon nanotubes could be synthesized by various forms today, among which the main routes are arc discharge, laser ablation and chemical vapor deposition methods (Dai 2002; Mohan et al. 2019).

Arc discharge first was used for CNT synthesis by Iijima in 1991. This method consists of a gas breakdown by electricity resulting in plasma for the CNT generation. Two graphite electrodes are used, the carbon precursor being placed in the anode and the catalytic material (pure graphite) in the cathode. The electrodes are separated from each other by 1–2 mm under an inert gas atmosphere at low pressure. Then, a potential difference of around 20 V and a DC electric current of about 50–100 A are applied. The electrical current generates and keeps a non-fluctuating arc to create the plasma which is a mix of carbon, rare gases and catalyst vapor. A fluctuating arc affects the stability of the plasma which reflects on the final CNT

TABLE 6.2

Fullerene Production Details for Different Proceedings

Process/ Precursor	Fullerene Yield in the Soot (%)	Production Rate (g/h)	C60/C70 Quantities (%)	References
Arc plasma/ Graphene	13.1	1.2	69/24/7	Marković et al. (2007)
Pyrolysis/Toluene	17.5	44	43/28.5	Alford et al. (2008)
Pyrolysis/Tetralin	30	70	39/30.5	Alford et al. (2008)

quality. The generated plasma achieves very high temperatures between 3,730°C and 5,730°C. Finally, the synthesized nanotubes condense in the cathode. The arc discharge method can have the parameters appropriately adjusted to improve an increase in productivity and a higher crystallinity CNT, which promotes better electrical and mechanical properties. The main disadvantages are the production of by-products, the limitation of the size of around 1 mm and the difficulty in controlling the chirality of the nanotubes (Gupta, Gupta, and Sharma 2019; Arora and Sharma 2014; Mohan et al. 2019).

The SWNT was first synthesized using laser ablation in 1995 by Guo and coworkers (1995). The experimental set-up consists of a tubular furnace that is heated at 1,200°C, under a constant flow of an inert gas (helium or argon) at ~0.67 bar (or 500 torr). Thus, the target solid vaporization occurs by laser. That solid compound is a mixture of graphite and a metal catalyst, such as Co, Ni, Fe or Y. Then the generated product accumulates in the water-cooled collector. This technique presents a high production yield between 70% and 90% and a CNT with lower metal impurities; however, this method requires high-purity graphite and a high-power laser (Gupta, Gupta, and Sharma 2019; Baddour and Briens 2005; Prasek et al. 2011).

Chemical vapor decomposition (CVD) was first employed for CNT synthesis in 1993 by José-Yacamán and coworkers (1993). In this process, a hydrocarbon, usually ethylene or acetylene, and inert gas are fed in a tubular reactor, which is heated up between 550°C and 1,000°C. Then, the precursor together with the catalysts (normally Fe, Ni and Co) are decomposed and subsequently precipitated as CNT (Gupta, Gupta, and Sharma 2019; Baddour and Briens 2005). The CVD is the most applied method to obtain CNTs due to its simplicity, low cost, mild temperatures and ambient pressures. This method allows for the deployment of different precursor materials and creates diverse formats of products. The main disadvantage is the large size growth which leads to cracking and shrinking, thus causing the lower quality of the CNT produced (Gupta, Gupta, and Sharma 2019).

6.2.3.3   Use as Hydrogen Storage Media

Theoretically, a fullerene molecule can hold 60 hydrogen atoms in C60H60 form, on both internal and external particle surfaces. That theoretical configuration corresponds to a hydrogen capacity of 7.7 wt%. However, high temperatures are required to decompose the fullerene hydrate, around 550°C–580°C (Niemann et al. 2008). A feasible alternative to improve the hydrogen storage capacity of fullerenes was reported by Woo and coworkers. These authors synthesized a clathrate, where the C60 unit is immobilized on a hydroquinone (HQ) host framework. It showed a hydrogen capacity of 49.5 g/L at −196°C and 80 bar (Woo et al., 2018), which corresponds to 2.9 wt% taking into account the crystal density of 1.678 g/cm^3 (equivalent to the C60) (Krätschmer et al. 1990).

On the other hand, carbon nanotubes present a higher adsorption capacity. Back in 1997, Dillon et al. reported the hydrogen storage in SWNTs, which present a capacity of around 5 and 10 wt% at 0°C and 0.4 bar (Dillon et al. 1997). Although this is very good uptake, the hydrogen storage on CNTs may be influenced by internal and external (experimental conditions that can affect the adsorption capacity, like the temperature and pressure) factors. Internal factors are related to the inherent

CNT properties which could affect the magnitude of H_2 adsorption. Nanotubes with a smaller diameter cause the adsorption potentials on the walls to overlap, which increases the interaction forces between the adsorbate and adsorbent. Materials that have a higher surface area show a larger capacity since the specific surface area is related to the size and pore distribution (Lyu, Kudiiarov, and Lider 2020).

Molecular simulation can be used to estimate the behavior of carbon nanotubes. A smaller tube diameter provides better hydrogen distribution in the vicinity of the tube walls. On the other hand, for multi-walled carbon nanotubes (MWCNTs), the hydrogen adsorption capacity decreases with an increase in the number of layers, when the wall-to-wall space is kept constant. When the number of layers is constant, the adsorption capacity improves with an increase in spacing. When SWCNTs and MWCNTs are compared for the same tube diameter, single-walled carbon nanotubes (SWCNTs) exhibit higher adsorption performance (Lyu, Kudiiarov, and Lider 2020).

6.2.4 Carbon Nanofibers

Carbon nanofibers (CNFs) are defined as fibers with 92 wt% minimum of carbon in composition (Hassan et al. 2020; Frank et al. 2014). The structure is composed of a hexagonal arrangement of carbon atoms in a two-dimensional plane, which could present the turbostratic, graphitic or hybrid form depending on the precursor material and the production process. In a plane, the carbon atoms present covalent bonds in sp^2 orbitals, while between the layers van der Waals interactions occur, which is a relatively weak force. In graphitic form, the layers are organized in parallel to each other in a regular fashion. This kind of material presents an interplanar distance of about 0.335 nm. On the other hand, in the turbostratic material, the graphene sheets are irregularly layered, bent, tilted or split. This disordered stacking and the presence of sp^3 bonds cause an interplanar space around 0.344 nm. Usually, in precursors like mesophase pitch (MP) and in vapor-grown carbon fibers, the graphitic crystallinity is observed and in other precursor materials, such as the polyacrylonitrile (PAN), the turbostratic structure is obtained (Huang 2009). Figure 6.3 presents a contrast between the crystal graphite and the turbostratic material.

6.2.4.1 Methods of Synthesis

PAN (polyacrylonitrile) is the most widely used precursor for the production of CNF (Newcomb 2016; Huang 2009). The first step is the acrylonitrile (AN) polymerization,

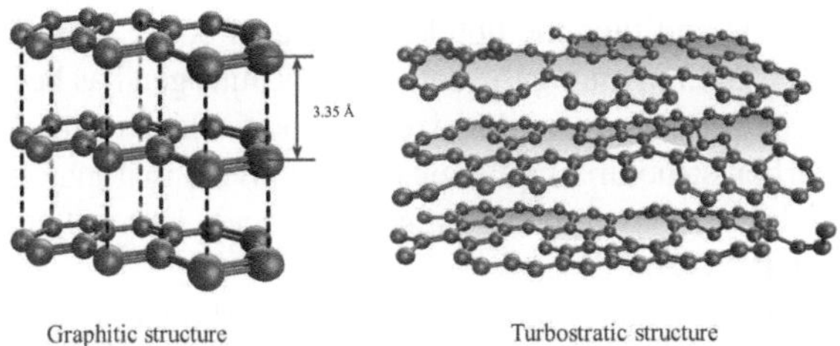

FIGURE 6.3 Comparison between a higher ordered crystalline graphitic structure and the turbostratic structure.

which is commonly carried out by solution polymerization or suspension polymerization. The copolymer used in CNF production usually has a typical molar mass value between 70,000 and 260,000 g/mol (Frank et al. 2014). The spinning process turns the organic precursor into a fiber by several techniques, such as dry spinning, melt spinning and wet spinning, which must be adequate to the nature of the starting material. Then the polymeric fiber is transformed into CNF by the stabilization and carbonization steps. In the stabilization step, the fiber passes through a thermal treatment in an oxidative atmosphere at temperatures between 200°C and 300°C, while being subject to a tensile strength. Subsequently, the material undergoes a carbonization step at a temperature range from 1,000°C to over 3,000°C in an inert atmosphere (Hassan et al. 2020; Newcomb 2016).

6.2.4.2   Use as Hydrogen Storage Media

Rodriguez's group reported higher adsorption capacities for carbon nanofibers in herring bone and platelet structural forms. These results were attributed to a mechanism of expansion of the gaps between graphite layers at higher pressures (Park et al. 1999; Chambers et al. 1998). Unfortunately, these results were not reproduced by any other laboratory. Hydrogen storage on nanotubes could be influenced by the synthesis, purification and post-synthesis treatments. These processes can create edges and separate adjacent layers in crystalline structure, thus forming preferential adsorption sites (Mohan et al. 2019; Ströbel et al. 2006).

6.3   MODIFICATIONS TO IMPROVE HYDROGEN UPTAKE

In general, high storage capacities of hydrogen on carbon materials are only achieved at cryogenic temperatures, due to the low adsorption enthalpy of hydrogen on these materials. To overcome this drawback, some strategies have been investigated to improve hydrogen storage at near room temperature (or realistic conditions), such as (i) doping with heteroatoms and (ii) doping with metals (Schaefer et al. 2016; Lozano-Castelló et al. 2013; Xia, Yang, and Zhu 2013).

6.3.1   DOPING WITH HETEROATOMS

This alternative involves the replacement of carbon atoms by light elements (e.g., N, F, P, B, Be) and modifies the electronic structure of the carbon matrix due to the electronegativity difference between carbon and heteroatoms introduced, increasing hydrogen interactions and thus the uptake (Allendorf et al. 2018; Zhao et al. 2013; Xia, Yang, and Zhu 2013). Among the heteroatoms, nitrogen has been the most studied due to some aspects such as similar size to carbon, which results in a better maintenance of the carbon structure; higher electronegativity, that may improve both the hydrogen uptake by carbons near nitrogen atoms; and chemical stability (He et al. 2021; Morandé et al. 2023).

The heteroatoms are usually introduced on the carbon matrix by post-synthesis treatment with chemicals or direct synthesis through the carbonization of precursors that contain the desired elements (Li et al. 2016). For N-doped carbons, e.g., ammonia and urea are common chemicals to post-synthesis treatment, whereas chitosan and EDTA are examples of N-rich precursors for direct carbonization (Wang et al. 2016; Das et al. 2018; He et al. 2021). Moreover, chemical vapor deposition has been pointed in the literature as an option for both post-synthesis treatment and direct synthesis. The former may minimize the reduction of the textural properties that carbons suffer owing to the doping process (Khan et al. 2020; Morandé et al. 2023).

6.3.2 DOPING WITH ALKALI AND TRANSITION METALS

Hydrogen storage improvement by metal doping (e.g., Co, Cu, Li, K, Ni, Pd, Pt) has been attributed to hydrogen spillover mechanism, which involves the split of molecular hydrogen by dissociative chemisorption on the metal particle, diffusion of hydrogen atoms toward the carbonaceous support and adsorption at surface sites (Schaefer et al. 2016; Gupta et al. 2021; Bader and Ouederni 2017).

Palladium and platinum have been extensively studied as carbon dopants, and their ability to enhance hydrogen uptake has been well established. However, their high cost has prompted research into more affordable metals such as cobalt (Schaefer et al. 2016). Wet impregnation and chemical vapor deposition are commonly employed methods for depositing metals onto carbon surfaces (Rossetti et al. 2015; Lazzarini et al. 2023; Cai et al. 2015).

Nevertheless, this alternative comprises the production of composite materials where both physisorption and chemisorption occur. The latter will affect overall adsorption reversibility and, thus, the use of this strategy should be carefully investigated (Rossetti et al. 2015).

6.4 HYDROGEN ADSORPTION

6.4.1 FUNDAMENTAL ASPECTS

6.4.1.1 Adsorption Capacity

Adsorption is defined as the interaction between a surface of a porous solid material and a target compound. In physisorption, relatively weak intermolecular forces are involved, whereas in chemisorption covalent bonds are formed between the adsorbent surface and the adsorbate (Ruthven 1984). For hydrogen, physisorption is based on van der Waals forces, where the interaction between the surface and adsorbate is often referred to as London Dispersion forces. It is proportional to the compounds' polarizability and decreases with $1/R^6$, where R is the distance between the adsorbate and adsorbent. The hydrogen molecules present a small quadrupole moment, which results in low adsorption energy (~4–5 kJ/mol) of hydrogen adsorbed on carbon materials (Lochan and Head-Gordon 2006; Ströbel et al. 2006; Mohan et al. 2019). Hydrogen chemisorption may happen in any carbon atom if the π bonds between carbons are available. Under high-pressure and high-temperature conditions, it is possible to break the hydrogen molecule bond, which requires about 4.52 eV, to create a new C–H bond. However, in the case of chemical bonds, high temperatures are necessary for hydrogen desorption, which is not desirable for hydrogen storage processing cyclic storage/delivery (Ströbel et al. 2006; Mohan et al. 2019).

Within the hydrogen sorption mechanisms on carbonaceous materials, the physisorption on adsorbent surface area plays a major role in hydrogen reversible retention in the material. Because of this, adsorbents with a higher specific surface area store more hydrogen (Li et al. 2007). It is considered that at least 85.917 m²/mol are necessary to adsorb 1 mol of hydrogen (Ströbel et al. 2006). This value was estimated considering the capacity of one graphene sheet (3 wt%), which presents a specific surface area of 1,315 m²/g, at very low temperatures. Therewith, it is possible to correlate the adsorption capacity with the specific surface area by multiplying it by a factor of 2.27×10^{-3} wt% per m²/g. Then, the U.S. Department of Energy (U.S. DOE) defined a target for onboard hydrogen storage of 6.5 wt%, which theoretically requires a specific surface area of around 3,000 m²/g (U.S. Department of Energy 2017; Mohan et al. 2019; Ströbel et al. 2006).

So, the adsorption capacity is defined as the H_2 being stored in a porous material at a given temperature and pressure. The literature reports it on a mass or volumetric basis. The storage data also can be expressed in terms of the excess or absolute amount adsorbed. The excess adsorbed amount corresponds to the hydrogen quantity present in the adsorbed phase that exceeds that on the bulk gas phase at a given temperature and pressure. While the absolute adsorbed concentration represents the sum of excess adsorption and the amount in the bulk gas at a given pressure and temperature. On the other hand, the total stored capacity is the adsorbed amount added to the quantity of hydrogen that may be compressed within the internal and external voids of a bed of solids packed in a storage vessel (Broom et al. 2016; Bastos-Neto et al. 2012).

6.4.1.2 Adsorption Kinetics

Adsorption kinetics is the rate at which hydrogen sorption takes place on the material surface (Ruthven 1984). The hydrogen physisorption is completely reversible and has a fast kinetics of adsorption (Broom et al. 2016; Bastos-Neto et al. 2012). That is an important feature since in hydrogen applications fast charge and discharge are usually required (Thomas 2007).

Carbon material structure strongly impacts the adsorption kinetics. Nanotubes and nanofibers with large diameters and interlayer distances favor diffusion. The optimal diameter for CNT is around 0.5 nm, which maintains higher diffusion coefficients. That value was estimated taking into account the Arrhenius law and the quantum Boltzmann kinetic equation (Ströbel et al. 2006).

6.4.1.3 Adsorption Enthalpy

The adsorption affinity between the solid surface and adsorbate can be measured by the adsorption enthalpy (Ruthven 1984). Physisorption, in general, is an exothermic process. When the adsorbate/adsorbent system presents a higher adsorption enthalpy, the gas molecules are strongly bound to the surface. Consequently, in the desorption process, it will be necessary to supply large amounts of energy to release the adsorbate. Hydrogen storage needs both higher capacities and a fast charging and discharging process without disrupting the solid structure. Due to the weak van der Waals forces, the hydrogen adsorption in most adsorbent materials presents low enthalpies (~4–10 kJ/mol) at cryogenic temperatures (Ströbel et al. 2006; Lang et al. 2022; Mohan et al. 2019).

6.4.2 Comparative Performance

6.4.2.1 Chemical and Textural Properties of Carbon Materials

Carbonaceous materials have properties such as low density, an extensive pore structure and good chemical stability. Moreover, they have a diverse range of structural forms that can be modified using various synthesis techniques, activation processes and modifications. Therefore, carbonaceous materials have been widely studied for hydrogen storage (Thomas 2007).

The adsorption capacity in a given pressure and temperature condition may be related to the sample porosity. Generally, the higher the surface area and pore volume the sample presents more adsorption sites and more space it provides for filling (Blankenship and Mokaya 2022). Figure 6.4 shows the hydrogen adsorption in wt% with respect to the BET area for various carbonaceous materials. Despite some degree of scattering, a linear trend can be observed.

Higher specific surface areas may be associated with a wide distribution of pores. At low temperatures, the pore-filling mechanism plays an essential role in hydrogen adsorption (Thomas 2007). Figure 6.5 illustrates the variation of hydrogen adsorption concerning the micropore volume. Similar to the BET area comparison, a linear correlation can be observed between H_2 adsorption and the micropore volume. The narrowest pores contribute to the increase of van der Waals forces due to the potential overlap of the pore walls, which actuates on the hydrogen molecules. Consequently, Henry's constant and the adsorption energy increase with the reduction in the average pore size (Thomas 2007; Ramirez-Vidal et al. 2021). However, the reduction in pore size is limited since the adsorption can only occur if the adsorbent can diffuse into the pore. By convention, the optimal pore size is approximately 0.6 nm (Blankenship and Mokaya 2022).

Different investigations regarding hydrogen adsorption on carbonaceous materials are presented in Table 6.3. Activated carbons offer several possibilities for precursor

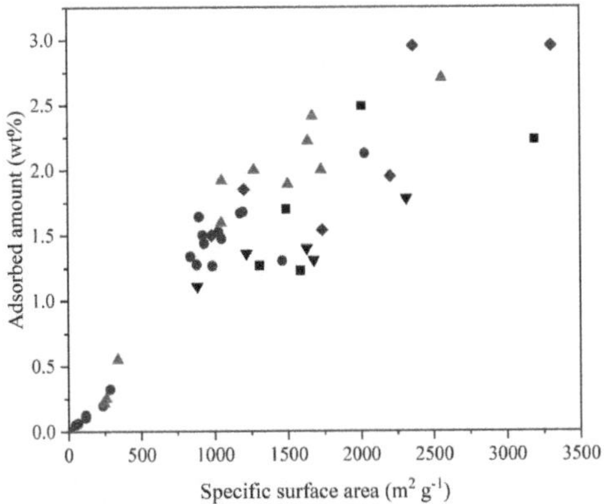

FIGURE 6.4 Correlation between the hydrogen adsorption capacity of carbon materials at 1 bar and −196°C and their specific surface areas: ■ (Wang, Gao, and Hu 2009), ● (Nijkamp et al. 2001), ▲ (Zubizarreta, Arenillas, and Pis 2009), ▼ (Pang et al. 2004), ◆ (Ramirez-Vidal et al. 2021) and ▶ (Baran, Buczek, and Zarębska 2022).

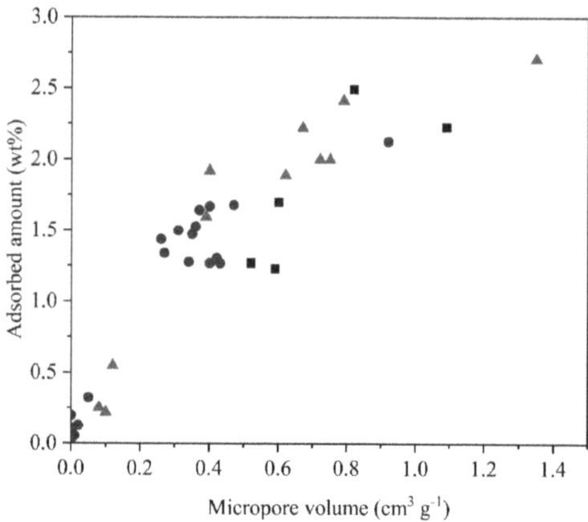

FIGURE 6.5 Correlation between the hydrogen adsorption capacity of carbon materials at 1 bar and −196°C and their micropore volume: ■ (Wang, Gao, and Hu 2009), ● (Nijkamp et al. 2001) and ▲ (Zubizarreta, Arenillas, and Pis 2009).

materials and production methodologies that could impact their physical and chemical properties. Among the activation techniques, the use of KOH and heating at 700°C–900°C creates materials with major specific surface areas (up to $4,000\,m^2g^{-1}$) while keeping the PSD narrow (Sevilla and Mokaya 2014). Most materials activated with KOH present areas between 1,500 and $3,300\,m^2/g$, except for a carbon prepared from empty fruit bunches that was carbonized at 500°C before physical activation with CO_2 at 900°C, as reported by Md Arshad et al. (2016). The material was then impregnated with different compositions of KOH at room temperature. Although the authors have increased the specific surface area with the increasing KOH concentration, the values obtained are not comparable with other chemically activated materials (Md Arshad et al. 2016).

At cryogenic temperatures, adsorption is strongly related to textural properties, especially the micropore volume and pore distribution. Heo and Park synthesized activated carbons from rice husk with different KOH/precursor ratios. The better materials were RHC1 and RHC2, which have capacities of 2.85 and 2.78 wt%, respectively. Although RHC1 presents a BET area 13% lower than RHC2, the micropore volume is $0.792\,cm^3/g$ for RHC1 and $0.775\,cm^3/g$ for RHC2 (Heo and Park 2015).

The BET area estimated for only one graphene sheet is around $2,600\,m^2/g$ (Ghosh et al. 2008). Table 6.3 presents some results of hydrogen material and BET area for graphene materials. The specific surface area obtained for synthetic materials did not reach this value because the prepared samples presented more than one layer or an agglomerated morphology (Srinivas et al. 2010; Ghosh et al. 2008). Guo and coworkers studied hierarchical graphene and found a hydrogen capacity of 4.01 wt% at −196°C and 1 bar, which is higher than other carbonaceous materials under the same conditions. The author attributed this performance to the physisorption on nanoporous structure, the presence of energetic sites, like edge and defect sites, and a minor contribution of functional groups (Guo, Wang, and Li 2013).

TABLE 6.3

Hydrogen Storage at Different Conditions Using Carbonaceous Materials

Sample	Precursor/Synthesis	S_{BET} (m^2/g)	H_2 Uptake (wt%)	Conditions	References
Activated Carbons					
AC3	Empty fruit bunch/CO_2 at 900°C and impregnated with KOH	687	1.97	−196°C, 1 bar	Md Arshad et al. (2016)
RHC1	Rice rusk/KOH/Rice rusk = 1, at 900°C	2,682	2.85	−196°C, 1 bar	Heo and Park (2015)
RHC2	Rice rusk/KOH/Rice rusk = 2, at 900°C	3,044	2.78	−196°C, 1 bar	Heo and Park (2015)
CAC3	Corncobs/chemistry activated with KOH at 800°C	3,245	2.95	−196°C, 1 bar	Liu et al. (2014)
MWSTC-50-3-700	Tamarind seeds/microwave carbonization and activation with KOH at 700°C	1,784	4.73	25°C, 40 bar	Ramesh et al. (2015)
Ch/700/700/3	Chitosan/carbonization at 700°C than activation with KOH at 700°C	3,066	2.90	−196°C, 1 bar	Wróbel-Iwaniec et al. (2015)
Maxsorb MSC30	Commercial supply by Kansai Coke and Chemicals	3,305	2.95	−196°C, 1 bar	Ramirez-Vidal et al. (2021)
AC – ultrasonic	Olive leaf/chemical activation with Zn_2Cl in an ultrasonic bath, then carbonized by microwave process	172	0.43	−196°C, 80 bar	Doğan et al. (2018)
AC – non-ultrasonic	Olive leaf/chemical activation with Zn_2Cl in a non-ultrasonic bath, then carbonized by microwave process	92	0.38	−196°C, 80 bar	Doğan et al. (2018)
Graphene					
Graphene (EG 1)	Graphite/exfoliation of graphite oxide	1,550	1.7	−196°C, 1 bar	Ghosh et al. (2008)
Hierarchical – Graphene	Graphite oxide/heating in steps up to 600°C	1,305	4.01	−196°C, 1 bar	Guo et al. (2013)
Graphene – nanosheets	Natural graphite/graphite exfoliation followed by reduction with hydrazine hydrate	640	0.64	−196°C, 1 bar	Srinivas et al. (2010)
Carbon Nanotubes					
CNT-acid	Carbon tetrachloride/chemical production, purification	470	–	–	Adeniran and Mokaya (2015)

(Continued)

TABLE 6.3 (*Continued*)

Hydrogen Storage at Different Conditions Using Carbonaceous Materials

Sample	Precursor/Synthesis	S_{BET} (m²/g)	H₂ Uptake (wt%)	Conditions	References
CN2800	Carbon tetrachloride/ chemical production, purification and activation with KOH/CNT = 2 at 600°C	2,925	4.80	−196°C, 20 bar	Adeniran and Mokaya (2015)
CN4800	Carbon tetrachloride/ chemical production, purification and activation with KOH/CNT = 4 at 600°C	3,802	5.40	−196°C, 20 bar	Adeniran and Mokaya (2015)
P-MWCNTs	Methane and hydrogen/ chemical deposition	481	0.67	25°C, 34 bar	Elyassi et al. (2017)
A-MWCNTs	Methane and hydrogen/ chemical deposition and activation with KOH at 900°C	995	1.24	25°C, 34 bar	Elyassi et al. (2017)
F-MWCNTs	Methane and hydrogen/ chemical deposition and be functionalized by sonication Process with H_2SO_4/HNO_3=3	23	0.40	25°C, 34 bar	Elyassi et al. (2017)
Carbon Nanofibers					
CNF	Methane/thermal decomposition	140	1.40	25°C, 120 bar	Hwang et al. (2002)

Adeniran and Mokaya (2015) studied carbon nanotubes that were activated with KOH. After the process, a sample that was treated with a KOH/CNT ratio of 2 and heated at 600°C presented an increase in specific surface area from 470 to 1,479 m²/g. This increase can be attributed to the creation of defects in the CNT structure, such as puckering and holes. This enhances the surface area and forms new nanopores. Table 6.3 presents a comparison between this material and two activated nanotubes, showing that this material has a higher surface area.

6.4.2.2 Surface Modification on Carbon Materials

As discussed earlier, optimizing textural properties improves hydrogen storage at cryogenic temperatures, but it alone cannot significantly improve hydrogen uptake at room temperature. Therefore, it is necessary to couple this with other strategies, such as doping with metals or heteroatoms, which may increase the adsorption enthalpy of hydrogen on carbon materials (target: ~15–25 kJ/mol), as this is a key parameter to achieve high storage capacities at room temperature (Sevilla and Mokaya 2014; Allendorf et al. 2018; Xia, Yang, and Zhu 2013). Several works have been developed in this scenario, evaluating different aspects, such as the variation of dopant concentration on carbon surface, application of different doping methods and the influence of surface functional groups on spillover process (Bader and Ouederni 2017; Chung et al. 2015; Conte et al. 2022; Zhao et al. 2013; Warczinski and Hättig 2021).

Regarding the dopant concentration variation, Conte and coworkers synthesized activated carbons from amorphous cellulose with different copper/carbon ratios (Cu/C: 0.2–5% in the starting material). As the amount of copper increased, the specific surface area improved, but reached a maximum for AC_Cu1%, and then decreased until it was very similar to the pristine sample. Likewise, the highest hydrogen storage capacity was observed for sample AC_Cu1% at both investigated temperatures, −196°C (3.91 wt%) and 25°C (0.34 wt%), indicating an enhancement of 14% and 21% with respect to the pristine sample, respectively. Above this proportion, there was no improvement, and regardless of the copper content, the increase in temperature led the capacities to drop dramatically (Conte et al. 2022).

On the other hand, Schaefer and colleagues synthesized activated carbon doped with different nickel contents (~7–32 wt%). Although the specific surface area of the samples was progressively reduced with the enhancement of Ni content, the sample doped with 32 wt% of Ni exhibited the highest hydrogen storage capacity below 5 bar (0.051 wt%) at 25°C. However, for the pressure range of 10–100 bar, the pristine sample exhibited higher hydrogen storage capacities. Moreover, high metal contents did not result in high capacities at room temperature, and the highest value observed was 0.52 wt% at 100 bar (Schaefer et al. 2016).

Furthermore, doping can be carried out by pre- or post-synthesis treatments. Bader and Ouederni evaluated the use of both strategies to incorporate Pd, Pt, Ni, Cu, Co and Ag onto activated carbons. Although incorporating metals during synthesis (after chemical activation and before the carbonization step) led to higher metal contents, it resulted in a decrease in hydrogen adsorption capacity. It was noted that metal particles would block access to micropores and could be present as agglomerates on the carbon matrix. On the other hand, post-synthesis addition of metals resulted in a small enhancement of H_2 adsorption capacity, which was attributed to the promotion of spillover kinetics, as these materials had lower textural properties compared to the pristine sample. The slight improvements were associated with the small amounts of metal incorporated (1.10–3.71 wt%) and their presence as oxides or hydroxides, as they were not previously reduced, which leads to poor catalytic activity (Bader and Ouederni 2017).

These reports exemplify the ongoing discussion regarding the occurrence of the spillover mechanism at room temperature and its actual impact on improving storage capacity, as well as the influence of varying metal content and different doping methods. Table 6.4 presents a selection of carbon materials doped with metals and their corresponding hydrogen storage capacities.

Regarding heteroatoms, their incorporation into the carbon matrix is usually achieved through post-treatment with chemicals or direct carbonization of heteroatom-containing precursors (Li et al. 2016). Nitrogen is one of the most studied elements, and its influence on the hydrogen storage capacity of carbons has been predicted theoretically. However, some experimental results reported in the literature are controversial. Some studies found that doped materials have higher capacities

TABLE 6.4

Hydrogen Uptake on Modified Carbons by Doping with Metals

Sample	Dopant/Method	H$_2$ Uptake (Before Doping)	H$_2$ Uptake (After Doping)	Conditions	References
AC-Ag (972 m^2/g)	Ag (3.71 wt%)/ WI	0.126 wt%	0.130 wt%	25°C, 25 bar	Bader and Ouederni (2017)
AC-Ni (781 m^2/g)	Ni (1.80 wt%)/ WI	0.126 wt%	0.129 wt%	25°C, 25 bar	Bader and Ouederni (2017)
AC-Cu (860 m^2/g)	Cu (1.75 wt%)/ WI	0.126 wt%	0.134 wt%	25°C, 25 bar	Bader and Ouederni (2017)
Cu-AC (604 m^2/g)	Cu (2.55 wt%)/ MWI	0.126 wt%	0.082 wt%	25°C, 25 bar	Bader and Ouederni (2017)
AC_Cu1% (1,459 m^2/g)	Cu (1 at%)/ MWI	3.42 wt%	3.91 wt%	−196°C, 80 bar	Conte et al. (2022)
AC_Cu5% (1,172 m^2/g)	Cu (5 at%)/ MWI	3.42 wt%	3.45 wt%	−196°C, 80 bar	Conte et al. (2022)
AC_Cu1% (1,459 m^2/g)	Cu (1 at%)/ MWI	0.28 wt%	0.34 wt%	25°C, 80 bar	Conte et al. (2022)
AC_Cu5% (1,172 m^2/g)	Cu (5 at%)/ MWI	0.28 wt%	0.27 wt%	25°C, 80 bar	Conte et al. (2022)
Ni-A15 (AC nanofiber) (1,468 m^2/g)	Ni/WI	1.17 wt%	1.63 wt%	30°C, 100 bar	Lee et al. (2007)
Ni-MWCNTs CR (70 m^2/g)	Ni (13 wt%)/ CR	1 wt %	0.6 wt%	30°C, 1.5 bar	Mehrabi et al. (2019)
AC-Pt0.5IMP (3,140 m^2/g)	Pt (0.46 wt%)/ WI	0.51 wt%	0.62 wt%	0°C, 40 bar	Rossetti et al. (2015)
AC-Pt0.5CVD (3,222 m^2/g)	Pt (0.28 wt%)/ CVD	0.51 wt%	0.47 wt%	0°C, 40 bar	Rossetti et al. (2015)
AC-Rh0.5IMP (3,382 m^2/g)	Rh (0.56 wt%)/ IMP	0.51 wt%	0.45 wt%	0°C, 40 bar	Rossetti et al. (2015)
AC-Rh0.5CVD (2,213 m^2/g)	Rh (0.15 wt%)/ CVD	0.51 wt%	0.52 wt%	0°C, 40 bar	Rossetti et al. (2015)
AC-Cu0.5IMP (3,262 m^2/g)	Cu (0.56 wt%)/ CVD	0.51 wt%	1.29 wt%	0°C, 40 bar	Rossetti et al. (2015)
AC-Cu0.5CVD (3,124 m^2/g)	Cu (0.61 wt%)/ CVD	0.51 wt%	1.36 wt%	0°C, 40 bar	Rossetti et al. (2015)
Pt-AX21 (2,452 m^2/g)	Pt (6 wt%)/WI	0.60 wt%	1.10 wt%	25°C, 100 bar	Wang and Kaskel (2012)
DWCNT	Pd (2 wt%)/WI	1.70 wt%	3 wt%	25°C, 30 bar	Wu et al. (2012)

CVD, Chemical vapor deposition; CR, Chemical reduction; MWI, Modified wet impregnation (during the synthesis); WI, Wet impregnation (post-synthesis).

TABLE 6.5

Hydrogen Uptake on Modified Carbons by Doping with Heteroatoms

Sample	Dopant	H$_2$ Uptake (Before Doping)	H$_2$ Uptake (After Doping)	Conditions	References
AC-N2 (1,247 m²/g)	N	2.25 wt%	2.52 wt%	−196°C, 4 bar	Li et al. (2016)
AC-P2 (1,807 m²/g)	P	2.25 wt%	3.00 wt%	−196°C, 4 bar	Li et al. (2016)
AC-S3 (943 m²/g)	S	2.25 wt%	2.72 wt%	−196°C, 4 bar	Li et al. (2016)
F1A15 (AC nanofiber) (1,053 m²/g)	F	1.17 wt%	1.32 wt%	30°C, 100 bar	Lee et al. (2007)
SCEMC (AC) (729 m²/g)	S (6.56 wt%)	–	2.02 wt%	−196°C, 20 bar	Xia et al. (2012)
SCEMC700 (AC) (1627 m²/g)	S (2.03 wt%)	–	4.43 wt%	−196°C, 20 bar	Xia et al. (2012)
N-HEG (graphene) (1,627 m²/g)	N	0.53 wt%	0.88 wt%	25°C, 20 bar	Parambhath et al. (2012)

AC, activated carbon.

than non-doped materials with similar areas, while others suggest that nitrogen doping does not affect hydrogen storage (Zhao et al. 2013; Sevilla and Mokaya 2014). Table 6.5 presents selected carbon materials doped with heteroatoms and their hydrogen storage capacities.

Moreover, although activated carbon typically has a certain oxygen content, the presence of specific oxygen-containing functional groups has been suggested to positively affect the hydrogen storage capacity of carbon materials. Blankenship and his colleagues investigated the impact of elevated levels of oxygen on the hydrogen storage capacity of carbon materials derived from cellulose acetate. The materials were activated with KOH (KOH/char = 4) at three different temperatures (600°C, 700°C and 800°C), producing AC-4600, AC-4700 and AC-4800, respectively. These materials exhibited high specific surface areas (3,771–2,000 m²/g) and micropore volumes (1.54–0.79 cm³/g), which coupled with oxygen functional groups (COOH, C–OH and O–C=O, confirmed by XPS and IR) resulted in high storage capacities. For AC-4700: (i) −196°C and 1 bar (3.9 wt%), (ii) −196°C and 20 bar (7 wt%) and (iii) 25°C and 30 bar (0.8 wt%). In addition, the isosteric heat of hydrogen adsorption was ~10 kJ/mol, which is significantly higher than the typical values reported for carbon materials (Blankenship, Balahmar, and Mokaya 2017).

Additionally, researchers have explored the impact of surface chemistry on the spillover mechanism in metal-doped carbons (Bader and Ouederni 2017; Li and Lueking 2011; Warczinski and Hättig 2021). Chung and collaborators, for instance, synthesized activated carbons with varying oxygen contents and doped them with Pd. They discovered that higher oxygen contents could enhance hydrogen uptake at room temperature by improving the spillover effect. Nevertheless, this effect is more pronounced for low pressures (<1 bar), possibly due to the short diffusion length (Chung et al. 2015).

6.4.2.3   Carbon Materials versus Other Adsorbents

The adsorption capacity of materials is strongly related to their textural properties. Silicas, zeolites and alumina are known to have lower hydrogen storage performance compared to other adsorbent materials (Bastos-Neto et al. 2012; Thomas 2007).

Metal-organic frameworks are a class of crystalline materials composed of secondary building units (SBUs) and organic linkers. MOFs exhibit great flexibility and adaptability in adsorption and catalysis. In hydrogen adsorption, van der Waals forces are responsible for the interaction between hydrogen and MOFs, and materials with a large specific surface area generally exhibit better adsorption performance (Langmi et al. 2014; Shet et al. 2021).

Some MOFs may also have accessible metal sites that can enhance hydrogen adsorption. Gedrich et al. studied the presence of open nickel sites in DUT-9 (DUT = Dresden University of Technology) and used CO_2 supercritical drying to remove the solvent while maintaining the framework structure and avoiding breakdown (Gedrich et al. 2010). This treatment improved the crystal structure and resulted in an increase in hydrogen adsorption from 1.33 to 1.66 wt% at 1 bar and −196°C (Langmi et al. 2003; Gehringer et al. 2020). Table 6.6 presents the hydrogen storage capacity of various non-carbonaceous adsorbents.

Carbonaceous materials have a relatively low adsorption enthalpy, ranging from 4 to 8 kJ/mol, whereas materials such as silica gels and zeolites have a slightly higher heat of adsorption, ranging from 5.4 to 7.7 kJ/mol and 5.9 to 7.9 kJ/mol, respectively (Thomas 2007). MOFs with open metal sites and modifications to carbonaceous materials can impact the hydrogen interactions with the material surface, thereby affecting the adsorption enthalpy. For instance, Sumida et al. studied Fe-BTT, which has Fe^{2+} coordination sites for adsorption. This material exhibits a zero-coverage isosteric heat of about 11.9 kJ/mol, which corresponds to the strong bonding energy of adsorption on the open metal sites. As coverage increases, the heat of adsorption decreases to 7 kJ/mol, which corresponds to the usual dispersive forces binding. However, these metallic sites only account for a small portion of the total available adsorption sites. Additionally, impurities such as water can compete with hydrogen and render these metals unavailable for adsorption (Thomas 2007; Sumida et al. 2010).

6.4.2.4   Gravimetric and Volumetric Capacity

The ultimate target to onboard hydrogen set by the US Office of Energy for system volumetric capacity is around 50 g/L (U.S. Department of Energy 2017). However, this capacity not only depends on the material properties but also takes into account the entire storage system, including thermal insulation, heat exchangers and other devices used to maintain cryogenic temperatures. Hence, the efficiency of a system projected to run at cryogenic temperatures can be more affected by the devices used to maintain these temperatures than a system projected to operate at ambient temperature (Allendorf et al. 2018).

TABLE 6.6

Hydrogen Storage in Various Conditions Using Different Adsorbent Materials

Sample	Precursor/Synthesis	S_{BET} (m²/g)	H₂ Uptake (wt%)	Conditions	Ref.
SBA-15	Synthetized	992	0.42	−196°C, 1 bar	Carraro et al. (2017)
SBA-15 (2.5)	Synthetized and impregnated with nickel	819	0.52	−196°C, 1 bar	Carraro et al. (2017)
NaY	Zeolite Y in Na-form	827	1.20	−196°C, 1 bar	Jhung et al. (2007)
KY	Zeolite Y/fully ion-exchanged into potassium form	777	1.41	−196°C, 1 bar	Jhung et al. (2007)
NaA	Synthetized by a hydrothermal route	–	1.54	−196°C, 15 bar	Langmi et al. (2003)
NaX	Synthetized by a hydrothermal route	622	1.79	−196°C, 15 bar	Langmi et al. (2003)
NaY	Commercial	725	1.81	−196°C, 15 bar	Langmi et al. (2003)
Al_2O_3 preshaped	Commercial	233	0.06	−196°C, 1 bar	Nijkamp et al. (2001)
Fe-BTT	Reaction between $FeCl_2$ and $H_3BTT \cdot 2HCl$ (where the BTT[3-] is the 1,3,5-benzenetristetrazolate)	2,191	2.30	−196°C, 1 bar	Sumida et al. (2010)
UiO-66 powder @80°C	Solvothermal method	1,413	1.90	−196°C, 1 bar	Bambalaza et al. (2020)
Cr-MIL101	Solvothermal methods	2,589	6.36	−196°C, 20 bar	Orcajo et al. (2018)
Ni-MOF-74	Solvothermal methods	1,286	3.18	−196°C, 20 bar	Orcajo et al. (2018)
Fe-MIL100	Commercial	1,350	2.36	−196°C, 20 bar	Orcajo et al. (2018)
MOF-5	Synthetized and activated by solvent exchange	3,512	4.50	−196°C, 100 bar	Ahmed et al. (2017)
IRMOF-20	Synthetized and activated by solvent exchange	4,073	5.70	−196°C, 100 bar	Ahmed et al. (2017)

The development of hydrogen adsorbents with the highest specific surface area and highest gravimetric basis capacity has highlighted that there is no clear correlation between these properties and volumetric capacity, which is an intrinsic material characteristic (Allendorf et al. 2018). Schlichtenmayer and Hirscher investigated hydrogen adsorption for several materials at different temperature and pressure conditions. Among the samples, the activated carbon AX-21_33 performed well in both gravimetric and volumetric capacity with, respectively, 5.71 wt% and 22 g/L at −196°C and 25 bar. This indicates that this material has a higher adsorption capacity in a compact volume, suggesting that adsorption contributes more than the hydrogen gas in the pores (Schlichtenmayer and Hirscher 2016).

However, most studies focus only on the gravimetric aspect of hydrogen adsorption, and there is no consensus on the methodology used to report the volumetric capacity. Various experimental processes can be employed to determine material density, which can impact the reported volumetric capacity. Juan-Juan et al. evaluated different density measurement methods, including true density, tap density, packing density and crystal density, for three materials (Maxsorb3000, AC-1 and MOF-5). The density values obtained increased in the order of tap density < packing density < crystal density < true density. Depending on the measurement technique and experimental conditions, the reported volumetric capacity can vary. Tap density and packing density are suitable for conversion to volumetric capacity, as true density provides only the volume of the atom, and crystal density does not account for inter-particle volume. As the ultimate goal is to fill a tank, tap density is a more appropriate measure (Juan-Juan et al. 2010).

6.5 CONCLUSION

The storage of hydrogen on carbon materials has been the subject of extensive research, but the weak interactions (i.e., low enthalpy of adsorption) between hydrogen and carbon surfaces remain a critical issue. Typically, carbon materials exhibit significant storage capacities only at cryogenic temperatures, and even then, the capacities are often small, requiring high pressures (~100 bar) to improve them. The practical application of hydrogen storage on a large scale depends on more realistic conditions of temperature and pressure.

To meet the targets for onboard storage set by the DOE, materials will need to store reasonable amounts of hydrogen at temperatures ranging from −40°C to 85°C. As a result, many efforts have been made to improve the storage capacity of carbon materials at or near ambient temperatures, including surface modifications by metal or heteroatom doping and functionalization, as well as the development of carbon composites with metal hydrides or even MOFs. The former is considered a strategy to enhance the enthalpy of adsorption of hydrogen on carbons, while the latter attempts to combine the favorable aspects of different classes of materials. Metal hydrides have been suggested as relevant materials in terms of gravimetric capacity, and this has led to investigations into the use of carbons as their scaffolds.

REFERENCES

Abbas, Q., P. A. Shinde, M. A. Abdelkareem, A. H. Alami, M. Mirzaeian, A. Yadav, and A. G. Olabi. 2022. Graphene synthesis techniques and environmental applications. *Materials* 15(21): 7804. https://doi.org/10.3390/ma15217804.

Adeniran, B., and R. Mokaya. 2015. Low temperature synthesized carbon nanotube superstructures with superior CO2 and hydrogen storage capacity. *Journal of Materials Chemistry A* 3(9): 5148–61. https://doi.org/10.1039/c4ta06539e.

Ahmed, A., Y. Liu, J. Purewal, L. D Tran, A. G Wong-Foy, M. Veenstra, J. Matzger, and D. J. Siegel. 2017. Balancing gravimetric and volumetric hydrogen density in MOFs. *Energy and Environmental Science* 2017(10): 2459–71. https://doi.org/10.1039/c7ee02477k.

Alford, J. M., C. Bernal, M. Cates, and M. D. Diener. 2008. Fullerene production in sooting flames from 1,2,3,4-tetrahydronaphthalene. *Carbon* 46(12): 1623–25. https://doi.org/10.1016/j.carbon.2008.07.004.

Allendorf, M. D., Z. Hulvey, T. Gennett, A. Ahmed, T. Autrey, J. Camp, E. S. Cho, et al. 2018. An assessment of strategies for the development of solid-state adsorbents for vehicular hydrogen storage. *Energy and Environmental Science* 11(10): 2784–812. https://doi.org/10.1039/c8ee01085d.

Amiri, A., M. Naraghi, G. Ahmadi, M. Soleymaniha, and M. Shanbedi. 2018. A review on liquid-phase exfoliation for scalable production of pure graphene, wrinkled, crumpled and functionalized graphene and challenges. *FlatChem* 8(January): 40–71. https://doi.org/10.1016/j.flatc.2018.03.004.

Anctil, A., C. W. Babbitt, R. P. Raffaelle, and B. J. Landi. 2011. Material and energy intensity of fullerene production. *Environmental Science and Technology* 45(6): 2353–59. https://doi.org/10.1021/es103860a.

Ania, C. O., and E. Raymundo-Piñero. 2019. Nanoporous carbons with tuned porosity. In *Green Energy and Technology*, edited by K. Kaneko and F. Rodríguez-Reinoso, 91–135. Singapore: Springer. https://doi.org/10.1007/978-981-13-3504-4_5.

Arora, N., and N. N. Sharma. 2014. Arc discharge synthesis of carbon nanotubes: comprehensive review. *Diamond and Related Materials* 50: 135–50. https://doi.org/10.1016/j.diamond.2014.10.001.

Baddour, C. E., and C. Briens. 2005. Carbon nanotube synthesis: a review. *International Journal of Chemical Reactor Engineering* 3(1): 1279. https://doi.org/10.2202/1542-6580.1279.

Bader, N., and A. Ouederni. 2017. Functionalized and metal-doped biomass-derived activated carbons for energy storage application. *Journal of Energy Storage* 13: 268–76. https://doi.org/10.1016/j.est.2017.07.013.

Bambalaza, S. E., H. W. Langmi, R. Mokaya, N. M. Musyoka, and L. E. Khotseng. 2020. Experimental demonstration of dynamic temperature-dependent behavior of UiO-66 metal – organic framework : compaction of hydroxylated and dehydroxylated forms of UiO-66 for high- pressure hydrogen storage. *ACS Applied Materials & Interfaces* 12(22): 24883–94. https://doi.org/10.1021/acsami.0c06080.

Baran, P., B. Buczek, and K. Zarębska. 2022. Modified activated carbon as an effective hydrogen adsorbent. *Energies* 15(17): 6122. https://doi.org/10.3390/en15176122.

Bastos-Neto, M., C. Patzschke, M. Lange, J. Möllmer, A. Möller, S. Fichtner, C. Schrage, et al. 2012. Assessment of hydrogen storage by physisorption in porous materials. *Energy and Environmental Science* 5(8): 8294–303. https://doi.org/10.1039/c2ee22037g.

Bedia, J., M. Peñas-Garzón, A. Gómez-Avilés, J. J. Rodriguez, and C. Belver. 2020. Review on activated carbons by chemical activation with FeCl3. *C - Journal of Carbon Research* 6(2): 21. https://doi.org/10.3390/c6020021.

Bicil, Z., and M. Doğan. 2021. Characterization of activated carbons prepared from almond shells and their hydrogen storage properties. *Energy and Fuels* 35(12): 10227–40. https://doi.org/10.1021/acs.energyfuels.1c00795.

Blankenship, L. S., and R. Mokaya. 2022. Modulating the porosity of carbons for improved adsorption of hydrogen, carbon dioxide, and methane: a review. *Materials Advances* 3(4): 1905–30. https://doi.org/10.1039/d1ma00911g.

Blankenship, T. S., N. Balahmar, and R. Mokaya. 2017. Oxygen-rich microporous carbons with exceptional hydrogen storage capacity. *Nature Communications* 8(1): 1545. https://doi.org/10.1038/s41467-017-01633-x.

Broom, D. P., C. J. Webb, K. E. Hurst, P. A. Parilla, T. Gennett, C. M. Brown, R. Zacharia, et al. 2016. Outlook and challenges for hydrogen storage in nanoporous materials. *Applied Physics A: Materials Science and Processing* 122(3): 1–21. https://doi.org/10.1007/s00339-016-9651-4.

Broom, D. P., C. J. Webb, G. S. Fanourgakis, G. E. Froudakis, P. N. Trikalitis, and M. Hirscher. 2019. Concepts for improving hydrogen storage in nanoporous materials. *International Journal of Hydrogen Energy* 44(15): 7768–79. https://doi.org/10.1016/j.ijhydene.2019.01.224.

Cai, J., S. Bennici, J. Shen, and A. Auroux. 2015. The influence of metal- and N-species addition in mesoporous carbons on the hydrogen adsorption capacity. *Materials Chemistry and Physics* 161: 142–52. https://doi.org/10.1016/j.matchemphys.2015.05.029

Carraro, P. M., A. A. García Blanco, C. Chanquía, K. Sapag, M. I. Oliva, and G. A. Eimer. 2017. Hydrogen adsorption of nickel-silica materials: role of the SBA-15 porosity. *Microporous and Mesoporous Materials* 248: 62–71. https://doi.org/10.1016/j.micromeso.2017.03.057.

Chambers, A., C. Park, R. Terry, K. Baker, and N. M. Rodriguez. 1998. Hydrogen storage in graphite nanofibers. *The Journal of Physical Chemistry B* 102(22): 4253–56. https://doi.org/10.1021/jp980114l.

Chung, T. Y., C. S. Tsao, H. P. Tseng, C. H. Chen, and M. S. Yu. 2015. Effects of oxygen functional groups on the enhancement of the hydrogen spillover of Pd-doped activated carbon. *Journal of Colloid and Interface Science* 441: 98–105. https://doi.org/10.1016/j.jcis.2014.10.062.

Conte, G., A. Policicchio, O. De Luca, P. Rudolf, G. Desiderio, and R. G. Agostino. 2022. Copper-doped activated carbon from amorphous cellulose for hydrogen, methane and carbon dioxide storage. *International Journal of Hydrogen Energy* 47(42): 18384–95. https://doi.org/10.1016/j.ijhydene.2022.04.029.

Dai, H. 2002. Carbon nanotubes: synthesis, integration, and properties. *Accounts of Chemical Research* 35(12): 1035–44. https://doi.org/10.1021/ar0101640.

Das, T. K., S. Banerjee, P. Sharma, V. Sudarsan, and P. U. Sastry. 2018. Nitrogen doped porous carbon derived from EDTA: effect of pores on hydrogen storage properties. *International Journal of Hydrogen Energy* 43(17): 8385–94. https://doi.org/10.1016/j.ijhydene.2018.03.081.

Dillon, A. C., K. M. Jones, T. A. Bekkedahl, C. H. Kiang, D. S. Bethune, and M. J. Heben. 1997. Storage of hydrogen in single-walled carbon nanotubes. *Nature* 386(6623): 377–79. https://doi.org/10.1038/386377a0.

Doğan, E. E., P. Tokcan, and B. K. Kizilduman. 2018. Storage of hydrogen in activated carbons and carbon nanotubes. *Advances in Materials Science* 18(4): 5–16. https://doi.org/10.1515/adms-2017-0045.

Elyassi, M., A. Rashidi, M. R. Hantehzadeh, and S. M. Elahi. 2017. Hydrogen storage behaviors by adsorption on multi-walled carbon nanotubes. *Journal of Inorganic and Organometallic Polymers and Materials* 27(1): 285–95. https://doi.org/10.1007/s10904-016-0471-y.

Epelle, E. I., K. S. Desongu, W. Obande, A. A. Adeleke, P. P. Ikubanni, J. A. Okolie, and B. Gunes. 2022. A comprehensive review of hydrogen production and storage: a focus on the role of nanomaterials. *International Journal of Hydrogen Energy* 47(47): 20398–431. https://doi.org/10.1016/j.ijhydene.2022.04.227.

Ferrari, A. C., F. Bonaccorso, V. Fal'ko, K. S. Novoselov, S. Roche, P. Bøggild, S. Borini, et al. 2015. Science and technology roadmap for graphene, related two-dimensional crystals, and hybrid systems. *Nanoscale* 7(11): 4598–810. https://doi.org/10.1039/c4nr01600a.

Fierro, V., A. Szczurek, C. Zlotea, J. F. Marêché, M. T. Izquierdo, A. Albiniak, M. Latroche, G. Furdin, and A. Celzard. 2010. Experimental evidence of an upper limit for hydrogen storage at 77 K on activated carbons. *Carbon* 48(7): 1902–11. https://doi.org/10.1016/j.carbon.2010.01.052.

Frank, E., L. M. Steudle, D. Ingildeev, J. M. Spörl, and M. R. Buchmeiser. 2014. Carbon fibers: precursor systems, processing, structure, and properties. *Angewandte Chemie - International Edition* 53(21): 5262–98. https://doi.org/10.1002/anie.201306129.

Gao, Y., Q. Yue, B. Gao, and A. Li. 2020. Insight into activated carbon from different kinds of chemical activating agents: a review. *Science of the Total Environment* 746: 141094. https://doi.org/10.1016/j.scitotenv.2020.141094.

Gedrich, K., I. Senkovska, N. Klein, U. Stoeck, A. Henschel, M. R. Lohe, I. A. Baburin, U. Mueller, and S. Kaskel. 2010. A highly porous metal-organic framework with open nickel sites. *Angewandte Chemie - International Edition* 49(45): 8489–92. https://doi.org/10.1002/anie.201001735.

Gehringer, D., T. Dengg, M. N. Popov, and D. Holec. 2020. Interactions between a H2 molecule and carbon nanostructures: a DFT study. *C - Journal of Carbon Research* 6(1): 16. https://doi.org/10.3390/c6010016.

Ghany, N. A. A., S. A. Elsherif, and H. T. Handal. 2017. Revolution of graphene for different applications: state-of-the-art. *Surfaces and Interfaces* 9(June): 93–106. https://doi.org/10.1016/j.surfin.2017.08.004.

Ghosh, A., K. S. Subrahmanyam, K. S. Krishna, S. Datta, A. Govindaraj, S. K. Pati, and C. N. R. Rao. 2008. Uptake of H_2 and CO_2 by graphene. *The Journal of Physical Chemistry C* 112(40): 15704–7. https://pubs.acs.org/doi/abs/10.1021/jp805802w.

González-García, P. 2018. Activated carbon from lignocellulosics precursors: a review of the synthesis methods, characterization techniques and applications. *Renewable and Sustainable Energy Reviews* 82(August 2017): 1393–414. https://doi.org/10.1016/j.rser.2017.04.117.

Guo, C. X., Y. Wang, and C. M. Li. 2013. Hierarchical graphene-based material for over 4.0 wt% physisorption hydrogen storage capacity. *ACS Sustainable Chemistry and Engineering* 1(1): 14–8. https://doi.org/10.1021/sc3000306.

Guo, T., P. Nikolaev, A. Thess, D. T. Colbert, and R. E. Smalley. 1995. Catalytic growth of single-walled manotubes by laser vaporization. *Chemical Physics Letters* 243(1–2): 49–54. https://doi.org/10.1016/0009-2614(95)00825-O.

Gupta, N., S. M. Gupta, and S. K. Sharma. 2019. Carbon nanotubes: synthesis, properties and engineering applications. *Carbon Letters* 29(5): 419–47. https://doi.org/10.1007/s42823-019-00068-2.

Gupta, A., G. V. Baron, P. Perreault, S. Lenaerts, R. G. Ciocarlan, P. Cool, P. G.M. Mileo, et al. 2021. Hydrogen clathrates: next generation hydrogen storage materials. *Energy Storage Materials* 41(May): 69–107. https://doi.org/10.1016/j.ensm.2021.05.044.

Gutiérrez-Cruz, A., A. R. Ruiz-Hernández, J. F. Vega-Clemente, D. G. Luna-Gazcón, and J. Campos-Delgado. 2022. A review of top-down and bottom-up synthesis methods for the production of graphene, graphene oxide and reduced graphene oxide. *Journal of Materials Science* 57(31): 14543–78. https://doi.org/10.1007/s10853-022-07514-z.

Hassan, M. F., M. A. Sabri, H. Fazal, A. Hafeez, N. Shezad, and M. Hussain. 2020. Recent trends in activated carbon fibers production from various precursors and applications-a comparative review. *Journal of Analytical and Applied Pyrolysis* 145(May 2019): 104715. https://doi.org/10.1016/j.jaap.2019.104715.

He, H., X. Xu, D. Liu, J. Li, Y. Wei, H. Tang, J. Li, X. Li, Z. Z. Xie, and D. Qu. 2021. The impacts of nitrogen doping on the electrochemical hydrogen storage in a carbon. *International Journal of Energy Research* 45(6): 9326–39. https://doi.org/10.1002/er.6463.

Heo, Y. J., and S. J. Park. 2015. Synthesis of activated carbon derived from rice husks for improving hydrogen storage capacity. *Journal of Industrial and Engineering Chemistry* 31: 330–4. https://doi.org/10.1016/j.jiec.2015.07.006.

Huang, X. 2009. Fabrication and properties of carbon fibers. *Materials* 2(4): 2369–403. https://doi.org/10.3390/ma2042369.

Hwang, J. Y., S. H. Lee, K. S. Sim, and J. W. Kim. 2002. Synthesis and hydrogen storage of carbon nanofibers. *Synthetic Metals* 126(1): 81–5. https://doi.org/10.1016/S0379-6779(01)00543-4.

Iijima, S. 1991. Helical microtubules of graphitic carbon. *Nature* 354: 56–8.

Jastrzębski, K., and P. Kula. 2021. Emerging technology for a green, sustainable energy promising materials for hydrogen storage, from nanotubes to graphene-a review. *Materials* 14(10): 2499. https://doi.org/10.3390/ma14102499.

Jhung, S. H., J. S. Lee, J. W. Yoon, D. P. Kim, and J. S. Chang. 2007. Low-temperature adsorption of hydrogen on ion-exchanged Y zeolites. *International Journal of Hydrogen Energy* 32(17): 4233–7. https://doi.org/10.1016/j.ijhydene.2007.05.014.

José-Yacamán, M., M. Miki-Yoshida, L. Rendón, and J. G. Santiesteban. 1993. Catalytic growth of carbon microtubules with fullerene structure. *Applied Physics Letters* 62(6): 657–9. https://doi.org/10.1063/1.108857.

Juan-Juan, J., J. P. Arco-Lozar, F. Suárez-García, D. Cazorla-Amorós, and A. Linares-Solano. 2010. a comparison of hydrogen storage in activated carbons and a metal - organic framework (MOF-5). *Carbon* 48: 2906–9. https://doi.org/10.1016/j.carbon.2010.04.025.

Khan, A., M. Goepel, J. C. Colmenares, and R. Gläser. 2020. Chitosan-based N-doped carbon materials for electrocatalytic and photocatalytic applications. *ACS Sustainable Chemistry & Engineering* 8(12): 4708–27. https://doi.org/10.1021/acssuschemeng.9b07522.

Kopac, T. 2021. Hydrogen storage characteristics of bio-based porous carbons of different origin: a comparative review. *International Journal of Energy Research* 45(15): 20497–523. https://doi.org/10.1002/er.7130.

Kostoglou, N., C. Koczwara, C. Prehal, V. Terziyska, B. Babic, B. Matovic, G. Constantinides, et al. 2017. Nanoporous activated carbon cloth as a versatile material for hydrogen adsorption, selective gas separation and electrochemical energy storage. *Nano Energy* 40(January): 49–64. https://doi.org/10.1016/j.nanoen.2017.07.056.

Krätschmer, W., L. D. Lamb, K. Fostiropoulos, and D. R. Huffman. 1990. Solid C60: a new form of carbon. *Nature* 347(6291): 354–8. https://doi.org/10.1038/347354a0.

Kroto, H. W., J. R. Heath, S. C. O'Brien, R. F. Curl, and R. E. Smalley. 1985. C60: buckminsterfullerene. *Nature* 318(6042): 162–3. https://doi.org/10.1038/318162a0.

Krueger, A. 2010. *Carbon Materials and Nanotechnology.* Padstow: Wiley https://doi.org/10.1002/9783527629602.

Kumar, N., R. Salehiyan, V. Chauke, O. J. Botlhoko, K. Setshedi, M. Scriba, M. Masukume, and S. S. Ray. 2021. Top-down synthesis of graphene: a comprehensive review. *FlatChem* 27(February): 100224. https://doi.org/10.1016/j.flatc.2021.100224.

Lang, C., Y. Jia, X. Yan, L. Ouyang, M. Zhu, and X. Yao. 2022. Molecular chemisorption: a new conceptual paradigm for hydrogen storage. *Chemical Synthesis* 2: 1–13. https://doi.org/10.20517/cs.2021.15.

Langmi, H.W., A Walton, M.M. Al-Mamouri, S.R. Johnson, D Book, J.D. Speight, P.P. Edwards, I Gameson, P.A. Anderson, and I.R. Harris. 2003. Hydrogen adsorption in zeolites A, X, Y and RHO. *Journal of Alloys and Compounds* 356–357(August): 710–5. https://doi.org/10.1016/S0925-8388(03)00368-2.

Langmi, H. W., J. Ren, B. North, M. Mathe, and D. Bessarabov. 2014. Hydrogen storage in metal-organic frameworks: a review. *Electrochimica Acta* 128(2014): 368–92. https://doi.org/10.1016/j.electacta.2013.10.190.

Lazzarini, A., A. Marino, R. Colaiezzi, O. De Luca, G. Conte, A. Policicchio, A. Aloise, and M. Crucianelli. 2023. Boronation of biomass-derived materials for hydrogen storage. *Compounds* 3(1): 244–79. https://doi.org/10.3390/compounds3010020.

Lee, S. Y., J. H. Lee, Y. H. Kim, J. W. Kim, K. J. Lee, and S. J. Park. 2022. Recent progress using solid-state materials for hydrogen storage: a short review. *Processes* 10(2): 1–19. https://doi.org/10.3390/pr10020304.

Li, Q., and A. D. Lueking. 2011. Effect of surface oxygen groups and water on hydrogen spillover in pt-doped activated carbon. *Journal of Physical Chemistry C* 115(10): 4273–82. https://doi.org/10.1021/jp105923a.

Li, Y., D. Zhao, Y. Wang, R. Xue, Z. Shen, and X. Li. 2007. The mechanism of hydrogen storage in carbon materials. *International Journal of Hydrogen Energy* 32(13): 2513–7. https://doi.org/10.1016/j.ijhydene.2006.11.010.

Li, F., J. Zhao, and Z. Chen. 2015. Carbon-based nanomaterials for H2 storage. In *Carbon Nanomaterials for Advanced Energy Systems: Advances in Materials Synthesis and Device Applications*, edited by W. Lu, J-B Baek and L. Dai, 407–37. Haboken, New Jersey: John Wiley & Sons. https://doi.org/10.1002/9781118980989.ch13.

Li, D., W. B. Li, J. S. Shi, and F. W. Xin. 2016. Influence of doping nitrogen, sulfur, and phosphorus on activated carbons for gas adsorption of H_2, CH_4 and CO_2. *RSC Advances* 6(55): 50138–43. https://doi.org/10.1039/c6ra06620h.

Liu, X., C. Zhang, Z. Geng, and M. Cai. 2014. High-pressure hydrogen storage and optimizing fabrication of corncob-derived activated carbon. *Microporous and Mesoporous Materials* 194: 60–5. https://doi.org/10.1016/j.micromeso.2014.04.005.

Llewellyn, P. L. 2019. Storage of hydrogen on nanoporous adsorbents. In *Green Energy and Technology*, edited by K. Kaneko and F. Rodríguez-Reinoso, 255–86. Singapore: Springer. https://doi.org/10.1007/978-981-13-3504-4_10.

Lochan, R C., and M. Head-Gordon. 2006. Computational studies of molecular hydrogen binding affinities: the role of dispersion forces, electrostatics, and orbital interactions. *Physical Chemistry Chemical Physics* 8(12): 1357–70. https://doi.org/10.1039/b515409j.

Lozano-Castello, D., F. Suarez-Garcia, Á. Linares-Solano, and D. Cazorla-Amoros. 2013. Advances in hydrogen storage in carbon materials. In *Renewable Hydrogen Technologies: Production, Purification, Storage, Applications and Safety*, edited by L. M. Gandía, G. Arzamendi and P. M. Diéguez, 269–91. Amsterdam: Elsevier. https://doi.org/10.1016/B978-0-444-56352-1.00012-X.

Lyu, J., V. Kudiiarov, and A. Lider. 2020. An overview of the recent progress in modifications of carbon nanotubes for hydrogen adsorption. *Nanomaterials* 10(2): 255. https://doi.org/10.3390/nano10020255.

Marković, Z., B. Todorović-Marković, I. Mohai, Z. Farkas, E. Kovats, J. Szepvolgyi, D. Otašević, P. Scheier, S. Feil, and N. Romčević. 2007. Comparative process analysis of fullerene production by the arc and the radio-frequency discharge methods. *Journal of Nanoscience and Nanotechnology* 7(4–5): 1357–69. https://doi.org/10.1166/jnn.2007.315.

Marsh, H., and F. Rodríguez-Reinoso. 2006. *Activated Carbon*. Oxford: Elsevier. https://doi.org/10.1016/B978-008044463-5/50018-2.

Md Arshad, S. H., N. Ngadi, A. A. Aziz, N. S. Amin, M. Jusoh, and S. Wong. 2016. Preparation of activated carbon from empty fruit bunch for hydrogen storage. *Journal of Energy Storage* 8(November): 257–61. https://doi.org/10.1016/j.est.2016.10.001.

Mehrabi, M., A. Reyhani, P. Parvin, and S. Z. Mortazavi. 2019. Surface structural alteration of multi-walled carbon nanotubes decorated by nickel nanoparticles based on laser ablation/chemical reduction methods to enhance hydrogen storage properties. *International Journal of Hydrogen Energy* 44(7): 3812–23. https://doi.org/10.1016/j.ijhydene.2018.12.122.

Mohan, M., V. K. Sharma, E. Anil Kumar, and V. Gayathri. 2019. Hydrogen storage in carbon materials-a review. *Energy Storage* 1(2): e35. https://doi.org/10.1002/est2.35.

Mojica, M., J. A. Alonso, and F. Méndez. 2013. Synthesis of fullerenes. *Journal of Physical Organic Chemistry* 26(7): 526–39. https://doi.org/10.1002/poc.3121.

Morandé, A., P. Lillo, E. Blanco, C. Pazo, A. B. Dongil, X. Zarate, M. Saavedra-Torres, et al. 2023. Modification of a commercial activated carbon with nitrogen and boron: hydrogen storage application. *Journal of Energy Storage* 64(January): 107193. https://doi.org/10.1016/j.est.2023.107193.

Newcomb, B. A. 2016. Processing, structure, and properties of carbon fibers. *Composites Part A: Applied Science and Manufacturing* 91: 262–82. https://doi.org/10.1016/j.compositesa.2016.10.018.

Niemann, M. U., S. S. Srinivasan, A. R. Phani, A. Kumar, D. Yogi Goswami, and E. K. Stefanakos. 2008. Nanomaterials for hydrogen storage applications: a review. *Journal of Nanomaterials* 2008(1): 950967. https://doi.org/10.1155/2008/950967.

Nijkamp, M. G., J. E. M. J. Raaymakers, A. J. Van Dillen, and K. P. De Jong. 2001. Hydrogen storage using physisorption-materials demands. *Applied Physics A: Materials Science and Processing* 72(5): 619–23. https://doi.org/10.1007/s003390100847.

Novoselov, K. S., A. K. Geim, S. V. Morozov, D. Jiang, Y. Zhang, S. V. Dubonos, I. V. Grigorieva, and A. A. Firsov. 2004. Electric field in atomically thin carbon films. *Science* 306(5696): 666–9. https://doi.org/10.1126/science.1102896.

Orcajo, G., H. Montes-Andr, A. Villajos, C. Martos, J. A. Botas, and G. Calleja. 2018. Li-crown ether complex inclusion in MOF materials for enhanced H2 volumetric storage capacity at room temperature. *International Journal of Hydrogen Energy* 44(35): 19285–93. https://doi.org/10.1016/j.ijhydene.2018.03.151.

Pan, Y., X. Liu, W. Zhang, Z. Liu, G. Zeng, B. Shao, Q. Liang, et al. 2020. Advances in photocatalysis based on fullerene C60 and its derivatives: properties, mechanism, synthesis, and applications. *Applied Catalysis B: Environmental* 265(October 2019): 118579. https://doi.org/10.1016/j.apcatb.2019.118579.

Pang, J., J. E. Hampsey, Z. Wu, Q. Hu, and Y. Lu. 2004. Hydrogen adsorption in mesoporous carbons. *Applied Physics Letters* 85(21): 4887–9. https://doi.org/10.1063/1.1827338.

Parambhath, V. B., R. Nagar, and S. Ramaprabhu. 2012. Effect of nitrogen doping on hydrogen storage capacity of palladium decorated graphene. *Langmuir* 28(20): 7826–33. https://doi.org/10.1021/la301232r.

Park, C., P. E. Anderson, A. Chambers, C. D. Tan, R. Hidalgo, and N. M. Rodriguez. 1999. Further studies of the interaction of hydrogen with graphite nanofibers. *Journal of Physical Chemistry* B 103(48): 10572–81. https://doi.org/10.1021/jp990500i.

Prasek, J., J. Drbohlavova, J. Chomoucka, J. Hubalek, O. Jasek, V. Adam, and R. Kizek. 2011. Methods for carbon nanotubes synthesis - review. *Journal of Materials Chemistry* 21(40): 15872–84. https://doi.org/10.1039/c1jm12254a.

Prauchner, M. J., and F. Rodríguez-Reinoso. 2008. Preparation of granular activated carbons for adsorption of natural gas. *Microporous and Mesoporous Materials* 109 (1–3): 581–4. https://doi.org/10.1016/j.micromeso.2007.04.046.

Prekodravac, J. R., D. P. Kepić, J. C. Colmenares, D. A. Giannakoudakis, and S. P. Jovanović. 2021. A comprehensive review on selected graphene synthesis methods: from electrochemical exfoliation through rapid thermal annealing towards biomass pyrolysis. *Journal of Materials Chemistry C* 9(21): 6722–48. https://doi.org/10.1039/d1tc01316e.

Ramesh, T., N. Rajalakshmi, and K. S. Dhathathreyan. 2015. Activated carbons derived from tamarind seeds for hydrogen storage. *Journal of Energy Storage* 4: 89–95. https://doi.org/10.1016/j.est.2015.09.005.

Ramirez-Vidal, P., R. L. S. Canevesi, G. Sdanghi, S. Schaefer, G. Maranzana, A. Celzard, and V. Fierro. 2021. A step forward in understanding the hydrogen adsorption and compression on activated carbons. *ACS Applied Materials and Interfaces* 13(10): 12562–74. https://doi.org/10.1021/acsami.0c22192.

Rodriguez-Reinoso, F. 2002. Production and applications of activated carbons. In *Handbook of Porous Solids*, edited by F. Schüth, K. S. W. Sing, and J. Weitkamp, 3:1766–1827. Weinheim: Wiley-VCH Verlag GmbH. https://doi.org/10.1002/9783527618286.ch24a.

Rosas, J. M., R. Berenguer, M. J. Valero-Romero, J. Rodríguez-Mirasol, and T. Cordero. 2014. Preparation of different carbon materials by thermochemical conversion of lignin. *Frontiers in Materials* 1(December): 1–17. https://doi.org/10.3389/fmats.2014.00029.

Rossetti, I., G. Ramis, A. Gallo, and A. Di Michele. 2015. Hydrogen storage over metal-doped activated carbon. *International Journal of Hydrogen Energy* 40(24): 7609–16. https://doi.org/10.1016/j.ijhydene.2015.04.064.

Ruthven, D. M. 1984. *Principles of Adsorption and Adsorption Processes*. New York: John Wiley & Sons.

Schaefer, S., V. Fierro, A. Szczurek, M. T. Izquierdo, and A. Celzard. 2016. Physisorption, chemisorption and spill-over contributions to hydrogen storage. *International Journal of Hydrogen Energy* 41(39): 17442–52. https://doi.org/10.1016/j.ijhydene.2016.07.262.

Schlichtenmayer, M., and M. Hirscher. 2016. The usable capacity of porous materials for hydrogen storage. *Applied Physics A: Materials Science and Processing* 122(4): 379. https://doi.org/10.1007/s00339-016-9864-6.

Sethia, G., and A. Sayari. 2016. Activated carbon with optimum pore size distribution for hydrogen storage. *Carbon* 99: 289–94. https://doi.org/10.1016/j.carbon.2015.12.032.

Sevilla, M., and R. Mokaya. 2014. Energy storage applications of activated carbons: supercapacitors and hydrogen storage. *Energy and Environmental Science* 7(4): 1250–80. https://doi.org/10.1039/C3EE43525C.

Shams, S. S., R. Zhang, and J. Zhu. 2015. Graphene synthesis: a review. *Materials Science-Poland* 33(3): 566–78. https://doi.org/10.1515/msp-2015-0079.

Shet, S. P., S. S. Priya, K. Sudhakar, and M. Tahir. 2021. A review on current trends in potential use of metal-organic framework for hydrogen storage. *International Journal of Hydrogen Energy* 46(21): 11782–803. https://doi.org/10.1016/j.ijhydene.2021.01.020.

Sing, K. S. W. 2013. Adsorption by active carbons. *Adsorption by Powders and Porous Solids: Principles, Methodology and Applications: 2nd ed.* Heidelberg: Elsevier. https://doi.org/10.1016/B978-0-08-097035-6.00010-3.

Srinivas, G., Y. Zhu, R. Piner, N. Skipper, M. Ellerby, and R. Ruoff. 2010. Synthesis of graphene-like nanosheets and their hydrogen adsorption capacity. *Carbon* 48(3): 630–5. https://doi.org/10.1016/j.carbon.2009.10.003.

Ströbel, R., J. Garche, P. T. Moseley, L. Jörissen, and G. Wolf. 2006. Hydrogen storage by carbon materials. *Journal of Power Sources* 159(2): 781–801. https://doi.org/10.1016/j.jpowsour.2006.03.047.

Sumida, K., S. Horike, S. S. Kaye, Z. R. Herm, W. L. Queen, C. M. Brown, F. Grandjean, G. J. Long, A. Dailly, and J. R. Long. 2010. Hydrogen storage and carbon dioxide capture in an iron-based sodalite-type metal-organic framework (Fe-BTT) discovered via high-throughput methods. *Chemical Science* 1(2): 184–91. https://doi.org/10.1039/c0sc00179a.

Thomas, K. M. 2007. Hydrogen adsorption and storage on porous materials. *Catalysis Today* 120 (3–4): 389–98. https://doi.org/10.1016/j.cattod.2006.09.015.

Titirici, M. M., R. J. White, N. Brun, V. L. Budarin, D. S. Su, F. D. Monte, J. H. Clark, and M. J. MacLachlan. 2015. Sustainable carbon materials. *Chemical Society Reviews* 44(1): 250–90. https://doi.org/10.1039/c4cs00232f.

U.S Department of Energy. 2017. DOE Technical Targets for Onboard Hydrogen Storage for Light-Duty Vehicles | Department of Energy. Office of Energy Efficiency & Renewable Energy. 2017. https://www.energy.gov/eere/fuelcells/doe-technical-targets-onboard-hydrogen-storage-light-duty-vehiclesIls/doe-technical-targets-onboard-hydrogen-storage-light-duty-vehicles.

Vergara-Rubio, A., L. Ribba, D. E. P. Borregales, K. Sapag, R. Candal, and S. Goyanes. 2022. Ultramicroporous carbon nanofibrous mats for hydrogen storage. *ACS Applied Nano Materials* 5(10): 15353–61. https://doi.org/10.1021/acsanm.2c03401.

Wang, J., and S. Kaskel. 2012. KOH activation of carbon-based materials for energy storage. *Journal of Materials Chemistry* 22 45): 23710–25. https://doi.org/10.1039/c2jm34066f.

Wang, Z., X. Ke, Z. Zhu, F. Zhu, M. Ruan, H. Chen, R. Huang, and L. Zheng. 2001. A new carbon solid made of the world's smallest caged fullerene C20. *Physics Letters, Section A: General, Atomic and Solid State Physics* 280(5–6): 351–6. https://doi.org/10.1016/S0375-9601(00)00847-1.

Wang, H., Q. Gao, and J. Hu. 2009. High hydrogen storage capacity of porous carbons prepared by using activated carbon. *Journal of the American Chemical Society* 131(20): 7016–22. https://doi.org/10.1021/ja8083225.

Wang, Z., L. Sun, F. Xu, H. Zhou, X. Peng, D. Sun, J. Wang, and Y. Du. 2016. Nitrogen-doped porous carbons with high performance for hydrogen storage. *International Journal of Hydrogen Energy* 41(20): 8489–97. https://doi.org/10.1016/j.ijhydene.2016.03.023.

Warczinski, L., and C. Hättig. 2021. How nitrogen doping affects hydrogen spillover on carbon-supported Pd nanoparticles: new insights from DFT. *Journal of Physical Chemistry C* 125(17): 9020–31. https://doi.org/10.1021/acs.jpcc.0c11412.

Woo, Y., B. S. Kim, J. W. Lee, J. Park, M. Cha, S. Takeya, J. Im, et al. 2018. Enhanced hydrogen-storage capacity and structural stability of an organic clathrate structure with fullerene (C60) guests and lithium doping. *Chemistry of Materials* 30(9): 3028–39. https://doi.org/10.1021/acs.chemmater.8b00749.

Wróbel-Iwaniec, I., N. Díez, and G. Gryglewicz. 2015. Chitosan-based highly activated carbons for hydrogen storage. *International Journal of Hydrogen Energy* 40(17): 5788–96. https://doi.org/10.1016/j.ijhydene.2015.03.034.

Wu, H., D. Wexler, and H. Liu. 2012. Effects of different palladium content loading on the hydrogen storage capacity of double-walled carbon nanotubes. *International Journal of Hydrogen Energy* 37(7): 5686–90. https://doi.org/10.1016/j.ijhydene.2011.12.120.

Xia, Y., Y. Zhu, and Y. Tang. 2012. Preparation of sulfur-doped microporous carbons for the storage of hydrogen and carbon dioxide. *Carbon* 50(15): 5543–53. https://doi.org/10.1016/j.carbon.2012.07.044.

Xia, Y., Z. Yang, and Y. Zhu. 2013. porous carbon-based materials for hydrogen storage: advancement and challenges. *Journal of Materials Chemistry A* 1(33): 9365–81. https://doi.org/10.1039/c3ta10583k.

Zhang, L., M. D. Allendorf, R. Balderas-Xicohténcatl, D. P. Broom, G. S. Fanourgakis, G. E. Froudakis, T. Gennett, et al. 2022. Fundamentals of hydrogen storage in nanoporous materials. *Progress in Energy* 4(4): ac8d44. https://doi.org/10.1088/2516-1083/ac8d44.

Zhao, W., V. Fierro, N. Fernández-Huerta, M. T. Izquierdo, and A. Celzard. 2013. Hydrogen uptake of high surface area-activated carbons doped with nitrogen. *International Journal of Hydrogen Energy* 38(25): 10453–60. https://doi.org/10.1016/j.ijhydene.2013.06.048.

Zheng, J., C.-G. Wang, H. Zhou, E. Ye, J. Xu, Z. Li, and X. J. Loh. 2021. Current research trends and perspectives on solid-state nanomaterials in hydrogen storage. *Research* 2021: 3750689. https://doi.org/10.34133/2021/3750689.

Zhu, S. E., F. Li, & G. W. Wang. 2013. Mechanochemistry of fullerenes and related materials. *Chemical Society Reviews* 42(18): 7535–70. https://doi.org/10.1039/c3cs35494f.

Zubizarreta, L., A. Arenillas, and J. J. Pis. 2009. Carbon materials for H2 storage. *International Journal of Hydrogen Energy* 34(10): 4575–81. https://doi.org/10.1016/j.ijhydene.2008.07.112.

7 Glass Microspheres for Hydrogen Storage

*Samuel Eshorame Sanni, Unwana Udoh Robert,
Emeka Emmanuel Okoro, Babalola Aisosa Oni,
and Christian Benedict*

7.1 INTRODUCTION

The environment is considerably affected by the generation, movement, and utilization of energy, resulting in numerous problems like solid waste management, air pollution, water pollution, thermal pollution, and climate change (European Environment Agency, 2017). Burning fossil fuels, one of the sources of energy today, causes the release of black soot also known as black carbon, and other greenhouse gases into the atmosphere, hence polluting the air. Water bodies are not left out due to oil spills during the extraction and transportation of fossil fuels. Coal mining also poses a threat to water quality as minerals are leached into water bodies (European Environment Agency, 2004).

Due to the problems mentioned above, the quest to find a better means of production and consumption of energy led to the discovery of hydrogen as a source of clean energy. The element hydrogen is the most prevalent throughout the universe. It is about 74% of all the atoms in the entire universe and is usually associated with other atoms to form compounds or molecules (ThinkTech Hawaii, 2016).

President Bush stated in his 2003 Union address that the utilization of hydrogen as fuel for a car could generate solely water, without the release of exhaust fumes. Research has shown that hydrogen is indeed a clean energy source. However, despite its benefits, efforts are still ongoing with respect to subscribing and transitioning to the use of hydrogen energy. One reason for this is that gaseous fuels have a low density, which necessitates a fuel storage method. This poses a significant obstacle for hydrogen, given its comparatively lower energy density relative to hydrocarbon fuels. Achieving greater than 300 miles of driving range requires storing a large amount of hydrogen in vehicles; this contributes an extra mass in bulk beyond that of the fuel. The storage system's size and safety are therefore critical concerns for hydrogen storage. To address these challenges, hydrogen is stored in several materials including balloons, cylinders, microspheres, etc. (MIT OpenCourseWare, 2012). The discourse of this chapter will focus on glass microspheres for hydrogen storage. Figure 7.1 is an illustration of different baloon systems used for hydrogen storage.

DOI: 10.1201/9781003382553-9

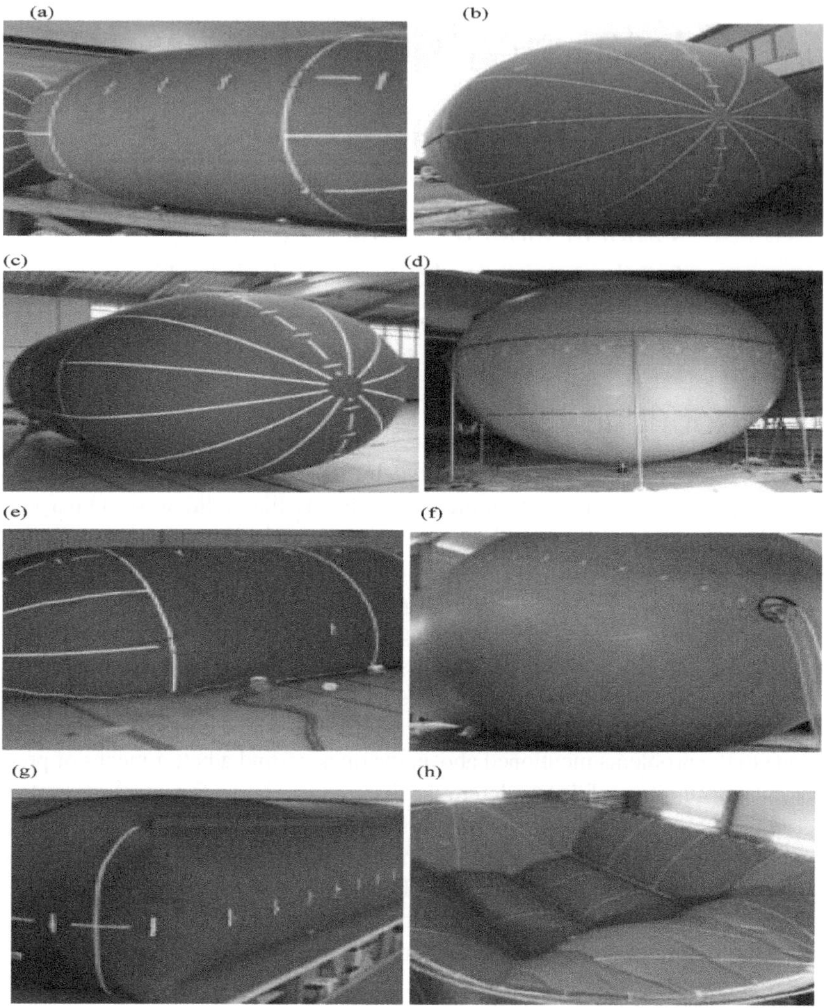

FIGURE 7.1 A fixed mass of hydrogen is stored in different hydrogen buffers (Sanni et al., 2001).

7.2 SOURCES OF HYDROGEN

It is quite overwhelming that the major source of hydrogen is fossil fuels (coal and natural gas) with about 60% or one-third of the world's hydrogen obtained via "dedicated" facilities for hydrogen synthesis. Only a small fraction of hydrogen can be produced from the electrolysis of water. However, synthetic hydrogen often requires dehydration or other forms of cleaning in order to enhance its purity. Presently, the majority of hydrogen is produced in close proximity to where it will be utilized (Jolly, 2019). A significant portion of the world's hydrogen (approximately 75%), is currently produced from natural gas. This amounts to 70 million tonnes (MtH_2) of hydrogen each year, which is derived from 6% or 205 billion cubic meters (bcm) of natural gas. Whereas, steam methane reformers use natural gas as raw material to produce the hydrogen in ammonia as well as the methanol industries and refineries.

The most prevalent technique for generating hydrogen from natural gas is methane steam reforming. It can assume any form including: the use of water as an oxidant and source of hydrogen, partial oxidation by oxygen in air as an oxidant, or a combination of both processes, known as auto-thermal reforming (ATR). Hydrogen can be obtained from natural gas, liquefied petroleum gas, and naphtha through steam reforming. On the other hand, hydrogen can be produced from heavy fuel oil and coal through partial oxidation (Birol, 2019).

7.3 THE NEED FOR HYDROGEN STORAGE

Hydrogen storage is deemed necessary because hydrogen is considered a promising component of the upcoming sustainable energy portfolio. Owing to the quest to curb global emissions by carbon-based fuels, the emergence of hydrogen for use in hydrogen fuel cells offers an eco-friendly alternative for generating low-power for various applications like small electronic gadgets (Schlapbach & Züttel, 2001), automobiles, airplanes, and infrastructure.

Hydrogen has the potential for long-term storage in significant amounts. It can be produced and conserved on an industrial scale, unlike batteries. The stored hydrogen can be utilized as an emergency energy source.

In the transportation sector, hydrogen can serve as complementary fuel alongside battery support. Efforts in research and infrastructure are still needed for the complete implementation of hydrogen as a key component of clean energy solutions, with storage being an important aspect of this process. The Hydrogen and Fuel Cell Technologies Office (HFTO) of the United States focuses on advancing hydrogen use through applied research, development, and innovation for various applications, including transportation. The Department of Energy (DOE) also supports the development of environmentally and economically feasible technologies for hydrogen production (Birol, 2019).

In hydrogen-based economies, prior to distribution and use, the key economic determinants for hydrogen storage include cost and safety after the gas has been produced. Hydrogen can be kept in the form of a gas or a liquid by storing it within solid materials (metal hydrides, cylinders, balloons, and glass microspheres). Using various storage methods for hydrogen entails different pros and cons, such as expenses, mass, stability, user-friendliness, and energy capacity. Compared to conventional methods, hydrogen containment in metal hydrides is regarded as one of the best means of storing hydrogen due to their high volumetric storage capacities and safety. However, metal hydrides for hydrogen storage employ hydrogen-containing materials that are reactive in the presence of water or other liquid compounds, like alcohols which may give rise to side reactions with evident by-products. Recent advances in research have also projected glass mesospheres as potentially viable storage materials for hydrogen, hence the need to exploit possible approaches to improve the technology for producing these materials.

7.3.1 Challenges Associated with Hydrogen Storage

Generally, every fuel has its own degree of risk, which is influenced by three primary factors: the fuel's quantity, the oxidant, and the source of ignition (Sanni et al., 2021). The adoption of apt engineering controls is a measure that can be taken in order to mitigate the risks associated with fuels, even hydrogen.

Due to its high diffusivity, low density, and flammability range that differ from those of hydrocarbons and low-molecular-weight alcohols, hydrogen storage poses distinct challenges and safety concerns. In lieu of its flammability, some basic properties (non-toxicity, low molecular weight, zero potential for emissions, etc.) of hydrogen make it safer than many other commonly used fuels (Sanni et al., 2020). Literature also has it that, in the event of an accident, hydrogen's quick diffusion in the air, as opposed to its concentration at a spot and possibly igniting, as seen in batteries or petroleum, is another benefit. However, there is still evidence of hydrogen-related hazards. Hydrogen has a broad flammable range in air, which makes leak detection and ventilation tests critical in handling hydrogen. Flame detectors designed for hydrogen combustion are also necessary due to the near-invisible flame produced when it burns. Additionally, it requires lower ignition energy compared to petrol and natural gas. Moreover, choosing the right materials for storing hydrogen is important because some metals may lose their strength and become fragile when in contact with hydrogen (Shelby et al., 2009). Since hydrogen containment in pressure vessels requires adequate training of personnel, it then becomes necessary to seek alternative approaches with low risks for storing the gas for safe dispensation as the need arises. Hydrogen gas can also be stored underground in disused oil and gas fields or aquifers, or salt caverns (Usman, 2022). Despite being viable, cavern storage is quite an expensive option. There are only a few instances of cavern-based hydrogen storage systems in operation, mostly in Europe and the USA. The majority of these systems utilize depleted underground natural gas stores, repurposed as hydrogen reservoirs to store excess renewable energy (Chen, & Zhu, 2008).

Storing hydrogen is a challenging endeavor because of its low energy density per volume, lightness (i.e., lighter than helium), simple nature, low boiling point (liquid hydrogen has a boiling point of −258°C), and is easily lost into the atmosphere owing to its reactive nature (Sanni et al., 2021). It has a density that is 3.2 times lower than that of natural gas and 2,700 times lower than that of gasoline. Therefore, storing hydrogen would require some form of compression (in high-pressure tanks within 350–700 bar or 5,000–10,000 psi), liquefaction, and chemical combination with other elements/carriers prior to storage or cryogenic separation.

7.3.2 Storage Media for Hydrogen with Emphasis on Glass Microspheres

According to current research, no material has yet met all the essential criteria for efficient hydrogen storage. One such material is palladium, which exhibits low-temperature reversible behavior, but its low storage capacity of less than 1 wt% and how expensive it is render it unsuitable for use as a storage material for hydrogen. Conversely, $Li_3Be_2H_7$, a material made of multiple components, has been found to be very effective for storing hydrogen (i.e., about 8.7 wt% of hydrogen that can be stored and released repeatedly) (Shelby and Hall, 2008), besides being toxic and adaptable to temperatures up to 300°C. AlH_3, a material that is highly efficient for storage and operates at relatively low temperatures (10.0 wt%), containing inexpensive Al metal priced at $1300 per tonne has nearly all of its hydrogen uptake in an irreversible form.

A highly efficient hydrogen storage system is present in $NaBH_4$, when dissolved in H_2O, can store approximately 9.2 wt% hydrogen, and with the right catalyst, H_2 evolution can be adequately controlled; however, it is difficult to restore the initial material. Water in its pure form contains hydrogen in a weight percentage of 11.1%, but its breakdown necessitates a substantial amount of thermal, electric, or chemical energy. Recent advancements in hydrogen storage technology involve nitrides and imides, which have storage capacities of about 6.5–7.0 wt% hydrogen at high temperatures (approximately 300°C).

Literature also has it that Hollow Glass Microspheres (HGMs) can store compressed hydrogen gas safely (Schmitt et al., 2006), although their poor thermal conductivity limits the release rate of hydrogen gas confined in these spheres when situated in a packed bed (Kohli et al., 2008). To address this limitation, monolithic glass plates doped with an optically active element like Fe_3O_4 can greatly enhance the rate of hydrogen release when exposed to infrared radiation (Palma et al., 2016). Researchers have shown that radiation emitted from an infrared lamp can accelerate the release of hydrogen from such a material (Shelby & Hall, 2008).

7.3.2.1 What Are Microspheres?

Microspheres are classified into two groups: solid and hollow. The latter is also referred to as microbubbles or hollow microspheres, while the former is referred to as microbeads or microspheres. These materials are utilized in various industries such as pharmaceuticals, food, cosmetics, chemicals, transportation, and construction, due to their distinctive physical and chemical properties. These features encompass being light in weight, possessing low thermal conductivity, exhibiting resistance to compressive stress, and almost complete chemical inertness.

In today's world, microspheres and microbubbles have found progressive applications, especially in areas such as optical science and technology. For instance, by serving as resonators for whispering gallery mode (WGM) devices, they have enabled the creation of advanced laser (Lewkowicz, 1974) and sensor micro-devices. This trend is expected to continue as the exploration for more condensed and resilient uses, specifically in the field of biosensing, is presently a research and development priority (Chiasera et al., 2010; Righini et al., 2011).

7.3.2.2 Types of Glass Microspheres

Microspheres and microbubbles can be produced using amorphous materials, specifically oxide and chalcogenide glasses. However, it should be noted that numerous other materials, both artificial and organic, can be utilized to fabricate microspheres for diverse uses.

Glass microspheres can be made from various materials, including amorphous polymers and stainless steel, which can serve as conductive spacers, shock absorbers, and micromotor bearings (Ghalichechian et al., 2008) hollow microspheres made of nickel metal with improved magnetic characteristics. On the other hand, Ni/Pt bimetallic microbubbles are capable of catalyzing portable hydrogen generation systems (Yi et al., 2009); ferrite microspheres have found potential uses in magnetic applications, ferrofluid technology, and biomedical fields, including cancer diagnosis and treatment, biomolecular separations, and magnetic resonance imaging (Deng et al., 2005). Single-crystal semiconductor microspheres are being used as active WGM resonators (Okamoto et al., 2014), while ceramic ZrO_2 hollow microspheres are suitable for thermal applications (Gulyaev, 2015).

Commercially, a range of microspheres made from glass, polymer, ceramic, solid metal, and hollow metal are readily accessible with varying qualities of uniformity, sphericity (the comparison of the surface area of a particle to that of a sphere with the same volume), particle size, and distribution, which can be tailored for specific applications.

7.3.2.2.1 Oxide-Based Glass Microspheres

These are microspheres created from pure silica or multiple oxide compounds and are commonly utilized in research and development as well as industries, and have found commercial applications. In laboratory settings, in cases where the unique properties of an individual microsphere are required to utilize the properties of a whispering gallery mode resonator (Righini, 2019) or to deploy a micro/nano Coordinate Measuring Machine (CMM) sensor (Fan et al., 2010), creating microspheres of high quality is a sensitive process that necessitates the expertise of a qualified specialist. The most frequently utilized fabrication method involves liquefying the tip of an optical fiber and relying on surface tension to form an almost flawless sphere. Various heat sources such as electric arcs (as seen in commercial fiber splicers), high-power lasers (like CO_2), or oxygen/butane torches can be employed (Ruan et al., 2014).

In the case of fibers pulled from multi-oxide glasses that possess a lower melting point than pure silica, a basic resistive micro-heater could potentially be adequate (Wang et al., 2012). A modified or commercially available optical fiber fusion splicer, such as the Furukawa FITEL S182K in Tokyo, Japan, provides precise control over the process. The splicer inserts a cleaved fiber tip into one arm and generates a series of arcs that partially melt the tip, resulting in a spherical shape due to surface tension forces. The illustration in Figure 7.2 shows an integrated optical fiber microsphere in a fusion splicer; the temperature generated by the electrode discharge can melt pure silica, reaching up to approximately 2,000°C.

According to Yu et al. (2014), an experimental apparatus was used to fabricate a compact microsphere made of optical fibers, measuring under 100 µm in diameter, with a 2D roundness deviation of no more than 0.70 µm, and an actual sphericity of roughly 0.5 µm. The production process involved utilizing the Taguchi approach in

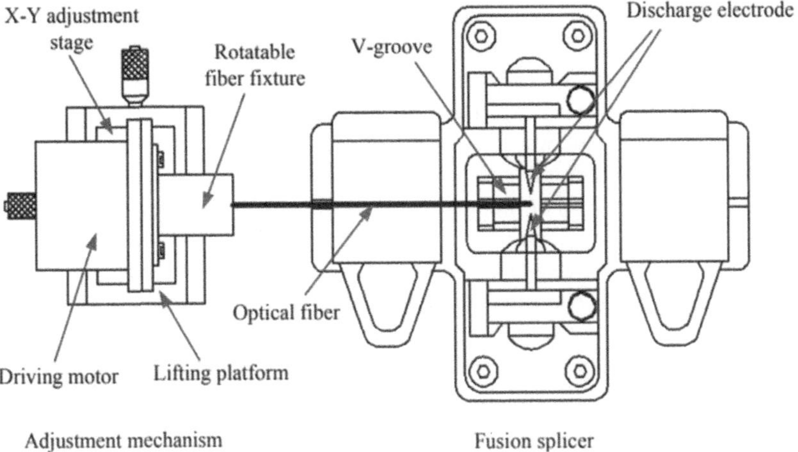

FIGURE 7.2   A microsphere of optical fiber integrated within a fusion splicer (Yu et al., 2014).

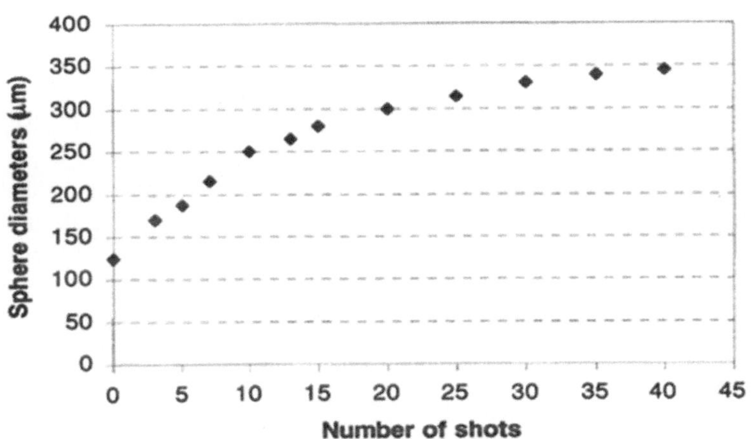

FIGURE 7.3   Different-sized microspheres form at the end of a 125 µm fiber by adjusting the arc shots in a fiber fusion splicer (Brenci et al., 2004).

combination with a fiber tapering method for statistical process optimization (Nair et al., 1992). Obtaining the true sphericity of a microsphere involves using a numerical method based on information obtained from a 2D cross-section photo. If the fiber is not tapered, the microsphere's size is greater than 125 µm, which is the standard cladding diameter of the telecom SMF-28 SM silica fiber, and it increases with the number of electric arc shots until it reaches a maximum diameter of about 350 µm as illustrated in Figure 7.3.

According to Yano et al. (2015), the conventional methods for producing microspheres result in the production of only a single microsphere, and its size is largely influenced by the size of the original fiber. Although the microspheres remain an integral part of the fiber stem, this approach may not be advantageous in certain applications. To obtain discrete particles, it is preferable to produce a fine glass powder that can be melted. This technique also allows the production of microspheres from various oxide glasses, utilizing the localized-laser-heating (LLH) technique, powered by a continuous wave Ti-sapphire laser operating at 810 nm and 200 mW. Alternative methods comprise utilizing a plasma torch powered by microwaves (Nunzi Conti et al., 2006) or melting the glass ingredients in a furnace and depositing the molten glass onto a rotating plate (Peng et al., 2003). However, one drawback of employing either of these methods is that it yields numerous unattached particles with a broad range of sizes, which requires sorting by size and checking their surface qualities.

In laboratory settings, it is desirable to create a microbubble by attaching a microbubble to a capillary to form a whispering gallery mode resonator that has integrated microfluidics for biomedical applications (Berneschi et al., 2016; Yang, 2014). To achieve this, a widely used method includes heating a pressurized silica capillary with a CO_2 laser or an electric arc discharge, resulting in single or double-pass structures, such as spherical shells with one or two openings (Yang, 2014). However, commercial uses require individual micro-balloons, and several manufacturing techniques have been developed recently (Veatch, 1961). It is possible to mass-produce microspheres and microbubbles with good size dispersion control via industrial processes that are suitable for the purpose. By heating and melting glass powders, solid glass microspheres can be created. On the other hand, glass microbubbles can be produced by adding a blowing agent to the glass powder (Berneschi et al., 2016; Yang, 2014).

The commercial importance of glass microspheres with either a solid or hollow is considerable, and the research market has projected that the commercial value of the glass microspheres market would reach around \$1993.36 million in 2019 and experience an annual growth rate of around 12.4% (Micromarketmonitor, 2018). Naturally, various patents exist for different types of glass microspheres which are intended for various applications.

7.3.2.2.2　Chalcogenide Glass Microspheres

Chalcogenide glass microspheres consist mainly of three chalcogen elements, namely sulfur, selenium, and tellurium with nonlinear properties, such as the ability to be affected by light, minimal vibrational energy, and the ability to allow infrared light to pass through them (Elliott, 2015). The fabrication of chalcogenide optical fibers can be achieved using the melting of the fiber tip method (Sanghera, 2002). However, the more typical process involves a vertical furnace purged with argon, which is an inert gas, which will be used to crush the chalcogenide glass due to its reactive nature when molten (Elliott et al., 2007). Solid chalcogenide microspheres have various applications, including the detection of biological molecules, measurement of temperature changes, amplification of light, and generation of laser light (Ahmad et al., 2012; Palma et al., 2013). Additionally, metal-chalcogenide nanocrystals that are binary, ternary, or quaternary, such as $CuInS_2$, CdSe, Cu_2ZnSnS_4, and PbTe, are also promising for renewable energy applications, as they have the potential to increase the energy conversion device's efficiency (Aldakov, 2013).

7.3.2.3 Application of Glass Microspheres in Hydrogen Storage

According to previous studies, applications for both solid and hollow microspheres depend on the characteristics of the material utilized as well as their sizes. Their utilization encompasses a broad spectrum of technologies and has attracted significant attention across numerous fields for many years (Amos & Yalcin, 2015), especially that of field energy, which has gained increased importance in recent times; three sub-areas can be identified, and these include reducing energy consumption, storing energy, and generating energy.

The quest for vehicles that have fewer negative impacts on the environment relative to vehicles powered by gasoline or diesel using an internal combustion engine has been driven by the need for environmental protection in recent years. Electric vehicles appear to be one of the best solutions, with fuel cells expected to offer significant advantages (O'Hayre et al., 2016). Electric vehicles powered by fuel cells can travel more than 500 km on a single fuel tank, and unlike battery vehicles, they can be refilled in a matter of minutes. However, hydrogen storage presents a significant obstacle to the advancement and sustainability of hydrogen-fueled vehicles (Lim et al., 2010; Durbin and Malardier-Jugroot, 2013). HGMs demonstrate their potential in addressing the challenge of hydrogen storage. HGMs, which have a size range of 1–100 μm, and a density between 1.0 and 2.0 g/cc, with porous walls containing openings of 1–100 nm, are regarded as a promising materials for hydrogen storage (Shetty, 2011; Schmid et al., 2017). Recent research and patents demonstrate advancements in using HGMs for hydrogen storage. HGMs can be utilized to trap hydrogen gas by exploiting its diffusion through the thin wall of the shperes at high temperatures and pressures and subsequently cooling it at room temperature. However, the inadequate heat conduction of HGMs has been a limitation to their commercial use owing to the unsuitable slow liberation of hydrogen gas. A potential solution to address this issue involves adding transition metals to the glass. For instance, cobalt incorporation into HGMs improved their thermal conductivity from 0.072 to 0.198 W/mK with increasing weight percentages up to 10 wt% (Dalai et al., 2017). However, the highest hydrogen adsorption capacity was achieved with a cobalt loading of 2 wt%; above this level, the storage capacity decreased due to the uneven deposition of cobalt oxide (CoO) on the microspheres' surface, leading to pore blockage (Dalai, 2017).

For applications requiring fast storage and release of gas, proper strategies must be developed, and one effective method utilizes an infrared light lamp to enhance the rate of gas release, known as photo-induced out-gassing. The method to increase thermal conductivity can be applied to improve the out-gassing process, which involves alloying the glass with optically responsive elements like cobalt, nickel, and iron. According to Rapp and Shelby (2004), 7,070 borosilicate glass doped with 0.5 wt% Fe_3O_4 was effective at promoting hydrogen release, which was observed to be directly proportional to the light intensity of the lamp. The Fe^{2+}/Fe^{3+} ratio was observed to increase as a result of the reaction between hydrogen and iron-doped glass, which boosted the amount of infrared light adsorbed and further enhanced the amount of hydrogen produced from the process of photo-induced out-gassing. Shetty and Hall (2011) improved out-gassing by spraying cobalt-doped HGM and found 3 wt% CoO to be more effective than 1.5 wt% CoO. They also developed a framework for predicting the parameters needed to create HGMs with specific geometric properties and ratios of wall thickness to diameter.

TABLE 7.1

Potential Glasses for Making Microspheres, Their Young Modulus, Compositions, and Tensile Strengths

Glass Molecular Formula	Constituent Ratio (mol%: mol%)	Young Modulus, $*10^{10}$ Pa	Tensile Strength $*10^{10}$ Pa
SiO_2 (Quartz)	–	7.3	0.69
Na_2-SiO_2	13:87	6.2	0.51
Na_2-SiO_2	16:84	6.0	0.46
Na_2-SiO_2	17:83	5.9	0.44
Na_2-SiO_2	20:80	5.6	0.40
SiO_2, $Na_2O + K_2O$, PbO, BaO	78.7:12.4:8.8:0.1	6.1	0.48
BK10	–	7.4	0.49

Source: Adapted from Kholi et al. (2008).

According to recent studies, hollow microspheres with a crystalline structure, including but not limited to vanadium pentoxide, as well as multi-layered NiO and TiO_2 microspheres, hold promise for use as anodes in lithium-ion batteries and supercapacitors for storing electrical energy as they may serve as safe and inexpensive solution for energy storage (Tian et al., 2018; Ren et al., 2014; Qi et al., 2016).

Employing hollow microspheres made of glass or polymer for the storage of pressurized gases is attractive since they allow gases to pass through and can securely retain them. In the domain of atomic power, HGMs have been studied since the late 1950s for use in fusion power reactors. Nuckolls et al. (1972) demonstrated the feasibility of laser thermonuclear burns using small pellets of deuterium and tritium (DT), thereby opening avenues for the creation of such reactors. Recent publications on the subject have also helped to present a summary of the difficulties and current advanced models for inertial confinement fusion (ICF) research (Craxton et al., 2015; Betti, 2016; Zohuri, 2017). Any successfully deployed laser-fusion system must be characterized by the economical fabrication of appropriate fuel capsules which gained attention for preparing hydrogen or isotope-filled nuclear fusion targets since the 1970s. HGMs with diameters ranging from approximately 50–500 µm and wall thicknesses of 1–20 µm have emerged as strong candidates for this purpose (Lewkowicz, 1974; Qi et al., 2013). Table 7.1 comprises of some glasses for microsphere construction and their properties while Table 7.2 consists of different gravimetric efficiencies of glass microbubbles/microspheres with varying aspect ratios.

7.4 CONCLUSION

In recent years, there has been consistent growth in the usage of glass microspheres, i.e., both solid and hollow types for scientific and commercial applications. This has been made possible by significant advancements in their production, which now allows for the creation of high-quality and large quantities of these systems. Glass microspheres are particularly advantageous as it is possible to easily integrate

TABLE 7.2

Gravimetric Efficiency for Glass Microbubbles/Microspheres for Varying Aspect Ratios

Aspect Ratio	Sphere Radius (µm)	Wall Thickness (µm)	Maximum Allowable Pressure *102 bar	Density of Hydrogen Gas Stored *101 (kg/m³)	Mass of Hydrogen *10^{-1} (µg)	Mass Ratio of Hydrogen to Sphere, M_H/M_S (w/w)	Volumetric Density of Gas *101 (kg/m³)	Gravimetric Efficiency (%)
50	100	2	6.0	3.46	1.4	0.21	3.26	17
	250	5	6.0	3.46	21.3	0.21	3.26	17
	500	10	6.0	3.46	170	0.21	3.26	17
100	100	1	3.0	2.06	0.8	0.25	1.99	19.7
	250	2.5	3.0	2.06	13.0	0.25	1.99	19.7
	500	5	3.0	2.06	104	0.25	1.99	19.7
150	100	0.67	2.0	1.40	0.6	0.26	1.37	20.3
	250	1.67	2.0	1.40	9.0	0.26	1.37	20.3
	500	3.33	2.0	1.40	71.3	0.26	1.37	20.3
200	100	0.5	1.5	1.09	0.5	0.27	1.07	21.0
	250	1.25	1.5	1.09	7.0	0.27	1.07	21.0
	500	2.5	1.5	1.09	56.3	0.27	1.07	21.0

Source: Adapted from Kholi et al. (2008).

chemical elements and compounds into them to improve their abilities regarding storing hydrogen. Since hydrogen is a flammable gas, it is also necessary to investigate the safest ways of deploying glass microspheres when used for hydrogen storage. As in the case of balloons, glass microspheres can have integrated features that can control issues related to gas explosion, especially safety mechanisms that will bring about a slow release/diffusion rate of the gas in case of an emergency.

To date, the accelerated discharge of reserved hydrogen gas upon exposure to high-intensity light in contrast to conventional heating methods is quite rapid and occurs in less than 1 second. MoSci Corporation has manufactured iron oxide-loaded hollow glass microspheres with varying concentrations. The company designed, calibrated, and constructed a device that utilizes pressure, volume, and temperature control to measure photo-enhanced diffusion. A residual gas analyzer (RGA) for taking kinetic data during out-gassing is also integrated with the system. Gas diffusion induced by photo and thermal effects has been investigated to analyze its impact on HGMs in relation to gas identity, pressure, temperature, and dopant concentration. It has been shown that there is a linear relationship between gas pressure and gas content in HGMs. The filling rate is affected by temperature, and the dopant concentration affects the photo-induced response time (Couvreur & Guiot, 2018). The base glass used by MoSci Corp for HGM production is made up of about 85% recycled glass and is mixed with various dopants, including iron oxide, to create a hybrid melt with improved properties. After processing, the HGMs exhibit better thermal stability and expansion coefficients, glass transformation temperatures, and potential for gas storage. Hence, MoSci Corp now intermittently delivers approximately 100 g samples of hydrogen gas.

Furthermore, reports also have it that studies involving various techniques, including electron paramagnetic resonance and magnetic susceptibility, are used to measure the extent of hydrogen interaction with the glass during filling, out-gassing/desorption, storage, as well as analysis of the microspheres to assess their size dispersion, wall width, and changes in composition, which are all under consideration.

REFERENCES

Ahmad, H., Aryanfar, I., Lim, K.S., Chong, W.Y., Harun, S.W. (2012). Thermal response of chalcogenide microsphere resonators, *Quantum Electronics*, 42(5), pp. 462–464. https://doi.org/10.1070/qe2012v042n05abeh014813.

Aldakov, D., Lefrancois, A., Reiss, P. (2013). Ternary and quaternary metal chalcogenide nanocrystals: synthesis, properties and applications, *Journal of Materials Chemistry. C, Materials for Optical and Electronic Devices*, 1(24), pp. 3756–3776. https://doi.org/10.1039/c3tc30273c.

Amos, S.E., Yalcin, B. (2015). *Hollow Glass Microspheres for Plastics, Elastomers, and Adhesives Compounds*, Elsevier, New York.

Berneschi, S., Baldini, F., Barucci, A., Cosci, A., Cosi, F., Farnesi, D., Nunzi Conti, G., Righini, G.C., Soria, S., Tombelli, S., et al. (2016). Localized biomolecules immobilization in optical microbubble resonators, In Proceedings of the 2016 SPIE, Laser and Applications, San Francisco, CA, USA, 16–18.

Betti, R., Hurricane, O.A. (2016). Inertial-confinement fusion with lasers, *Nature Physics*, 12(5), pp. 435–448. https://doi.org/10.1038/nphys3736.

Birol, F. (2019). The Future of Hydrogen: Seizing Today's Opportunities, the IEA for the G20, Japan.

Brenci, M., Calzolai, R., Cosi, F., Nunzi Conti, G., Pelli, S., Righini, G.C. (2004). Microspherical resonators for biophotonic sensors, In: *Proceedings of the International Society for Optical Engineering (SPIE)*, Warsaw, Poland, pp. 20–22.

Chiasera, A., Dumeige, Y., Féron, P., Ferrari, M., Jestin, Y., Nunzi Conti, G., Pelli, S., Soria, S., Righini, G.C. (2010). Spherical whispering-gallery-mode microresonators: Spherical WGW microresonators, *Laser & Photonics reviews*, 4(3), pp. 457–482. https://doi. org/10.1002/lpor.200910016.

Chen, P., Zhu, M. (2008). Recent progress in hydrogen storage, *Materials Today*, 11(12), pp. 38–46.

Couvreur, P., Guiot, P. (2018). Biodegradable polymeric nanoparticles as drug carrier for anti-tumor agents, In *Polymeric Nanoparticles and Microspheres*. CRC Press, Boca Raton, FL, pp. 27–94.

Craxton, R.S., Anderson, K.S., Boehly, T.R., Goncharov, V.N., Harding, D.R., Knauer, J.P., McCrory, R.L., McKenty, P.W., Meyerhofer, D.D., Myatt, J.F., et al. (2015). Direct-drive inertial confinement fusion: a review, *Physics of Plasmas*, 22(11), p. 110501. https://doi. org/10.1063/1.4934714.

Dalai, S., Savithri, V., Sharma, P. (2017). Investigating the effect of cobalt loading on thermal conductivity and hydrogen storage capacity of hollow glass microspheres (HGMs), *Materials Today Proceedings*, 4(11), pp. 11608–11616. https://doi.org/10.1016/j.matpr.2017.09.072.

Deng, H., Li, X., Peng, Q., Wang, X., Chen, J., Li, Y. (2005). Monodisperse magnetic single-crystal ferrite microspheres, *Angewandte Chemie* (International ed. in *English*), 44(18), pp. 2782–2785. https://doi.org/10.1002/anie.200462551.

Durbin, D.J., Malardier-Jugroot, C. (2013). Review of hydrogen storage techniques for on board vehicle applications, *International Journal of Hydrogen Energy*, 38, pp. 14595–14617. https://doi.org/10.1016/j.ijhydene.2013.07.058.

Elliott, G.R., Hewak, D.W., Murugan, G.S., Wilkinson, J.S. (2007). Chalcogenide glass microspheres; their production, characterization and potential, *Optics Express*, 15(26), pp. 17542–17553. https://doi.org/10.1364/oe.15017542.

Elliott, S.R. (2015). Chalcogenide phase-change materials: past and future, *International Journal of Applied Glass Science*, 6(1), pp. 15–18. https://doi.org/10.1111/ijag.12107.

European Environment Agency. (2004). Environmental Impact of Energy - European Environment Agency. [online] www.eea.europa.eu. Available at: https://www.eea.europa. eu/help/glossary/eea-glossary/environmental-impact-of-energy#:~:text=The%20 environmental%20problems%20directly%20related.

Fan, K.C., Cheng, F., Wang, W., Chen, Y., Lin, J.Y. (2010). A scanning contact probe for a micro-coordinate measuring machine (CMM), *Measurement Science & Technology*, 21(5), p. 054002. https://doi.org/10.1088/0957-0233/21/5/054002.

Ghalichechian, N., Modafe, A., Beyaz, M.I., Ghodssi, R. (2008). Design, fabrication, and characterization of a rotary micromotor supported on microball bearings, *Journal of Microelectromechanical Systems: A Joint IEEE and ASME Publication on Microstructures, Microactuators, Microsensors, and Microsystems*, 17(3), pp. 632–642. https://doi.org/10.1109/jmems.2008.916346.

Gulyaev, I. (2015). Experience in plasma production of hollow ceramic microspheres with required wall thickness, *Ceramics International*, 41(1), pp. 101–107. https://doi. org/10.1016/j.ceramint.2014.08.040.

Jolly, W.L. (2019). Hydrogen properties, uses, & facts. *Encyclopedia Britannica*, www.britan-nica.com/science/hydrogen

Kohli, E.K., Khardekar, R.K., Singh, R., Gupta, P.K. (2008). Glass micro-container based hydrogen storage scheme, *International Journal of Hydrogen Energy*, 33, pp. 417–422. https://doi.org/10.1016/j.ijhydene.2007.07.044.

Lewkowicz, I. (1974). Spherical hydrogen targets for laser-produced fusion, *Journal of Physics D: Applied Physics*, 7(4), pp. L61–L62. https://doi.org/10.1088/0022-3727/7/4/101.

Lim, K.L., Kazemian, H., Yaakob, Z., Daud, W.R.W. (2010). Solid-state materials and methods for hydrogen storage: a critical review, *Chemical Engineering Technology*, 33(2), pp. 213–226. https://doi.org/10.1002/ceat.200900376.

Micromarketmonitor. (2014-2019). Glass Microspheres Market Report Trends, Analysis, Forecast - Micro Market Monitor. [online] Available at: https://www.micromarketmonitor.com/market-report/glassmicrospheres-reports-6570147015.html [Accessed 10 March 2023].

MIT OpenCourseWare. (2012). Lecture 6: Hydrogen Storage, and Atoms to Molecules | Introduction to Modeling and Simulation | Materials Science and Engineering. [online] Available at: https://ocw.mit.edu/courses/3-021j-introduction-to-modeling-and-simulation-spring-2012/resources/lecture-6/.

Nair, V.N., Abraham, B., MacKay, J., Box, G., Kacker, R.N., Lorenzen, T.J., Lucas, J.M., Myers, R.H., Vining, G.G., Nelder, J.A., et al. (1992) Taguchi's parameter design: a panel discussion, *Technometrics: A Journal of Statistics for the Physical, Chemical, and Engineering Sciences*, 34(2), pp. 127–161. https://doi.org/10.1080/00401706.1992.10484904.

Nuckolls, J., Wood, L., Thiessen, A., Zimmerman, G. (1972). Laser compression of matter to super-high densities: thermonuclear (CTR) applications, *Nature*, 239(5368), pp. 139–142. https://doi.org/10.1038/239139a0.

Nunzi Conti, G., Chiasera, A., Ghisa, L., Berneschi, S., Brenci, M., Dumeige, Y., Pelli, S., Sebastiani, S., Féron, P., Ferrari, M., et al. (2006). Spectroscopic and lasing properties of Er3+-doped glass microspheres, *Journal of Non-Crystalline Solids*, 352(23–25), pp. 2360–2363. https://doi.org/10.1016/j.jnoncrysol.2006.01.089.

O'Hayre, R., Cha, S.W., Prinz, F.B., Colella, W. (2016). *In Fuel Cell Fundamentals*, John Wiley & Sons, Hoboken, NJ.

Okamoto, S., Inaba, K., Iida, T., Ishihara, H., Ichikawa, S., Ashida, M. (2014). Fabrication of single-crystalline microspheres with high sphericity from anisotropic materials, *Scientific Reports*, 4(1), p. 5186. https://doi.org/10.1038/srep05186.

Palma, G., Bia, P., Mescia, L., Yano, T., Nazabal, V., Taguchi, J., Moréac, A., Prudenzano, F. (2013). Design of fiber coupled Er3+: chalcogenide microsphere amplifier via particle swarm optimization algorithm, *Optical Engineering (Redondo Beach, California)*, 53(7), p. 071805. https://doi.org/10.1117/1.oe.53.7.071805.

Palma, G., Falconi, M.C., Starecki, F., Nazabal, V., Yano, T., Kishi, T., Kumagai, T., Prudenzano, F. (2016). Novel double step approach for optical sensing via microsphere WGM resonance, *Optics Express*, 24(23), pp. 26956–26971. https://doi.org/10.1364/oe.24.026956.

Peng, X., Song, F., Jiang, S., Peyghambarian, N., Kuwata-Gonokami, M., Xu, L. (2003). Fiber-taper-coupled L-band Er3+-doped tellurite glass microsphere laser, *Applied Physics Letters*, 82(10), pp. 1497–1499. https://doi.org/10.1063/1.1559653.

Qi, X., Gao, C., Zhang, Z., Chen, S., Li, B., Wei, S. (2013). Fabrication and characterization of millimeter-sized glass shells for inertial confinement fusion targets, *Chemical Engineering Research & Design: Transactions of the Institution of Chemical Engineers*, 91(12), pp. 2497–2508. https://doi.org/10.1016/j.cherd.2013.03.010.

Qi, X., Zheng, W., Li, X., He, G. (2016). Multishelled NiO hollow microspheres for high-performance supercapacitors with ultrahigh energy density and robust cycle life, *Scientific Reports*, 6(1), p. 33241. https://doi.org/10.1038/srep33241.

Rapp, D.B., Shelby, J.E. (2004). Photo-induced hydrogen outgassing of glass, *Journal of Non-Crystalline Solids*, 349, pp. 254–259. https://doi.org/10.1016/j.jnoncrysol.2004.08.151.

Ren, H., Yu, R., Wang, J., Jin, Q., Yang, M., Mao, D., Kisailus, D., Zhao, H., Wang, D. (2014). Multishelled TiO2 hollow microspheres as anodes with superior reversible capacity for lithium ion batteries, *Nano Letters*, 14(11), pp. 6679–6684. https://doi.org/10.1016/s1631-0748(02)01450-9.

Righini, G.C. (2019). *Glass Micro- and Nanospheres. Physics and Applications*. Pan Stanford Publishing, Singapore.

Righini, G.C., Dumeige, Y., Féron, P., Ferrari, M., Nunzi Conti, G., Ristic, D., Soria, S. (2011). Whispering gallery mode microresonators: fundamentals and applications, *Riv. Nuovo Cimento*, 34, pp. 435–488.

Ruan, Y., Boyd, K., Ji, H., Francois, A., Ebendorff-Heidepriem, H., Munch, J., Monro, T.M. (2014). Tellurite microspheres for nanoparticle sensing and novel light sources, *Optics Express*, 22(10), pp. 11995–12006. https://doi.org/10.1364/OE.22.011995.

Sanghera, J., Shaw, L.B., Aggarwal, I.D. (2002). Applications of chalcogenide glass optical fibers, *Comptes Rendus Chimie*, 5(12), pp. 873–883. https://doi.org/10.1016/s1631-0748(02)01450-9.

Sanni, S.E., Alade, T.A., Agboola, O., Alaba, P.A. (2020). Catalytic dehydrogenation of formic acid-triethanolamine mixture using copper nanoparticles, *International Journal of Hydrogen Energy*, 45, pp. 4606–4624. https://doi.org/10.1016/j.ijhydene.2019.12.121.

Sanni, S.E., Alaba, P.A., Okoro, E., Emetere, M., Oni, B., Agboola, O., Ndubuisi, A.O. (2021). Strategic examination of the classical catalysis of formic acid decomposition for intermittent hydrogen production, storage and supply: a review, *Sustainable Energy Technologies and Assessments*, 45, p. 101078. https://doi.org/10.1016/j.seta.2021.101078.

Schlapbach, L., Züttel, A. (2001). Hydrogen-storage materials for mobile applications, *Nature*, 414(6861), pp. 353–358. https://doi.org/10.1038/35104634.

Schmid, G.H.S., Bauer, J., Eder, A., Eisenmenger-Sittner, C. (2017). A hybrid hydrolytic hydrogen storage system based on catalyst-coated hollow glass microspheres: hybrid hydrolytic hydrogen storage system, *International Journal of Energy Research*, 41(2), pp. 297–314. https://doi.org/10.1002/er.3659.

Schmitt, M.L., Shelby, J.E., Hall, M.M. (2006). Preparation of hollow glass microspheres from sol-gel derived glass for application in hydrogen gas storage, *Journal of Non-Crystalline Solids*, 352, pp. 626–631. https://doi.org/10.1016/j.jnoncrysol.2005.11.057.

Shelby, J., Hall, M. (2008). Glass Microspheres for Hydrogen Storage. Alfred University, PDF presentation, Project ID: STP47. Available at: https://www.hydrogen.energy.gov/pdfs/review05/stp_47_hall.pdf [Accessed 9 March 2023].

Shelby, J.E., Hall, M.M., Snyder, M.J., Wachtel, P.B. (2009). A radically new method for hydrogen storage in hollow glass microspheres, Technical Report, Alfred University, United States. https://doi.org/10.2172/958673.

Shetty, S., Hall, M. (2011). Facile production of optically active hollow glass microspheres for photo-induced outgassing of stored hydrogen, *International Journal Hydrogen Energy*, 36(16), pp. 9694–9701. https://doi.org/10.1016/j.ijhydene.2011.04.195.

ThinkTech Hawaii. (2016). *Safely Making, Storing and Dispensing Hydrogen*. Available at: https://www.youtube.com/watch?v=g32BLZssoNo&t=5s [Accessed 9 March 2023].

Tian, P., Song, Q., Pang, H., Ning, G. (2018). Hollow microspherical vanadium pentoxide fabricated via non-hydrothermal route for lithium ion batteries, *Materials Letters*, 227, pp. 13–16. https://doi.org/10.1016/j.matlet.2018.04.121.

Usman, M.R. (2022). Hydrogen storage methods: review and current status, *Renewable and Sustainable Energy Reviews*, 167, p. 112743. https://doi.org/10.1016/j.rser.2022.112743.

Veatch, F., Alford, H.E., Croft, R.D. (1961). Method of producing hollow glass spheres, U.S. Patent 2,978,339, 4.

Wang, P., Murugan, G.S., Lee, T., Ding, M., Brambilla, G., Semenova, Y., Wu, Q., Koizumi, F., Farrell, G. (2012). High-Q bismuth-silicate nonlinear glass microsphere resonators, *IEEE Photonics Journal*, 4(3), pp. 1013–1020. https://doi.org/10.1109/jphot.2012.2202385.

Yang, Y., Ward, J., Chormaic, S.N. (2014). Quasi-droplet microbubbles for high resolution sensing applications, *Optics Express*, 22(6), pp. 6881–6898. https://doi.org/10.1364/OE.22.006881.

Yano, T., Kishi, T., Kumagai, T. (2015). Glass microspheres for optics, *International Journal of Applied Glass Science*, 6(4), pp. 375–386. https://doi.org/10.1111/ijag.12145.

Yi, R., Shi, R., Gao, G., Zhang, N., Cui, X., He, Y., Liu, X. (2009). Hollow metallic micro-spheres: fabrication and characterization, *The Journal of Physical Chemistry. C, Nanomaterials*, 113, pp. 1222–1226.

Yu, H., Huang, Q., Zhao, J. (2014). Fabrication of an optical fiber micro-sphere with a diameter of several tens of micrometers, *Materials*, 7(7), pp. 4878–4895. https://doi.org/10.3390/ma7074878.

Zohuri, B. (2017). Inertial confinement fusion (ICF), In: *Inertial Confinement Fusion Driven Thermonuclear Energy*. Springer International Publishing, Cham, pp. 193–238.

8 Metal-Organic Frameworks (MOF) for Hydrogen Storage

Fatemeh Zarei-Jelyani, Mohammad Zarei-Jelyani,
Fatemeh Salahi, and Mohammad Reza Rahimpour

8.1 INTRODUCTION

The phenomenon of rapid industrialization coupled with overpopulation has resulted in the extensive utilization of fossil fuels for the purpose of energy production. Over the span of a century, there has been an eightfold increase in energy consumption, predominantly fueled by non-renewable sources such as fossil fuels. Fossil fuels are responsible for producing significant amounts of greenhouse gases, primarily carbon dioxide (CO_2), as well as other detrimental environmental byproducts. It is projected that global utilization of non-renewable energy sources will experience a 56% surge by the year 2040 (Hochreiter et al., 2003).

The utilization of non-renewable energy sources has two significant consequences: the hastened depletion of energy reserves and the negative effects on public health and the environment (Zacharia and Rather, 2015).

Research is currently being conducted on materials that can selectively extract greenhouse gases from the atmosphere. This involves the optimization of low-energy mixtures to efficiently manage waste materials and safely store potentially dangerous fuels such as hydrogen, methane, and ammonia. The utilization of hydrogen energy, which has an energy density of 140 MJ/kg, can potentially address the challenges posed by the scarcity of primary reserves, population expansion, and the adverse effects of industrialization. This is due to the fact that hydrogen energy surpasses all fossil fuels, such as natural gas, coal, and methane, in terms of significance. Consequently, hydrogen gas undergoes combustion as a source of energy, producing water vapor as a by-product that is deemed harmless and lacking in toxicity. Hydrogen energy has been identified as a promising potential energy resource for the future due to its low environmental impact and high energy yield (Suksaengrat et al., 2016). Currently, established methods for hydrogen storage, including compressed gas stored in pressurized tanks and liquefied hydrogen stored in cryogenic tanks, are being utilized in fuel cell vehicles (Eberle et al., 2012). Nevertheless, both these technologies exhibit significant limitations when it comes to mobile applications. Hydrogen gas exhibits a low density of $0.08988\,kg/m^3$ under standard atmospheric pressure. In order to attain high storage capacities, it is imperative to subject the gas to considerably high pressures (Schlapbach and Züttel, 2001). The necessity for

DOI: 10.1201/9781003382553-10

the storage system leads to the utilization of weighty and bulky storage containers. Hydrogen exhibits a low boiling point of 20 K, necessitating significant cooling measures for the liquefaction of this element. The aforementioned result in a system of considerable complexity, encompassing boil-off losses and a reduction in the energy density of hydrogen (Schlapbach and Züttel, 2001).

The adsorption of H_2 through physisorption on metal-organic frameworks (MOFs) has been observed to be a more viable and enduring process in comparison to alternative phase conversion methods, such as cryo-compressed storage and liquefied storage. MOFs are defined by their crystalline formations, high porosity, and a substantial surface area that render them capable of effectively trapping a vast quantity of gas molecules within their voids. The desorption of adsorbed hydrogen on the pores of MOFs, which is held by weak van der Waals force, can be rapidly achieved by applying heat or pressure. The present procedure is capable of producing hydrogen gas on a significant scale under atmospheric pressure and temperature conditions (Singh et al., 2020). The production of hydrogen gas can be achieved through various methods, including steam reforming of methane, coal gasification, and water electrolysis. The utilization of water electrolysis is favored due to its ability to generate environmentally friendly hydrogen and its potential to decrease carbon emissions (Gao et al., 2018).

MOFs are consist of secondary building units (SBUs) and organic linkers, and their alterations are appropriate for both catalytic and storage purposes (Liao et al., 2018). When compared to alternative hydrogen molecule storage materials, such as CNT, hydrides, zeolites, and clathrates, MOFs exhibit a relatively elevated BET surface area and porosity. MOF-based materials have garnered significant attention in the areas of energy storage, energy conversion, and gas separation (Lee et al., 2019). The fundamental factors that determine H_2 adsorption are the volumetric storage capacity and storage gravimetric (Pęska et al., 2020; Liao et al., 2020). Cryogenic pressure vessels that are coated with MOFs have the potential to serve as a viable option for the storage of cryogenic H_2. Vessels under cryogenic pressure exhibit cost-effectiveness and reduced evaporation loss in contrast to compressed storage; however, they remain inadequate in fulfilling the Department of Energy's (DOE) stipulated objective. DOE has set a performance objective of 30 g/L and 4.5 wt% of usable H_2 storage capacity at 5–12 bars and 233–358 K for 2020. DOE has established a target of 6.5 wt% and 50 g/L as the ultimate goal. The aims are intended to give a 500 km refueling distance for light motor vehicles. The storage device is expected to cost roughly \$10/kWh (Jörissen, 2011). Several results toward the development of H_2 energy and its storage utilizing MOFs have been discussed. Hydrogen has the potential to serve as a viable medium for storage, specifically for utilization in fuel cells. At present, H_2 gas is stored through various methods such as liquefaction, utilization of pressurized vessels, employment of MOFs, carbon adsorption, and deployment of chemical H_2 energy storage materials. Commercially available technologies exhibit a low volumetric storage capacity and necessitate a substantial amount of energy for storage. MOFs have high adsorption capabilities; however, further research is needed to meet the DOE objective (Olabi et al., 2021; Chanchetti et al., 2020).

In this chapter, the recent developments in hydrogen storage utilizing modified MOF materials are described, starting with a study of the H_2 storage mechanism, followed by a discussion of the H_2 storage capacity of MOFs by the doping process, metal centers, and MOF hybrids. Some of these advancements must be resolved by transforming them into real-time applications.

8.2   A SUMMARY OF HYDROGEN STORAGE

At room temperature, hydrogen gas exhibits a relatively large volume in comparison to other gases possessing equivalent energy levels. Therefore, it is imperative that hydrogen be subjected to liquefaction, compression, or other suitable techniques for its utilization in vehicular and mobile contexts. The utilization of compressed gas necessitates the utilization of vessels that are pressurized, which tend to be weighty and voluminous owing to the elevated pressures involved. Hydrogen liquefaction is achieved at a temperature of 253°C, after which it is subsequently stored in a thermally insulated tank. The process of evaporation results in a persistent depletion of hydrogen, necessitating a substantial expenditure of energy to sustain the liquid phase. The process of carbon adsorption is currently in the phase of research and development. The high specific surface area and porosity of metal-organic frameworks make them a viable option for hydrogen storage.

8.3   ENERGY STORAGE MECHANISM OF MOFs

The hydrogen adsorption has weak interaction, which may be overcome by hydrogen spillover adsorption on porous carbon materials (PCM). The utilization of the spillover mechanism technique is deemed more advantageous for hydrogen storage due to its operability at ambient temperature, and its efficacy can be further augmented by the incorporation of metal-based catalysts. H_2 molecules move from the catalyst to the substrate in order to improve hydrogen storage characteristics (Zhou et al., 2015). The mechanisms of spillover are described as: surface chemisorption, catalytic dissociation on a metal surface, the transference of hydrogen from a metallic catalyst to the PCM surface, and substrate surface desorption and diffusion. Hydrogen chemisorption can be accomplished through a process of phase nucleation. In cases where chemisorption of H_2 onto the catalyst demands 0.8–1.8 eV of energy, the catalyst must have a very high capacity for adsorbing hydrogen. The migration of the hydrogen atom to the substrate requires overcoming a significant energy barrier of 2.45–3.2 eV. Through the use of hole doping, the energy barrier may be lowered. The H_2 atom diffusion within the substrate is hindered by the robust C-H bond, while the MOF surface presents a 1.05–2.16 eV energy barrier. Figure 8.1 illustrates the process of hydrogen migration through the MOF.

8.4   MOF ADVANCEMENTS FOR HYDROGEN STORAGE IN RECENT YEARS

8.4.1   MOF STRUCTURAL CHARACTERISTICS

MOFs are composed of two primary constituents, SBUs and organic linkers. Linkers and SBU combine to produce various MOFs. Various MOFs can be produced through the amalgamation of an organic ligand and distinct SBUs, as depicted in Figure 8.2.

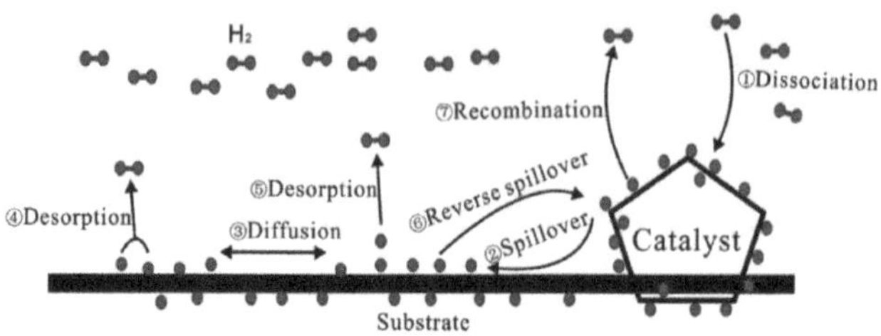

FIGURE 8.1 Illustration of hydrogen movement paths in H_2 storage by spillover (Shet et al., 2021; Guo et al., 2020).

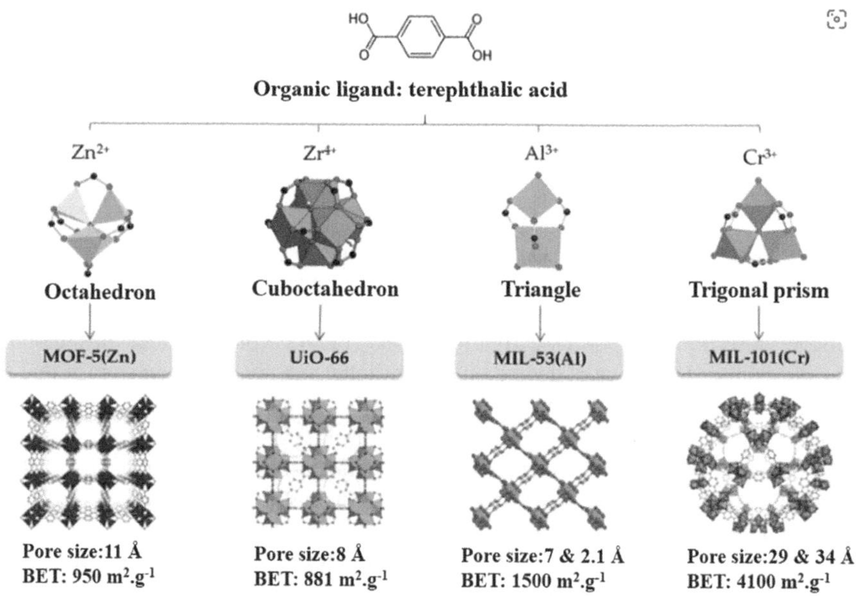

FIGURE 8.2 The node-and-connector method to prepare MOFs (Shet et al., 2021; Rocío-Bautista et al., 2019).

Consequently, a substantial quantity of MOFs can be synthesized utilizing different combinations. The pore diameters of connectors made from carbon chains may be modified while still retaining their structure and symmetry. The determination of pore diameter in the MOF is contingent upon the carbon chain linker's length. Furthermore, the linker is fortified with substitutes to enhance chemical selectivity and characteristics (Butova et al., 2016). The incorporation of linkers into SBUs can induce modifications in the unit cell parameters, either by elongating the carbon chain or by altering the symmetry of the SBU through changes in the relative positioning of its functionalities.

The introduction of a linker in SBUs can lead to modifications in the unit cell parameters, either through the elongation of the carbon chain or the alteration of symmetry resulting from changes in the relative positioning of functionalities. Hence, it can be inferred that the enhancement of MOFs' functionality can be achieved through the alteration of their oxidation states (Song et al., 2019). The coordination number of SBUs exhibits a range of variability from 3 to 66. Post-synthetic modification (PSM) is a technique employed on SBUs to create innovative MOFs that can regulate the capacity and features of MOFs. The introduction of functional groups and guest ions into SBUs has been observed to have a substantial impact on the proton conductivity of MOFs. The employment of the PSM approach enables the synthesis of strategic business units that were previously unattainable through conventional means (Ha et al., 2020).

The orientation of organic linkers has the potential to dictate the structural topologies. The linker functionalized with Zr-DMBD (dimercapto-1,4-benzenedicarboxylate) exhibited the capability of reducing the concentration of Hg^{2+} in water to levels below 0.01 ppm (Bon et al., 2019). The compound $ZrCl_4$ undergoes a reaction with the tpdc linker, specifically the 9,10-bis(triisopropylsilyloxy)phenanthrene-2,7-dicarboxylate variant, resulting in the formation of a Zr-MOF. This material exhibits stable functionalized pores and well-preserved frameworks. The anionic MOF was converted to a neutral MOF through the reaction of $CuCl_2$ and ttpm (tetra-kis (4-tetrazolylphenyl) methane) linker. The results indicated a greater surface area and increased binding affinity (Lu et al., 2014).

8.4.2 MOFs with High Surface Areas

MOFs are widely utilized owing to their high surface area, which enables them to adsorb gases within their metallic frameworks. The relationship between the surface area and H_2 adsorption is positively correlated, and in accordance with Chahine's principle, a 1 wt% H_2 uptake is observed for every $500\,m^2g^{-1}$ (Bakuru et al., 2019; Chen et al., 2018b). The process of heat treatment serves to augment the quantity of microporous surface area and pores (Broom et al., 2019). Composite materials based on a matrix of polymer of intrinsic microporosity (PIM-1) and a filler of porous aromatic framework (PAF-1) have been synthesized with the aim of enhancing the surface area available for porous adsorption and improving the efficiency of hydrogen adsorption. The PAF-1 material exhibited a surface area measurement of $3,787\,m^2/g$ and a hydrogen uptake of 1.47 wt%. The hydrogen storage capacity of PIM-1 was observed to increase by 84% when exposed to 37.5% PAF-1 (Rochat et al., 2017). GCMC simulations were performed at 77 and 298 K for MOF-519 $(Al_8(OH)_8(BTB)_4(H_2BTB)_4)$ and MOF-520 $(Al_8(OH)_8(BTB)_4(-HCOO)_4)$. The MOF-519 exhibited a surface area measuring $2400\,m^2/g$, and demonstrated H_2 adsorption of 2.13 wt% (1 bar) and 3.49 wt% (130 bar) at 77 K (Xia and Liu, 2017).

The Chahine rule was employed to forecast the hydrogen storage capacity of various MOFs, and it exhibited reliable and consistent outcomes. The MOF-5 exhibits a surface area of $3,512\,m^2g^{-1}$, a hydrogen capacity of 6.8%, and a MOF-5 crossing rate of 0.9% among MOFs. In comparison to MOF-5, it was observed that IRMOF-20 exhibited significantly greater usable volumetric and gravimetric capacities, with values of 9.1 wt% and 51 g H_2/L, respectively (Ahmed et al., 2017).

A novel MOF denoted as $Zn_4O(bpdc)(btctb)_{4/3}$ (DUT-32) has been synthesized utilizing both ditopic (bpdc-4,4'-biphenylendicarboxylic acid) and tritopic (btctb-4,4',

4′-[benzene-1,3,5-triyltris(carbonylimino)]trisbenzoate) linkers. The material exhibited a surface area of 6,411 m²/g and demonstrated a 7.8 wt% hydrogen capacity at 77 K and 53 bar, as well as 14.21 wt% at 77 K and 80 bar. The linkers resulted in an increase in both the surface area and the capacity for H_2 uptake. The properties of Zr-based MOF H_2 adsorption, specifically NU-1101, NU-1102, and NU-1103, were evaluated using both gravimetric and volumetric methods. The experimental results indicate that the respective capacities of NU-1101, NU-1102, and NU-1103 were determined to be 46.6 g/L (9.2 wt%), 43.7 g/L (9.6 wt%), and 43.2 g/L (12.6 wt%), respectively. The enhancement in gravimetric performance can be attributed to the augmentation in pore volume, whereas the alteration in void fraction accounts for the improvement in volumetric performance (Gómez-Gualdrón et al., 2017).

8.4.3 MOFs with High Storage Capacity at Room Temperature

MOFs exhibit a notable capacity for hydrogen storage at cryogenic temperatures (77 K) that is in proximity to the hydrogen storage objective set by the United States DOE. However, from a practical standpoint, the storage of hydrogen at a temperature of 77 K isn't a feasible choice. At standard temperature and pressure, hydrogen exists in a van der Waals state as a gas. MOFs typically exhibit a high isosteric adsorption heat, which results in notable hydrogen preservation at ambient temperature. The introduction of a Pt/C catalyst resulted in a 3.5% increase in hydrogen adsorption for microporous CuBTC at a temperature of 298 K and a pressure of 20 bar (Zacharia and Rather, 2015). Wang et al. (2019b) employed a deposition-reduction approach to synthesize magnesium-based MOFs including ZIF-8, ZIF-67, and MOF-74. $MgCl_2$ was solubilized within MOFs to enhance the efficacy of hydrogen adsorption. MOFs generally demonstrate a considerable isosteric adsorption heat, leading to significant preservation of hydrogen at room temperature. Adding Mg resulted in notable enhancements in the dehydrogenation characteristics, with the Mg-doped ZIF-67 MOF exhibiting the minimum temperature requirement (289°C) for hydrogen liberation. Following 100 cycles of hydrogen absorption and desorption, the Mg/ZIF-67 material exhibited a surface area of 3,357 m²/g and was capable of adsorbing 5.3% of H_2 at a temperature of 350°C, while retaining 100% of its adsorption rate (Wang et al., 2019b; Chen et al., 2018a). Sun et al. have chosen HKUST-1 [$Cu_3(BTC)_2(H_2O)_3$] on account of its narrow pore apertures and high thermal endurance of up to 523 K. The Cu-MOF material is employed as a loading template for $LiBH_4$ in the manufacture of H_2 storage material. In comparison to pristine $LiBH_4$ (380°C), $LiBH_4$@Cu-MOFs started to dehydrogenate at about 60°C. At ambient temperature, Cu-MOFs react with $LiBH_4$ to provide improved dehydrogenation characteristics (Sun et al., 2011).

8.4.4 Stability of MOFs

The assessment of the stability of MOFs is imperative due to the potential variability in hydrogen reactivity across varying temperatures. An unstable MOF can cause problems detecting the crystal structure or can collapse when the solvent is removed. Highly stable MOFs are produced when high valence metals are included into MOFs with low ligand exchange rates (Al Amery et al., 2020). Cryogenic cycling of MOF materials is a reliable method of gauging their long-term stability as hydrogen storage media. In order to carry out the examination, the MOFs are placed within cylindrical steel holders that are subsequently inserted into tubular pressure vessels. The range of pressures utilized in this process varies from 0.5 to 10 MPa (Wang et al., 2019a). The pressure cycling of MOF-5 was conducted over 300 cycles at a temperature of 77 K. Upon the introduction of various impurity gases, it was observed that MOF-5 exhibited a retention capacity of 97% after undergoing 300 cycles. The results of static exposure experiments indicate that the hydrogen storage efficiency of MOFs remained stable at concentrations of 8 ppm of H_2O and 0.004 ppm of H_2S. The H_2 storage capacity experienced a decrease from 6 to 5.8 wt% solely in the presence of HCl mixtures, while no significant effects of other gases were noted (Ming et al., 2016).

8.4.5 Loading Methods

The diverse loading techniques employed for MOFs can exert an influence on the catalyst characteristics and potentially affect the stability of MOFs. The wetness impregnation method was employed to load MPC-ZrO_2 with UIO-66-NH_2, resulting in the formation of Ru/MPC-ZrO_2. This method was utilized to generate Ru active sites that were well dispersed in MOF (Liao et al., 2018). The storage capacity of MOFs can be improved and their loading capabilities can be controlled by using a solvent-free loading approach (Seth et al., 2019). Nanomaterials are created by the solid grinding process and then encased inside the MOFs. The utilization of MOFs for the confinement of active metals at the nanoscale has resulted in enhanced hydrogen absorption capacity. The augmentation of hydrogen uptake capacity can be achieved through the doping of high valence metals with MOFs. The H_2 storage capacity at room temperature was observed to be 30% higher in ZIF-8 material impregnated with liquid Pd nanoparticles (Villajos et al., 2016). The selection of a loading method is a critical task as each method entails its own set of benefits and limitations.

8.5 MOF MODIFICATION FOR HYDROGEN STORAGE

8.5.1 Pristine MOFs

The pristine MOFs have garnered attention owing to their notable characteristics such as high surface area, porosity, pore size, and catalytic properties. Pristine MOFs exhibit a cooperative mechanism wherein the metal sites within the frameworks initiate hydrogen adsorption. Approximately 500,000 hypothetical MOFs were subjected to computational analysis, resulting in the identification of three MOFs, namely PCN-610/NU-100, UMCM-9, and SNU-70, which exhibit hydrogen storage capabilities surpassing that of IRMOF-20. The three MOFs exhibited a significantly high volumetric capacity and total gravimetric capacity of H_2 when compared to MOF-5 and IRMOF-20 across different pressure swings (Villajos et al., 2016). Among the three MOFs considered, PCN-610/NU-100 exhibited superior performance in both gravimetric and volumetric measurements. Enhancing the volume fraction, pore volume, gravimetric surface area, and minimizing the volumetric surface area and diameter resulted in an augmentation of the usable capacities. The findings suggest that there is no MOF that exhibits exceptional performance in both volumetric and gravimetric H_2 capacity for temperature and pressure swing (TPS). TPS, UMCM-9, and SNU-70 demonstrated a compelling amalgamation of volumetric and gravimetric H_2 capacities in the context of hydrogen storage.

8.5.2 MOF with Metal Centers

MOFs that are synthesized utilizing metal oxide clusters give rise to a crystalline framework that exhibits permanent porosity and generates robust binding energy, which is contingent upon the metal type. The adsorption mechanism observed in these MOFs is contingent upon the metal oxides present, and may manifest as either physisorption or chemisorption. The study conducted by Garcíasanchez et al. involved the utilization of a bismuth-based IEF-5 MOF in conjunction with DTTDC (dithieno[3,2-b:2′,3′-d]thiophene-2,6-dicarboxylicacid) as a hole transport linker. The characterization suggests that the presence of a ligand facilitates the charge transfer process, thereby contributing to the generation of molecular hydrogen. The process of H_2 production is significantly influenced by the selection of ligands (García-Sánchez et al., 2020). Suksaengrat and coworkers conducted a synthesis of M-MOF-525 and conducted an investigation into the impact of various metal ions, including V, Hf, Ti, and Zr, on the hydrogen storage capacity of the material. The hydrogen's site binding energy within the metal oxide region was comparatively greater than that of the organic linker binding sites. Additionally, the MOFs exhibited van der Waals force of attraction. The hydrogen-binding energies of Zr and Hf-MOF-525 are reported to be 0.04–0.15 eV/H_2 and 0.06–016 eV/H_2, respectively. The Hf-MOF-525 and Zr exhibited high dipole moments, thereby enabling the trapping of hydrogen molecules on their surfaces (Suksaengrat et al., 2020).

8.5.3 Doping in MOF

Metal ions may be added to MOFs to increase their capacity for storing hydrogen. The evaluation of the DFT research for the doping of borophene with Li, Na, and K metal atoms reveals significant insights. In MOFs with doped elements with high hydrogen adsorption capacities, spillover mechanisms are seen. On the basis of various periodic hole patterns, the materials are designated as S_1, S_2, and S_3. For S_1, S_2, and S_3, the Li-doped borophene exhibited gravimetric hydrogen capacities of 8.36 wt% for S_1, 11.49 wt% for S_2, and 13.96 wt% for S_3, compared to 10.39 wt% for S_2 and 9.06 wt% for S_3 that were doped with Na. The desorption temperature of H_2 was observed to be within the range of 200–220 K. However, it was found that K-doped borophene and Na-doped S_1 were not suitable for this purpose due to their unfavorable adsorption energy (Wang et al., 2018). The synthesis of MOF-5 (Zn(OAc)$_2$.0.5H$_2$O/H$_2$BDC) and PANI-ES (H$_2$O) was carried out, and PANI-EB was obtained through the process of deprotonation of the PANI-ES (H$_2$O) powder. The synthesis of MOF-5/PANI composites was conducted in DMF, and it was observed that 75% MOF-EB yielded the most notable results in terms of specific surface area (2,737 m^2/g), micropore volume (1.49 cm^3/g), and micropore area (2,193.6 m^2/g).

The electrical conductivity of PANI-ES (H_2O) was observed to be significant when combined with 25% MOF-ES (H_2O), resulting in a conductivity value of 2.6×10^{-5} Scm^{-1}. MOF-5 has been suggested as a potential solution to address the issue of low specific surface area (Biserčić et al., 2020).

8.5.4 HYBRID MOFs

The enhancement of MOF efficiencies can be achieved through the coupling with complementary materials such as graphene oxide. The hybrid MOFs exhibited diverse mechanisms, including dissociation, chemisorption, and physisorption, contingent upon the utilized composites. The hybrid MOFs exhibited diverse mechanisms, including dissociation, chemisorption, and physisorption, contingent upon the utilized composites. The UiO-66(Zr) MOF composite was synthesized utilizing Zeolite Templated Carbons (ZTC) and PIM-1 as an adsorbent. The interaction between the hydroxyl group of UiO-66(Zr) and PIM-1 was found to be effective in mitigating pore blockage. The effectiveness of the PIM-1 polymer in alleviating pore blockage was observed through its interaction with the hydroxyl group of UiO-66(Zr). The materials based on ZTC exhibited greater stability and demonstrated superior capacity for hydrogen uptake. The present study employs a thermal exfoliation methodology for the production of Zr-MOF utilizing reduced graphene oxide (rGO) as a substrate. The incorporation of rGO into the MOF did not result in any alteration of the crystal structure (Musyoka et al., 2017).

8.6 MOF OPPORTUNITIES IN HYDROGEN STORAGE

MOFs have the potential to be employed as an alternative to storing clean energy. The utilization of MOFs has been observed to exhibit a higher hydrogen capacity compared to compressed hydrogen gas, thereby presenting a potential avenue for reducing the cost of fuel for hydrogen-powered vehicles. MOFs exhibiting sufficient hydrogen adsorption capacity under ambient conditions may be deemed suitable for utilization in the automotive industry. The 3D structured MOFs exhibit a high degree of hydrogen gas molecule adsorption capacity within their pores, and are capable of gas molecule condensation under ambient temperature conditions. MOF 3D printed composites may be utilized to prevent moisture damage. Proton exchange membrane (PEM) fuel cells have the capability to facilitate the conversion of stored hydrogen into electrical energy (Kapelewski et al., 2018).

8.7 CONCLUSION

Hydrogen storage utilizing MOFs with high surface area materials is being studied as a means to reduce reliance on non-renewable energy sources. The relationship between hydrogen adsorption and BET surface area, pore volume, and pore size is one of direct proportionality. The hydrogen uptake capacity of inert metals such as Pd and Ru can be enhanced through the process of doping with MOFs. Various techniques, including the introduction of nanoparticles, utilization of metal ions, and the creation of composites, have been demonstrated to enhance the hydrogen storage capabilities of MOFs. Once the effects of these strategies are comprehended, the process of optimizing metal-organic frameworks will become straightforward. The conductivity of MOFs can be improved through the process of guest doping, thereby improving them suitable for application in fuel cells. The chapter posits that the development of novel, economical materials and techniques for the secure storage of hydrogen remains a pressing requirement.

REFERENCES

Ahmed, A., Liu, Y., Purewal, J., Tran, L. D., Wong-Foy, A. G., Veenstra, M., Matzger, A. J. & Siegel, D. J. 2017. Balancing gravimetric and volumetric hydrogen density in MOFs. *Energy & Environmental Science*, 10, 2459–2471.

Al Amery, N., Abid, H. R., Al-Saadi, S., Wang, S. & Liu, S. 2020. Facile directions for synthesis, modification and activation of MOFs. *Materials Today Chemistry*, 17, 100343.

Bakuru, V. R., Dmello, M. E. & Kalidindi, S. B. 2019. Metal-organic frameworks for hydrogen energy applications: advances and challenges. *ChemPhysChem*, 20, 1177–1215.

Biserčić, M. S., Marjanović, B., Zasońska, B. A., Stojadinović, S. & Ćirić-Marjanović, G. 2020. Novel microporous composites of MOF-5 and polyaniline with high specific surface area. *Synthetic Metals*, 262, 116348.

Bon, V., Senkovska, I. & Kaskel, S. 2019. Metal-organic frameworks. In: Kaneko, K., Rodríguez-Reinoso, F. (eds) Nanoporous Materials for Gas Storage. Springer, Singapore, pp. 137–172.

Broom, D. P., Webb, C., Fanourgakis, G. S., Froudakis, G. E., Trikalitis, P. N. & Hirscher, M. 2019. Concepts for improving hydrogen storage in nanoporous materials. *International Journal of Hydrogen Energy*, 44, 7768–7779.

Butova, V. V. E., Soldatov, M. A., Guda, A. A., Lomachenko, K. A. & Lamberti, C. 2016. Metal-organic frameworks: structure, properties, methods of synthesis and characterization. *Russian Chemical Reviews*, 85, 280.

Chanchetti, L. F., Leiva, D. R., De Faria, L. I. L. & Ishikawa, T. T. 2020. A scientometric review of research in hydrogen storage materials. *International Journal of Hydrogen Energy*, 45, 5356–5366.

Chen, C., Wang, J., Wang, H., Liu, T., Xu, L. & Li, X. 2018a. Improved kinetics of nanoparticle-decorated Mg-Ti-Zr nanocomposite for hydrogen storage at moderate temperatures. *Materials Chemistry and Physics*, 206, 21–28.

Chen, Y., Sakata, O., Nanba, Y., Kumara, L. S. R., Yang, A., Song, C., Koyama, M., Li, G., Kobayashi, H. & Kitagawa, H. 2018b. Electronic origin of hydrogen storage in MOF-covered palladium nanocubes investigated by synchrotron X-rays. *Communications Chemistry*, 1, 61.

Eberle, U., Müller, B. & Von Helmolt, R. 2012. Fuel cell electric vehicles and hydrogen infrastructure: status 2012. *Energy & Environmental Science*, 5, 8780–8798.

Gao, M., Shih, C.-C., Pan, S.-Y., Chueh, C.-C. & Chen, W.-C. 2018. Advances and challenges of green materials for electronics and energy storage applications: from design to end-of-life recovery. *Journal of Materials Chemistry A*, 6, 20546–20563.

García-Sánchez, A., Gomez-Mendoza, M., Barawi, M., Villar-Garcia, I. J., Liras, M., Gándara, F. & De La Peña O' Shea, V. A. 2020. Fundamental insights into photoelectrocatalytic hydrogen production with a hole-transport bismuth metal-organic framework. *Journal of the American Chemical Society*, 142, 318–326.

Gómez-Gualdrón, D. A., Wang, T. C., García-Holley, P., Sawelewa, R. M., Argueta, E., Snurr, R. Q., Hupp, J. T., Yildirim, T. & Farha, O. K. 2017. Understanding volumetric and gravimetric hydrogen adsorption trade-off in metal-organic frameworks. *ACS applied materials & interfaces*, 9, 33419–33428.

Guo, J.-H., Li, S.-J., Su, Y. & Chen, G. 2020. Theoretical study of hydrogen storage by spillover on porous carbon materials. *International Journal of Hydrogen Energy*, 45, 25900–25911.

Ha, J., Lee, J. H. & Moon, H. R. 2020. Alterations to secondary building units of metal-organic frameworks for the development of new functions. *Inorganic Chemistry Frontiers*, 7, 12–27.

Hochreiter, E., Schmidt-Hebbel, K. & Winckler, G. 2003. Monetary union and dollarization: theory and regional experience. *International Journal of Finance & Economics*, 8, 277.

Jörissen, L. 2011. Hydrogen and fuel cells. fundamentals, technologies and applications edited by Detlev Stolten. *Wiley Online Library*, 50, 9787.

Kapelewski, M. T., Runčevski, T., Tarver, J. D., Jiang, H. Z. H., Hurst, K. E., Parilla, P. A., Ayala, A., Gennett, T., Fitzgerald, S. A., Brown, C. M. & Long, J. R. 2018. Record high hydrogen storage capacity in the metal-organic framework Ni2(m-dobdc) at near-ambient temperatures. *Chemistry of Materials*, 30, 8179–8189.

Lee, C. C., Chen, C. I., Liao, Y. T., Wu, K. C. W. & Chueh, C. C. 2019. Enhancing efficiency and stability of photovoltaic cells by using perovskite/Zr-MOF heterojunction including bilayer and hybrid structures. *Advanced Science*, 6, 1801715.

Liao, Y.-T., Matsagar, B. M. & Wu, K. C.-W. 2018. Metal-organic framework (MOF)-derived effective solid catalysts for valorization of lignocellulosic biomass. *ACS Sustainable Chemistry & Engineering*, 6, 13628–13643.

Liao, Y.-T., Ishiguro, N., Young, A. P., Tsung, C.-K. & Wu, K. C.-W. 2020. Engineering a homogeneous alloy-oxide interface derived from metal-organic frameworks for selective oxidation of 5-hydroxymethylfurfural to 2, 5-furandicarboxylic acid. *Applied Catalysis B: Environmental*, 270, 118805.

Lu, W., Wei, Z., Gu, Z.-Y., Liu, T.-F., Park, J., Park, J., Tian, J., Zhang, M., Zhang, Q. & Gentle III, T. 2014. Tuning the structure and function of metal-organic frameworks via linker design. *Chemical Society Reviews*, 43, 5561–5593.

Ming, Y., Purewal, J., Yang, J., Xu, C., Veenstra, M., Gaab, M., Müller, U. & Siegel, D. J. 2016. Stability of MOF-5 in a hydrogen gas environment containing fueling station impurities. *International Journal of Hydrogen Energy*, 41, 9374–9382.

Musyoka, N. M., Ren, J., Langmi, H. W., North, B. C., Mathe, M. & Bessarabov, D. 2017. Synthesis of rGO/Zr-MOF composite for hydrogen storage application. *Journal of Alloys and Compounds*, 724, 450–455.

Olabi, A., Abdelghafar, A. A., Baroutaji, A., Sayed, E. T., Alami, A. H., Rezk, H. & Abdelkareem, M. A. 2021. Large-vscale hydrogen production and storage technologies: current status and future directions. *International Journal of Hydrogen Energy*, 46, 23498–23528.

Pęska, M., Dworecka-Wójcik, J., Płociński, T. & Polański, M. 2020. The influence of cerium on the hydrogen storage properties of La1-xCexNi5 alloys. *Energies*, 13, 1437.

Rochat, S., Polak-Kraśna, K., Tian, M., Holyfield, L. T., Mays, T. J., Bowen, C. R. & Burrows, A. D. 2017. Hydrogen storage in polymer-based processable microporous composites. *Journal of Materials Chemistry A*, 5, 18752–18761.

Rocío-Bautista, P., Taima-Mancera, I., Pasán, J. & Pino, V. 2019. Metal-organic frameworks in green analytical chemistry. *Separations*, 6, 33.

Schlapbach, L. & Züttel, A. 2001. Hydrogen-storage materials for mobile applications. *Nature*, 414, 353–358.

Seth, S., Vaid, T. P. & Matzger, A. J. 2019. Salt loading in MOFs: solvent-free and solvent-assisted loading of NH4NO3 and LiNO3 in UiO-66. *Dalton Transactions*, 48, 13483–13490.

Shet, S. P., Shanmuga Priya, S., Sudhakar, K. & Tahir, M. 2021. A review on current trends in potential use of metal-organic framework for hydrogen storage. *International Journal of Hydrogen Energy*, 46, 11782–11803.

Singh, R., Altaee, A. & Gautam, S. 2020. Nanomaterials in the advancement of hydrogen energy storage. *Heliyon*, 6, e04487.

Song, D., Bae, J., Ji, H., Kim, M.-B., Bae, Y.-S., Park, K. S., Moon, D. & Jeong, N. C. 2019. Coordinative reduction of metal nodes enhances the hydrolytic stability of a paddlewheel metal-organic framework. *Journal of the American Chemical Society*, 141, 7853–7864.

Suksaengrat, P., Amornkitbamrung, V., Srepusharawoot, P. & Ahuja, R. 2016. Density functional theory study of hydrogen adsorption in a Ti-decorated Mg-based metal-organic framework-74. *ChemPhysChem*, 17, 879–884.

Suksaengrat, P., Amornkitbamrung, V. & Srepusharawoot, P. 2020. Enhancements of hydrogen adsorption energy in M-MOF-525 (M= Ti, V, Zr and Hf): a DFT study. *Chinese Journal of Physics*, 64, 326–332.

Sun, W., Li, S., Mao, J., Guo, Z., Liu, H., Dou, S. & Yu, X. 2011. Nanoconfinement of lithium borohydride in Cu-MOFs towards low temperature dehydrogenation. *Dalton Transactions*, 40, 5673–5676.

Villajos, J. A., Orcajo, G., Calleja, G., Botas, J. A. & Martos, C. 2016. Beneficial cooperative effect between Pd nanoparticles and ZIF-8 material for hydrogen storage. *International Journal of Hydrogen Energy*, 41, 19439–19446.

Wang, L., Chen, X., Du, H., Yuan, Y., Qu, H. & Zou, M. 2018. First-principles investigation on hydrogen storage performance of Li, Na and K decorated borophene. *Applied Surface Science*, 427, 1030–1037.

Wang, T. C., White, J. L., Bie, B., Deng, H., Edgington, J., Sugar, J. D., Stavila, V. & Allendorf, M. D. 2019a. Design rules for metal-organic framework stability in high-pressure hydrogen environments. *ChemPhysChem*, 20, 1305–1310.

Wang, Y., Lan, Z., Huang, X., Liu, H. & Guo, J. 2019b. Study on catalytic effect and mechanism of MOF (MOF= ZIF-8, ZIF-67, MOF-74) on hydrogen storage properties of magnesium. *International Journal of Hydrogen Energy*, 44, 28863–28873.

Xia, L. & Liu, Q. 2017. Adsorption of H2 on aluminum-based metal-organic frameworks: a computational study. *Computational Materials Science*, 126, 176–181.

Zacharia, R. & Rather, S. U. 2015. Review of solid state hydrogen storage methods adopting different kinds of novel materials. *Journal of Nanomaterials*, 2015, 1–18.

Zhou, H., Zhang, J., Zhang, J., Yan, X.-F., Shen, X.-P. & Yuan, A.-H. 2015. Spillover enhanced hydrogen storage in Pt-doped MOF/graphene oxide composite produced via an impregnation method. *Inorganic Chemistry Communications*, 54, 54–56.

9 Pipelines for Hydrogen Transportation

Yimin Zeng and Minkang Liu

9.1 INTRODUCTION

Over the past centuries, the rapid growth of global population and economic activity has led to a significant increase in energy demand to meet our daily needs (Gielen et al., 2019). Although intensive efforts have been made to explore clean and renewable energy alternatives (e.g., bioenergy, solar, hydro, wind, geothermal, and tidal energy), our energy supply still primarily relies on the extensive combustion of fossil fuels, including coal, petroleum, and natural gas. This has resulted in their gradual depletion and has caused serious environmental damage, such as greenhouse gas emissions, global warming, smog, and acid rain (Ganda, 2019, Crippa et al., 2021, Kuramochi et al., 2020). To achieve the target of limiting global average surface temperature increases to less than 2°C, worldwide CO_2 emissions from the combustion of fossil fuels need to be substantially reduced—by up to 85% compared to emission levels in 2000 (Agency, 2017, Onyebuchi et al., 2018). Unfortunately, the global reduction in CO_2 emissions is lagging far behind the targets set by the UN Paris Agreement (Januta, 2021). Therefore, a swift transition from carbon-based energy sources to hydrogen energy has become an urgent task. Hydrogen is an ideal carbon-free energy carrier with unique features that offer a potential solution to the problems of fossil fuel depletion and environmental pollution (McQueen et al., 2020).

Hydrogen is a completely renewable fuel that can be produced from various methods including natural gas reforming, thermochemical conversion of biomass, electrolysis of water, or a combination of these methods (McQueen et al., 2020, Abe et al., 2019, Acar and Dincer, 2020). Currently, approximately 76% of global hydrogen production comes from natural gas via steam-methane reforming, 22% from coal gasification, and 2% from water electrolysis (DOE, 2020). Hydrogen holds long-term application promise in various industrial sectors, such as petroleum refining, iron and steel production, transportation, power generation, and energy storage. Safe and cost-competitive transportation of large amounts of hydrogen or hydrogen/natural gas blends to these end-use points is essential for establishing an economically feasible, hydrogen-fueled society. Among the possible transportation methods—road and railway transportation using hydrogen tube trailers, ocean transportation with liquefied hydrogen carriers, and existing natural gas pipeline transportation—the latter is considered the most technologically mature and scalable option (Singla et al., 2021). If successful, more than 300,000 miles of existing natural gas pipelines could be utilized for short-term transportation of H_2/natural gas mixtures and long-term

DOI: 10.1201/9781003382553-11

distribution of 100% hydrogen, due to the expected rising demand (Russo, 2020). Additionally, over 1,600 miles of new gas pipelines have been constructed specifically for hydrogen transportation (Agency, 2017).

Most gas pipelines are made of mid-strength microalloyed steels, such as API grade 5L X65, due to their low cost, acceptable mechanical strength, wide availability, and good durability. It is worth noting that existing gas pipelines were designed for the distribution of natural gas at low-pressure ranges of 0.0017–0.69 bar, and only a few pipelines operate at higher pressures up to 27.6 bar (Melaina et al., 2013, Somerday and San Marchi, 2006). These pressures are much lower than those described for hydrogen transportation in the following sections (Melaina et al., 2013, Somerday and San Marchi, 2006, Boggess et al., 2013, Matsumoto et al., 2018, Briottet et al., 2012). The primary challenge in high-pressure hydrogen transportation lies in the hydrogen uptake and susceptibility to hydrogen embrittlement (also known as hydrogen-induced cracking) of the pipelines (Brilz et al., 2022, Atrens et al., 2018a, Cheng, 2007, Djukic et al., 2019). In fact, the hydrogen embrittlement problem is one of the oldest issues in material degradation and was first reported in 1875 (Johnson, 1875). The potential consequences of such catastrophic pipe failure, considering the amount and flammability of hydrogen, could significantly hinder the current transition to clean and sustainable hydrogen energy, especially in densely populated regions (Erdener et al., 2023). Unfortunately, most previous studies have focused on investigating hydrogen embrittlement and pipeline steel integrity in aqueous solutions under corrosion and cathodic proton reduction states, as well as in low-pressure gaseous hydrogen environments. These conditions differ substantially from the planned hydrogen pipeline transportation settings (Li et al., 2022b, Wulf et al., 2018, Timmerberg and Kaltschmitt, 2019). To fill significant knowledge gaps and realize full-scale hydrogen pipeline deployment, intensive research efforts have been directed at addressing the challenges of hydrogen embrittlement in high-pressure hydrogen environments in recent years (Trautmann et al., 2020, Barrera et al., 2018b, Nanninga et al., 2012b). For instance, a recent study based on Density Functional Theory (DFT) and thermodynamic calculations indicates that pipeline steels may face a high risk of hydrogen embrittlement during the transportation of high-pressure hydrogen (Li et al., 2022b).

Therefore, this chapter aims to introduce planned high-pressure hydrogen transportation, review previously proposed hydrogen embrittlement models, and analyze the influence of operating environmental factors and steel properties on the susceptibility of pipeline steels to hydrogen embrittlement. Potential strategies for preventing hydrogen embrittlement are also presented and discussed.

9.2 BACKGROUND AND HE MECHANISMS

9.2.1 Designed Hydrogen Pipeline Transportation Conditions

The short-term transportation of H_2 and natural gas blends is designed to operate at a maximum hydrogen partial pressure of 2 MPa and a maximum temperature of 50°C (Eames et al., 2022, Mahajan et al., 2022). Long-term, near-pure hydrogen

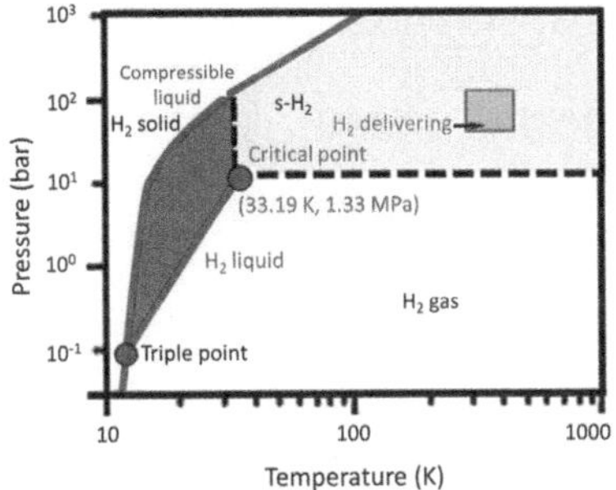

FIGURE 9.1 Hydrogen phase diagram constructed based on the database in the reference (Züttel, 2004).

transportation may operate in a pressure range of 5–20 MPa at near room temperature (Briottet et al., 2012). As shown in Figure 9.1, the supercritical point of hydrogen is located at approximately 33.19 K and 1.33 MPa. Therefore, the delivered hydrogen fluid should be in a supercritical state (s-H$_2$), where hydrogen uptake into and hydrogen embrittlement of pipeline steel could differ significantly from those exposed to low-pressure gaseous hydrogen. Unlike gaseous hydrogen, s-H$_2$ exhibits unique properties such as high compressibility, high density, and high fluidity. These properties are due to specific microstructures related to hydrogen bonding, ionic solvation, cluster formation, and ion association (Li et al., 2022b; Abdulagatov and Skripov, 2020; Yigzawe and Sadus, 2013).

9.2.2 Hydrogen Uptake Process

Hydrogen uptake (H$_2$-uptake) is the first and one of the most critical steps in determining the risk of hydrogen embrittlement (HE) in steel. Effectively controlling H-uptake is crucial for reducing the susceptibility of metals to HE. Since 1875, extensive studies have been conducted to investigate the generation, absorption, adsorption, and diffusion processes of hydrogen during the corrosion and cathodic protection of various materials in aqueous solutions. In these contexts, the hydrogen evolution reactions (Equations 9.1–9.6) could occur (Zhang et al., 2021; Jayabal et al., 2017; Zheng et al., 2015):

H generation and adsorption:

$$H_3O^+ + M + e^- \rightarrow MH_{ads} + H_2O \text{ (Acid Volmer Reaction)} \tag{9.1}$$

$$H_2O + M + e^- \rightarrow MH_{ads} + OH^- \text{ (Neutral or Alkaline Volmer Reaction)} \tag{9.2}$$

H$_2$ production:

$$MH_{ads} + H_3O^+ + e^- \rightarrow M + H_2 + H_2O \text{ (Acid Heyrovsky Reaction)} \tag{9.3}$$

$$MH_{ads} + H_2O + e^- \rightarrow M + H_2 + OH^- \text{ (Neutral or Alkaline Heyrovsky Reaction)} \quad (9.4)$$

$$2\ MH_{ads} \rightarrow M + H_2 \text{ (Tafel Reaction)} \quad (9.5)$$

H Absorption:

$$MH_{ads} \rightarrow MH_{abs} \quad (9.6)$$

where M represents a metal surface, MH_{ads} refers to hydrogen atoms adsorbed on the metal surface, and MH_{abs} is hydrogen atoms dissolved into the surface.

Unfortunately, very limited efforts have been directed toward studying H-uptake into metals in gaseous hydrogen environments, particularly under high-pressure s-H_2 conditions. In gaseous environments, the H-uptake by pipeline steels possibly involves the following distinct steps: (i) the physical and chemical adsorption and dissociation of H_2 molecules; and (ii) the absorption and permeation of dissociated hydrogen atoms into the steels (Pourazizi et al., 2020, Michler and Naumann, 2010, Dwivedi and Vishwakarma, 2018). Note that in the first step, physical adsorption/dissociation is a reversible process due to van der Waals forces between H_2 molecules and the steel surface, and it will quickly reach an equilibrium state. In contrast, chemisorption involves the dissociation of hydrogen molecules followed by the chemical binding of hydrogen atoms to the steel surface, as schematically shown in Figure 9.2 (Michler and Naumann, 2010; Pourazizi et al., 2020). Table 9.1 provides a summary of the hydrogen equilibrium solubilities in iron and commercial pipeline steels under ambient conditions. The solubility increases with steel strength, possibly due to variations in microstructure (Jo'm et al., 1971, Mohtadi-Bonab et al., 2013a, Jack et al., 2020, Li et al., 2019a, Han et al., 2019). Additionally, the concentration of dissolved hydrogen (C_H) in pipeline steel is also

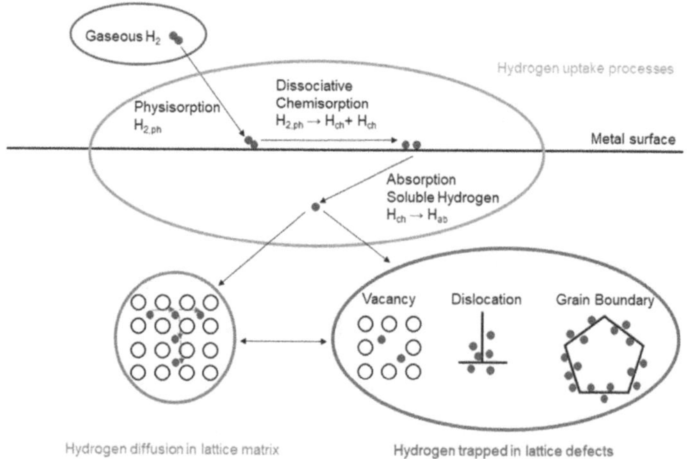

FIGURE 9.2 Schematic illustration of hydrogen adsorption, dissociation, absorption, permeation, and trapping processes, reprinted with permission from Li et al. (2022a).

TABLE 9.1

Equilibrium Hydrogen Solubility in Iron and Pipeline Steels at Room Temperature and Atmosphere Pressure

Materials	Hydrogen Solubility (mol/cm³)	References
Iron	1.5×10^{-7}	Jo'm et al. (1971)
X60	8.76×10^{-6}	Mohtadi-Bonab et al. (2013a)
X65	8.92×10^{-6}	Jack et al. (2020)
X70	8.91×10^{-6}	Mohtadi-Bonab et al. (2013a)
X80	1.07×10^{-5}	Li et al. (2019a)
X100	1.88×10^{-5}	Han et al. (2019)

related to hydrogen partial pressure, often referred to as fugacity (f_H), which is estimated from the Abel-Noble equation (Elhoud et al., 2010; Liu et al., 2014; San Marchi et al., 2007):

$$f_{H_2} = P_{H_2} * \exp * \left(\frac{P * b}{R * T} \right) \tag{9.7}$$

where P_{H_2} is the hydrogen partial pressure, b is co-volume constant, P is the total gas pressure, R is the gas constant, and T is the temperature.

Based on Sievert's law, the C_H in the steel can be calculated by Equation 9.8:

$$C_H = S * \sqrt{f_H} \tag{9.8}$$

where S is the hydrogen solubility constant related to operating temperature. For a pipeline steel, there is a C_H threshold above which HE will occur (Lovicu et al., 2012, Lynch, 2012a).

After being absorbed into steel, hydrogen atoms will either diffuse through the steel driven by a chemical potential gradient or become localized, trapped at defective sites. Due to their small size, the diffusion of hydrogen atoms is dominated by interstitial jumps, that is, hydrogen atoms occupy interstitial sites and overcome an energy barrier to move to neighboring sites (Mehrer, 2007, Porter et al., 2021). As most pipeline steels have a body-centered cubic (BCC) structure, hydrogen atoms will occupy either tetrahedral or both tetrahedral and octahedral interstitial sites under gaseous and high-pressure s-H_2 conditions (Hickel et al., 2014, Li et al., 2022b). At room temperature, the hydrogen diffusivity in pure BCC iron ranges from approximately 10^{-5} to $10^{-4} cm^2 s^{-1}$ (Johnson, 1988). In contrast, octahedral interstitial sites in face-centered cubic (FCC) alloys tend to be more favorable for hydrogen diffusion (Song et al., 2013, Turnbull, 2012). This distinction in interstitial solute sites between BCC and FCC structures contributes to higher diffusivity of hydrogen in BCC steels compared to FCC alloys, as the diffusion distance between the nearest BCC tetrahedral interstitial sites is obviously shorter than the distance between two FCC octahedral sites (Barrera et al., 2018a, Hickel et al., 2014).

Absorbed hydrogen atoms can also be trapped at material defects, such as cracks, grain boundaries, dislocations, and vacancies (Olden et al., 2008). The solubility of hydrogen in a perfect crystal is typically low but can increase in real steels and alloys by several orders of magnitude due to the presence of these hydrogen traps. Depending on their trap binding energy, these defects can either be reversible traps (<60 kJ/mol activation energy) or irreversible (>60 kJ/mol) traps (Svoboda et al., 2014, Turnbull, 2015, Liu et al., 2016). The most common reversible traps include interstitial sites, dislocations, lath boundaries, low-angle grain boundaries, coherent precipitates, and micro-cracks, while carbides, inclusions, and inconsistent precipitates serve as examples of irreversible traps. Table 9.2 lists the trap binding energies of different defects in steels and alloys (Choo and Lee, 1982, Yagodzinskyy et al., 2012, Novak et al., 2010). Moreover, as the hydrogen trapping site is a key factor in evaluating HE susceptibility, the trap site density in pipeline steels has also been identified in previous studies, as shown in Table 9.3 (Araújo et al., 2014b, Haq et al., 2013, Dong et al., 2009, 2010). The trap site density can be estimated from hydrogen permeation curves obtained through Devanathan–Stachurski cell testing (Vecchi et al., 2019). However, it's worth noting that even for the same steel, reported trap site densities can vary considerably due to the incorrect use of simplified formulas or when using the electrochemical charging method with extrapolation (Araújo et al., 2014a). Based on previous research, it is generally agreed that high-strength pipeline steels tend to have a higher trap site

---

TABLE 9.2

Selected Trap Binding Energies for Hydrogen in Steel

Trap Site	E_b (kJ/mol)	Lattice Configuration	References
Dislocation	24–26.8	Ferritic	Choo and Lee (1982)
	10–20	Austenitic	Yagodzinskyy et al. (2012)
Elastic dislocation	0–20.2	/	Novak et al. (2010)
Screw dislocation core	20–30	/	Hirth (1980a)
Grain boundary	17.2–59	/	Novak et al. (2010)
Austenite-ferrite interface	52	Duplex	Turnbull and Hutchings (1994)
Carbide interface	67–94	/	Novak et al. (2010), Li et al. (2004)

TABLE 9.3

Trap Density of Selected Pipeline Steels as Measured by Hydrogen Permeation

Pipeline Steels	Trap Site Density (cm^{-3})	References
X65	2.35×10^{20}	Araújo et al. (2014b)
X70	2.37×10^{24}	Haq et al. (2013)
X80	1.61×10^{27}	Dong et al. (2010)
X100	9.25×10^{27}	Dong et al. (2009)

density compared to lower-strength steels. Furthermore, the presence of a greater number of carbide interfaces and lattice dislocations in high-strength steels results in higher hydrogen trapping capacity, ultimately leading to increased susceptibility to hydrogen embrittlement (HE).

9.2.3 Proposed HE Mechanisms

HE can occur through a variety of mechanisms, including hydride formation, hydrogen-enhanced decohesion (HEDE), hydrogen-enhanced local plasticity (HELP), adsorption-induced dislocation emission (AIDE), hydrogen-enhanced macroscopic plasticity (HEMP), hydrogen-assisted micro void coalescence (HDMC), and mixed fracture (MF) (Venezuela et al., 2015, Zaferani et al., 2017, Moli-Sanchez et al., 2014, Lynch, 2012a, Koyama et al., 2014). These mechanisms can act individually or in concert, leading to hydrogen-induced cracking in materials. Extensive studies have been performed on the HE mechanisms in a wide array of metals and alloys—such as pipeline steels, stainless steels, Ni-based alloys, Ti, and Zr alloys—in aqueous environments using a variety of experimental methodologies (Mohtadi-Bonab et al., 2015; LAN et al., 2021; Nagumo & Takai, 2019; Nanninga et al., 2012a; Melaina et al., 2013; Qin et al., 2015; Tal-Gutelmacher & Eliezer, 2005). However, no general theory has yet been developed to comprehensively describe hydrogen embrittlement. This is partially because the phenomenon is highly dependent on not just the chemical composition and microstructure of the material under investigation, but also the in-service environments. Therefore, it is essential to identify the primary causes of HE embrittlement on a case-by-case basis.

9.2.3.1 Hydrogen-Enhanced Decohesion Mechanism

The Hydrogen-Enhanced Decohesion Mechanism (HEDE) was first proposed by Pfeil et al. (1926) and remains one of the foundational theories applied to describe HE in pipeline steels. Simply put, this mechanism posits that absorbed hydrogen can weaken the cohesive interatomic bonds in the material, making atoms more likely to separate even under low tensile stresses (as illustrated in Figure 9.3). This decrease in cohesive interactions is thought to occur when the hydrogen 1s electron occupies the 3d cell of an iron atom (Zaferani et al., 2017, Ramamurthy

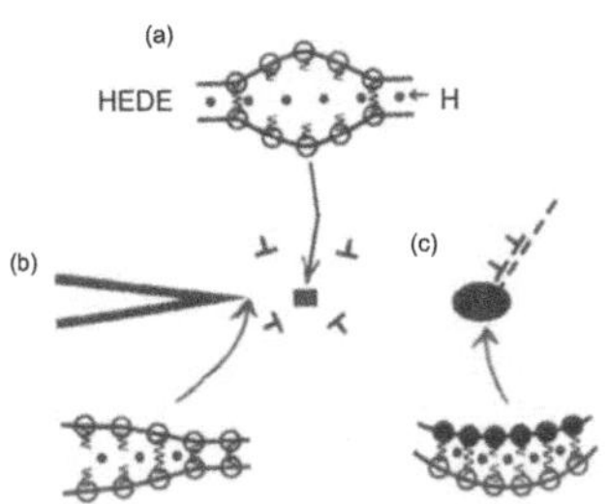

FIGURE 9.3 Scheme of the HEDE mechanism at atomistic scale, illustrating (a) hydrogen in the lattice, (b) adsorbed hydrogen, and (c) hydrogen at particle-matrix interfaces, reprinted with permission from Lynch (2012a).

and Atrens, 2013, Lynch, 2012b). As a result, fractures can occur at stress levels well below what would normally be considered permissible (Kappes et al., 2013). Support for the HEDE mechanism comes from observations of intergranular fracture where high hydrogen concentrations are present, as well as from quantum mechanical calculations (Martin et al., 2012, Wei et al., 2016). However, one of the major challenges facing the HEDE model is that the weakening of interatomic interactions in a metal by hydrogen has not yet been directly confirmed by experimental measurements. To date, no direct method or equipment exists that can accurately and precisely measure the cohesive forces at metal crack tips on the atomic level.

9.2.3.2   Hydrogen-Enhanced Local Plasticity Model (HELP)

The HELP mechanism was proposed by Beachem in 1972, based on detailed examination of hydrogen-assisted fracture surfaces (Beachem, 1972). This mechanism suggests that diffusible hydrogen facilitates dislocation mobility by reducing its resistance. This leads to accelerated dislocation movement and localized dislocation pileups, resulting in premature failure (Gangloff and Somerday, 2012, Kappes et al., 2013, Song and Curtin, 2013), as illustrated in Figure 9.4. Unlike HEDE, this model focuses on the interaction between the hydrogen atmosphere and dislocations and is mainly applied to understand plastic traces on the fracture surfaces. The observed fracture modes at crack tips—such as intergranular, transgranular, or quasi-cleavage—depend on several factors, including hydrogen concentration, microstructure, and stress intensity. Previous fractographic examinations have indicated apparent local plastic deformation accompanied by a reduction in macroscopic ductility. Moreover, the activation energy required for dislocation motion is reduced due to the presence of hydrogen atoms, and the activation area may also decrease correspondingly (Venezuela et al., 2016, 2018a, Ramamurthy and Atrens, 2013). Previous in situ TEM observations have shown that the presence of hydrogen can increase not only dislocation mobility but also the rate of dislocation nucleation (Robertson, 2001, Robertson et al., 2015). Furthermore, the HELP model is supported by both elasticity theory and atomistic calculations (Lynch, 2012a, Laadel et al., 2022). Although widely adopted to explain the HE behavior of materials, the HELP mechanism cannot account for certain reported phenomena, such as the suppression of dislocation motion by hydrogen in Alloy 718 (Li et al., 2019b) and pure aluminum (Xie et al., 2016).

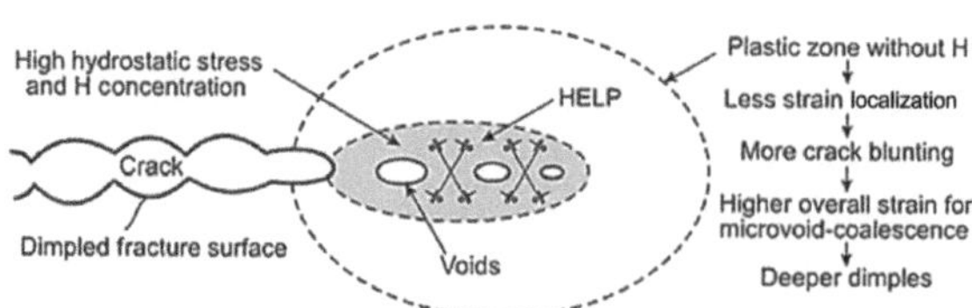

FIGURE 9.4   Schematic diagram of hydrogen-enhanced localized plasticity mechanism, reprinted with permission from Lynch (2012a).

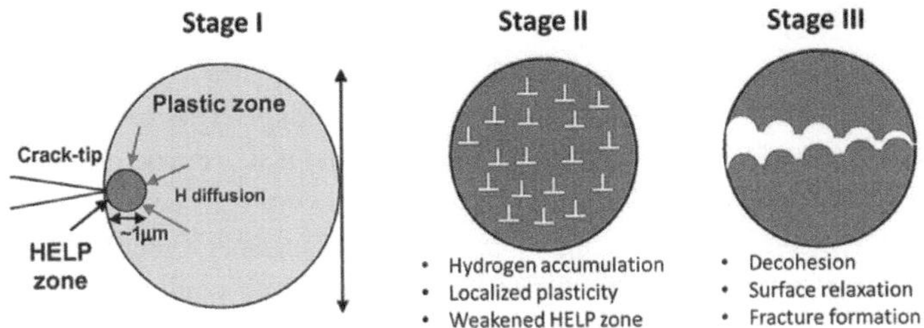

FIGURE 9.5 Schematic diagram of hydrogen-enhanced localized plasticity mechanism, reprinted with permission from Neeraj et al. (2012).

9.2.3.3 Adsorption-Induced Dislocation Emission (AIDE)

The AIDE mechanism was proposed by Lynch, based on his observations of striking similarities between HE-related fracture surfaces and those found in liquid metal embrittlement and stress corrosion cracking (Lynch, 1979). This mechanism can be considered a combination of both the HEDE and HELP models, as illustrated in Figure 9.5. In this approach, hydrogen atoms adsorb onto the material's surface, particularly in stress-concentrated regions like crack tips. These atoms then either break the interatomic bonds or reduce the cohesive strength of the material, subsequently increasing dislocation injection and activity from the crack tip. Crack growth then occurs through slip planes and the formation of micro-voids, following the principles of the HELP mechanism (Kappes et al., 2013, Lynch, 2003). The nucleation and growth of the crack are attributed to both decohesion and dislocation emission at the crack tip. Experimental observations, which include detecting high levels of adsorbed hydrogen on the surfaces of materials like iron, nickel, and titanium, support the existence of the AIDE mechanism (Lynch, 2012a, Ramamurthy and Atrens, 2013, Venezuela et al., 2018a, Nibur et al., 2006). The resulting cracks can be either intergranular or transgranular, depending on where dislocation activity and void formation most readily occur in the metals, such as at grain cores or grain boundaries. It is worth noting that this model has difficulty explaining the enhanced growth of micro-voids ahead of a crack.

9.2.3.4 Hydrogen-Enhanced Macroscopic Plasticity (HEMP)

Compared to the aforementioned three mechanisms, HEMP is a lesser-known mechanism that seeks to explain the influence of hydrogen on the mechanical properties of steel. This mechanism is applicable when hydrogen is present in large volumes (or concentrations) and interacts with the entire exposure area (Liu et al., 2018). In this model, the diffusion of hydrogen and its solid solution softening effect on the material can result in a reduction of yield strength. This, in turn, allows plasticization to occur throughout the exposure area, leading to a macroscopic enhancement of plasticity. Dislocation motion in areas with high hydrogen concentrations may facilitate the macroscopic movement of hydrogen at the yield stress level (Venezuela et al., 2018b, Atrens et al., 2018b). A limitation of this model is that it primarily focuses on the macroscopic aspects of plasticity, while detailed microscopic examination is lacking.

9.2.3.5 Hydrogen-Assisted Micro Void Coalescence (HDMC)

Unlike the previous models, the HDMC model focuses on micro void coalescence (MVC) and is considered a ductile fracture mechanism. This model was first proposed by John et al. in 1980 (Hirth, 1980b). In HDMC, crack propagation likely occurs in several stages, including void nucleation, void growth, void coalescence, crack extension, and the shearing of remaining ligaments (Venezuela, 2017). The presence of hydrogen facilitates dislocation motion and localized plastic deformation. Crack growth follows a zigzag pattern, with voids linking along the direction of crack propagation. MVC dimples, which result from the effect of hydrogen, show limited ductility. The final fracture occurs due to shear stress, resulting in shallow, parabolic shear dimples (Pradhan et al., 2012, Venezuela et al., 2018a). Unlike other models, this mechanism emphasizes the impact of hydrogen on the ductile MVC process and the role of localized shear stress. However, a limitation of this model is that it cannot account for other key failure factors such as dislocation movement or local dislocation pileups.

9.2.3.6 Mixed Fracture (MF)

Unlike the previous models, each of which focuses on a single type of micro-fracture—either brittle or ductile—some cases have shown through fracto-graphic examinations using scanning electron microscope (SEM) or transmission electron microscope (TEM) that the final fracture surface features both brittle ("fisheye"-shaped) and ductile (micro void coalescence) elements (Venezuela et al., 2018ab). Fisheyes typically appear in the central area and are often associated with inclusions (Atrens et al., 2018a). The growth and fracture of these fisheyes progress radially until they intersect with the ductile region or micro void coalescence. The fisheyes are then enclosed by a micro void coalescence fracture. This results in a mixed fracture, where both fracture modes (the hydrogen microfracture mechanism for "fisheyes" and MVC) coexist and compete with each other. In general, HE is a complex phenomenon that can lead to brittle or mixed fractures in pipeline steels. The HE processes involve various mechanisms, including the reduction of cohesive strength near the crack tip, enhanced local plasticity, adsorption-induced dislocation emission, and shifts in microfracture modes. The actual embrittlement mechanism and the specific combination of mechanisms responsible for HE in a material have yet to be completely determined. Ongoing research aims to identify the primary causes. The complexities and variations in HE process make it challenging to establish a direct link between proposed models and actual embrittlement phenomena. Moreover, HE processes in high-pressure H_2 transportation environments may be even more complicated due to the unique properties of supercritical hydrogen (s-H_2). Therefore, HE remains a significant and critical new area in H_2 pipeline research, and it urgently requires comprehensive analysis through fractography and other in situ and ex situ examination techniques to determine crack initiation, propagation, and failure mechanisms, and to explore cost-effective solutions before the deployment of specialized gaseous or supercritical hydrogen pipelines.

9.3 EFFECT OF PIPELINE STEEL VARIABLES AND ENVIRONMENTAL FACTORS ON HE

Numerous factors contribute to the vulnerability of pipeline steels to hydrogen-related embrittlement and degradation. Four general requirements trigger HE: (i) sufficient diffusible hydrogen in the steels; (ii) applied stress close to the steel's yield strength; (iii) the presence of HE-susceptible microstructures, such as micro-cracks; and (iv) ample time for hydrogen to move to susceptible sites (Wu et al., 2022; Lynch, 2012a). Unfortunately, long-term service pipes transporting high-pressure hydrogen can barely meet all these conditions. Moreover, variables and environmental factors related to pipeline service, such as operating pressure and temperature, local stress concentration, cyclic load, and hydrogen dissociation and permeation processes, could intensify the pipeline steel's susceptibility to HE. In general, a wide range of factors affect HE. This section aims to provide an overview of these variables and factors, particularly the material prerequisites for pipeline steels.

9.3.1 MICROSTRUCTURE

Microstructure is one of the most dominant factors affecting HE susceptibility, as thermomechanical fabrication processes lead to different microstructures and anisotropy of mechanical properties in mid- and high-strength pipeline steels. These steels usually consist of various microstructural phases, including ferritic pearlite, acicular ferrite, bainitic ferrite, bainite, and martensite (Nayak et al., 2008, Hwang et al., 2005). The formation of these microstructures, either individually or in combination, is heavily dependent on fabrication process parameters. For instance, rapid cooling rates and lower transformation temperatures lead to the development of bainitic ferrite, followed by acicular ferrites in diminishing order (Bae et al., 2015). In these microstructures, hydrogen exhibits distinct diffusion and trapping behaviors, thus leading to different HE susceptibilies. For example, acicular ferrite has a high dislocation density due to the separation of grains by low-energy boundaries and dispersed retained austenite and/or martensite-austenite islands (Wang et al., 2009). A previous study indicated that the presence of retained austenite was more helpful in resisting HE compared to a cementite structure (Mori et al., 2002). On the other hand, banded microstructural phases, such as ferrite-pearlite, are commonly present in lower-grade pipeline steels like X52, X60, and X65, which may lead to high HE susceptibility (Dunne et al., 2016, Domizzi et al., 2001, Mohtadi-Bonab et al., 2013b, Omale et al., 2017). One possible reason could be due to the high hardness around the pearlite band stabilizing crack propagation within that region (Xu et al., 2013). Another study (Ramirez et al., 2008) also investigated the roles of primarily ferrite, bainitic ferrite, and martensite, and found that the presence of a martensitic phase increased the susceptibility to hydrogen uptake and cracking due to the highly strained nature of martensite, which traps a significant amount of interstitial carbon within its lattice. It is also likely that the distribution of austenite and carbide within the martensitic laths may enhance hydrogen diffusion in the steel. In general, the martensitic phase has been found to be most susceptible to hydrogen uptake, while the ranking of other structures seems to depend on the specific operating conditions.

9.3.2 Mechanical Properties

The required mechanical properties of a pipeline steel in a designated service environment can be achieved through the formation of specific microstructures. It has been generally accepted that the presence of a high density of acicular and bainitic ferrite in the microstructure is beneficial in terms of grain refinement, strengthening, and increasing dislocation density (Lan et al., 2011). These microstructural features contribute to excellent toughness in high-grade pipeline steels. However, increasing steel strength often compromises its resistance to HE. For example, pipeline steel with a bainitic microstructure typically exhibits good mechanical properties but is more susceptible to hydrogen-related degradation (such as loss of ductility) (Arafin and Szpunar, 2011). Such phenomenon has been found in high-strength pipeline steels, such as X60, X80, and X100 steel (Hardie et al., 2006). Another comparative study on ferritic-bainitic steel, dual-phase steel, and transformation-induced plasticity steel also indicated a close relationship between hydrogen embrittlement susceptibility and strength (Depover et al., 2014). The embrittlement index of transformation-induced plasticity steel with the highest strength was the highest compared to the other two steels. This could be attributed to the presence of a crack-prone martensitic microstructure in the steel. Therefore, particular attention is seriously needed when using high-strength pipeline steels (such as X100 steel) for high-pressure hydrogen transportation.

9.3.3 Grain Boundaries

Grain boundaries (GBs) are the most common and easiest-to-characterize defects intrinsic in steels and alloys. As described above, they are reversible hydrogen traps and are preferred sites for the accumulation of hydrogen atoms. Embrittlement induced by hydrogen segregation at grain boundaries is thus one of the fundamental problems and has been extensively investigated by different groups (Huang et al., 2017, Song and Curtin, 2013, Zhou et al., 2016, Yamaguchi, 2011). For instance, a previous study revealed that a high concentration of hydrogen could decrease the strength of a grain boundary in fast grain boundary fracture by about 10% and become more severe in the case of slow grain boundary fracture. Therefore, the distribution of grain boundaries is a critical factor for pipeline steels used for gaseous and supercritical hydrogen pipeline transportation.

Depending on fabrication processes, high-angle grain boundaries, low-angle grain boundaries, and/or special coincidence site lattice boundaries are present in the steels. Compared to high-angle grain boundaries, low-angle and special coincidence site lattice have lower energy and are relatively less susceptible to hydrogen attack (Arafin and Szpunar, 2009). For example, a recent Density Functional Theory (DFT) modeling study indicated that high-angle grain boundaries (HAGBs) would exhibit the most negative free energy change and the lowest hydrogen adsorption energy compared to crystalline lattice sites (Sun et al., 2022). It is worth mentioning that the fraction of special coincidence site lattice boundaries found in pipeline steel is often low due to reduced stacking fault (Masoumi et al., 2017). Thus, it is helpful to control the fraction of high-angle grain boundaries during the manufacturing and heat treatment process for reducing the HE susceptibility along boundaries (Ohaeri et al., 2018, Ishikawa et al., 2016).

9.3.4 Inclusions and Precipitates

In addition to microstructures and crystallographic properties, inclusions and precipitates are also important factors affecting the HE susceptibility of pipeline steels. When the solubility of a metal is exceeded due to the addition of an excessive alloying element, a second phase (either an inclusion or precipitate) will form. Compared to inclusions, the second phase with smaller amounts and sizes is termed "precipitate" or "dispersoid" since it can retard the slip of dislocations within the steel matrix. The role of inclusions and precipitates in crack initiation and propagation is challenging to fully reveal due to the limited techniques available for in situ study of hydrogen-induced cracking nucleation around microscopic inclusions. However, it is generally accepted that non-metallic inclusions and incoherent precipitates can aid hydrogen attack on pipeline steel, as they serve as irreversible traps to accumulate hydrogen atoms and thus enhance HE through various processes (Mohtadi-Bonab et al., 2017, Qin and Szpunar, 2017, Depover et al., 2015). For instance, MnS is particularly susceptible to hydrogen embrittlement and hydrogen-induced cracking (Venkatasubramanian and Baker, 1984, Inés and Mansilla, 2013).

Inclusions are always present in produced pipeline steels due to dissolved oxygen and other exogenous materials, and they are usually harder than the steel matrix, creating incoherent interfaces that increase hydrogen trapping and lead to hydrogen-induced cracking (Depover et al., 2015). Previous studies have found that increased hydrogen trapping within incoherent inclusion particles can be attributed to a higher concentration of octahedral carbon vacancies and voids, where hydrogen atoms tend to congregate since stress concentration is generally higher at these spots than in the steel matrix (Mohtadi-Bonab and Eskandari, 2017a). To date, the prevalent theory for crack initiation around inclusions in pipeline steel involves the migration of hydrogen into interfacial voids. Under high-strain conditions, these voids become pressurized, leading to the formation of cracks (Mohtadi-Bonab and Eskandari, 2017b). Table 9.4 summarizes the types of inclusions in steels and their contributions to HE. Based on their compositions, non-metallic inclusions are classified into oxides, sulfides, silicates, nitrides, or various other non-metallic compounds. Note that an inclusion may contain several different phases. The most common inclusions in pipeline steels include MnS, Al_2O_3, complex (FeMn)S, silicates, and $FeO\text{-}Al_2O_3$ double oxides, among others. Regarding their impacts on HE, interesting observations are that the presence of Si-enriched and Al-Si-enriched inclusions enhance HE susceptibility (Liu et al., 2009), while Si-O inclusions are unlikely to facilitate hydrogen-induced cracking due to their coherency with the steel matrix (Li et al., 2016).

Various types of precipitates are also found in pipeline steels. The main types encountered in microalloyed steels are complex carbonitrides (Nb, Ti, V, Mo) (C, N). These precipitates come in different shapes (e.g., cubic for TiN and spherical for NbC), compositions, and sizes. The behaviors of hydrogen absorption and desorption are significantly influenced by the size of the precipitates. Notably, larger precipitates tend to trap more hydrogen compared to their smaller counterparts (Allam et al., 2020). However, all precipitates, regardless of their sizes, act as irreversible hydrogen trapping sites, and the mechanical properties of steels are unlikely to deteriorate.

TABLE 9.4

Summary on Inclusion Types and Their Contributions toward Hydrogen Degradation in Pipeline Steel

Pipeline Steel	Inclusion Characterization Technique	Inclusion Types	Findings	References
X70	SEM and energy-dispersive X-ray analysis (EDX) after slow strain rate tensile test	Si- and Al-enriched inclusions	Inclusions with morphology and composition that encourage hydrogen trapping and pressure build-up at the interface are more likely to initiate cracks. Si-enriched inclusions, which are coherent with the pipeline steel matrix, typically do not initiate cracks. On the other hand, cracks can initiate at Al-enriched inclusions due to their incoherency with the pipeline matrix.	Liu et al. (2009), Wang et al. (2018)
X100	SEM, EDX, and localized electrochemical impedance spectroscopy (LEIS)	Si-O enriched	Inclusions of Si-O type, which are coherent with the steel matrix, do not result in any obvious localized corrosion. In contrast, other inclusions containing relatively less Si-O experience localized corrosion attack after the complete dissolution or fall-out of the inclusions with increasing exposure time.	Li et al. (2016), Jin et al. (2010b)
X100	SEM and EDX	Mixture of Ca-Al-S, MnS, and oxides of Al and Si inclusions	HE occurred at a critical hydrogen charging concentration of 3.24 ppm, primarily at Si and Al oxide-enriched inclusions, rather than at elongated MnS and Ca-S-Al-O inclusions. This can be attributed to the higher degree of non-cohesiveness between Si and Al inclusions, which makes them more susceptible to hydrogen-induced embrittlement.	Jin et al. (2010a,b)

(Continued)

TABLE 9.4 (*Continued*)

Summary on Inclusion Types and Their Contributions toward Hydrogen Degradation in Pipeline Steel

Pipeline Steel	Inclusion Characterization Technique	Inclusion Types	Findings	References
X80	Electrochemical hydrogen permeation test with SEM and EDX inclusion characterization	Si- and Al-Si-enriched inclusions	The high presence of irreversible hydrogen traps (inclusions) at the weld heat-affected zone made the pipeline steel more susceptible to preferential anodic dissolution. Analysis of the inclusions revealed that hydrogen-related localized corrosion primarily occurred around those enriched with Si and Al-Si. A great amount of hydrogen trapped at these inclusion sites will enhance the local stress concentration.	Xue and Cheng (2011, 2013)

9.3.5 Microalloying Elements in Pipeline Steels

Microalloying is the process of adding small amounts of different alloying elements to steels to achieve desired mechanical and corrosion-resistant properties. In the case of pipeline steel, elements such as Nb, Ti, and V are added to restrict excessive grain growth, while other elements (e.g., C, N, S, and P) are intended to be removed during the hot-rolling process (Hong et al., 2003). Previous studies on Fe surfaces with different alloying elements suggested that the presence of W, Nb, Mo, Co, and Ti could help trap hydrogen atoms on the surface and prevent their inward diffusion (Cremaschi et al., 1995, Jiang and Carter, 2003, Muritala et al., 2020, Johnson and Carter, 2010, Xiong et al., 2023a). Aside from the surface, elements like Mo, V, Nb, Cr, and Ti can lead to the formation of nitrides and carbides, improving mechanical properties through solid solution strengthening and reducing hydrogen diffusion within the steel (Luo et al., 2022). Several previous studies indicated that the presence of Mo and Cr might enhance localized hydrogen trapping within pipeline steels (Shirband et al., 2011), and the addition of Al and Cu could decrease hydrogen diffusion (Shirband et al., 2011, Baba et al., 2016). Density Functional Theory (DFT) calculations have also been carried out to understand the adsorption and permeation processes of hydrogen on the steel surface, revealing that hydrogen atoms permeate faster through Cu, Ni, S, and Al-doped surfaces that have lower surface permeation barriers and higher permeation rate constants than Fe surfaces (Xiong et al., 2023b). On the other hand, W, Nb, Mo, Ti, Co, Mn, Cr, V, and C-doped surfaces have lower permeation rate constants and higher surface permeation barriers. More importantly, Mn, V, and Cr-doped surfaces decrease the adsorption stability and hydrogen coverage, making them ideal alloying elements for inhibiting hydrogen permeation in steels. However, the influence of alloying elements on hydrogen absorption, trapping, and diffusion is not yet fully understood due to their complex presence in steels. Further theoretical modeling and experimental testing are urgently needed to support the pipeline industry in designing new pipeline steels for long-term, safe, high-pressure hydrogen transportation in a cost-effective manner.

9.3.6 Environmental Factors—Temperature and Pressure

In addition to the properties of pipeline steels, operating or environmental factors during high-pressure hydrogen transportation can also affect HE susceptibility. At low or moderate hydrogen pressure and room temperature, previous studies found that atomic hydrogen has low mobility within the steel lattice, thus reducing HE susceptibility (Shirband et al., 2011). With increasing temperature, the mobility of hydrogen atoms increases, reducing the likelihood of hydrogen recombination and enhancing their diffusion to and interaction with defects, leading to a higher risk of HE (Liang et al., 2009). At elevated temperatures (>200°C), hydrogen can also interact with carbides to form CH_4, increasing HE susceptibility. The influence of temperature on HE in pipeline steels has not yet been explored due to limited testing facilities. A recent simulation suggested that increasing temperature could decrease hydrogen association and uptake into pipeline steels (Li et al., 2022b).

Compared with temperature, pressure plays a more important role in HE during high-pressure, supercritical hydrogen pipeline transportation. Recent studies indicate that increasing hydrogen pressure leads to higher hydrogen uptake into pipeline steels (Sun and Cheng, 2021, Li et al., 2022b). As stated above, increasing hydrogen pressure could also result in an increase in dissolvable hydrogen content (C_H) in the steels, thereby leading to higher HE risks. Interestingly, HE susceptibility does not linearly increase with pressure. For instance, Barthélémy et al. conducted rupture tests on AISI 321 steel and found that maximum embrittlement occurred at hydrogen partial pressures between 2 and 4 MPa (Barthélémy, 2007). Similar tests on X42 steel showed a rapid increase in HE susceptibility until the pressure reached ~7.5 MPa (Gutierrez-Solana and Elices, 1982). Further increasing pressure to 35 MPa had only a marginal effect on HE susceptibility. Thus, there may be a pressure threshold below which HE is severely intensified. More work is needed to validate this observation. Additionally, it's worth noting that the influence of pressure on steels and alloys may vary with their composition and structure. For instance, carbon steels, ferritic steels, and martensitic steels generally exhibit significant ductility losses, while high-alloy austenitic steels show varying responses to extensive hydrogen exposure (Liu et al., 2023).

Apart from temperature and pressure, the influence of CH_4 and other impurities (such as CO, H_2O, N_2, and H_2S) on hydrogen uptake and subsequently on HE of pipeline steels needs to be addressed. At present, the most economical hydrogen production occurs via steam reforming and water-gas shifting reactions of fossil fuels

(Balat, 2008, Haryanto et al., 2005, Nikolaidis and Poullikkas, 2017). Consequently, the produced hydrogen streams contain CO, H_2O, N_2, and H_2S impurities. As mentioned above, blending high-pressure hydrogen into existing natural gas pipelines is a short-term hydrogen transportation method to efficiently deliver hydrogen to end-users. Since methane (CH_4) is the main component of natural gas, understanding the effect of CH_4 presence on the adsorption and dissociation of hydrogen on the steel surface is essential for safe transportation. Unfortunately, little public information is available to describe the role of CH_4 in hydrogen uptake and HE under blending transportation conditions or high-pressure near-pure hydrogen transportation. Therefore, it is necessary to clarify the interaction of CO, H_2O, CH_4, and H_2S with supercritical hydrogen and their roles on hydrogen uptake and subsequent HE. These impurities may significantly degrade hydrogen quality. For example, it was reported that even a trace of CO (i.e., <10 ppm) in the H_2 stream can significantly degrade fuel cell performance (Abdollahi et al., 2012). According to a literature survey, further work is needed to address these knowledge gaps.

9.4 ALTERNATIVE PIPELINE MATERIALS FOR HYDROGEN TRANSPORTATION

To address the impact of hydrogen on pipeline steel materials, one alternative approach is to consider the use of Fiber-Reinforced Polymer (FRP) materials. FRP is commonly employed in the upstream oil and gas industry. FRP pipelines typically consist of an inner non-permeable liner, a protective layer, an interface layer, multiple reinforcement layers made of glass or carbon fibers, an outer pressure barrier layer, and an outer protective layer (Smith et al., 2005). These pipelines exhibit favorable mechanical properties and can be unspooled and trenched for approximately 0.5 miles of continuous pipe. More importantly, adjacent segments of FRP pipeline can be easily joined using simple connection techniques without welding, which eliminates concerns over improper welding treatment and mismatches in the microstructures of pipeline materials. FRP was also accepted into the ASME B31.12 (Hydrogen Piping and Pipelines Code) standard in 2006 (Hayden and Stalheim, 2009). Nevertheless, the lack of long-term, real-life assessments of these new pipeline materials restricts their application on a large scale, and further investigation is needed to fully explore their potential in this field. Overall, understanding the background and mechanisms of hydrogen embrittlement, as well as considering the effects of pipeline steel variables and environmental factors, is crucial for the prevention and mitigation of HE in pipeline transportation. Continued research and comprehensive analysis are necessary to understand the exact mechanism responsible for embrittlement in specific materials and to ensure the safe and reliable operation of specialized gaseous or supercritical hydrogen pipelines.

9.5 CONCLUSION

In conclusion, hydrogen is one of the most important energy sources nowadays, and hydrogen uptake or hydrogen embrittlement (HE) is a complex phenomenon that affects the integrity and safety of pipeline steels used in gaseous or supercritical hydrogen transmission. The process of hydrogen uptake in pipeline steels involves three distinct steps: dissociation, diffusion, and trapping. The concentration of absorbed hydrogen is influenced by environmental hydrogen pressure, and HE generally requires high hydrogen pressures during transportation. The susceptibility to HE increases with higher strength and more intricate microstructures of pipeline steels. Several well-known and widely accepted mechanisms of HE include hydrogen-enhanced decohesion, hydrogen-enhanced local plasticity, adsorption-induced dislocation emission, hydrogen-assisted micro void coalescence, and mixed fracture. These mechanisms can act individually or in combination, leading to the degradation and embrittlement of the material. Several measures have been taken to mitigate the susceptibility to HE of high-strength pipeline steels to a certain extent in low-pressure hydrogen transportation, such as design considerations to avoid notches, sharp variations, and regular variations in the material, baking operations, alloying, protective coatings, cadmium, nickel, and ion plating, and controlling traps and localized hydrogen concentration. However, most of the mitigation strategies have been generally studied under low-pressure/mixed natural gas hydrogen transportation; their capabilities under high-pressure pure hydrogen transportation are still unknown and need to be fully studied before large-scale deployment.

Several critical factors will also influence the vulnerability of pipeline steels to HE. Microstructure plays a crucial role, with microstructural features such as acicular ferrite and bainitic ferrite exhibiting better resistance to HE compared to ferrite-pearlite banded microstructures. Grain boundaries also play a significant role, with low-angle and special coincidence site lattice boundaries being less susceptible to hydrogen attack. Inclusions and secondary phases, such as non-metallic inclusions and precipitates, can act as hydrogen trap sites and contribute to hydrogen-induced cracking. It has also been found that higher-strength steels tend to be more susceptible to hydrogen-related degradation. The presence of microalloying elements can affect hydrogen diffusion, trapping, and precipitation strengthening, and the balance of microalloying elements needs to be carefully considered. In addition, environmental factors, such as temperature and pressure, also affect HE susceptibility. Higher temperatures increase atomic hydrogen mobility, while operating pressures can affect the degree of embrittlement, with a critical pressure level often observed.

ACKNOWLEDGMENTS

The authors gratefully acknowledge the financial support provided by the Department of Natural Resources Canada (NRCan) OERD H_2 Codes and Standards program.

REFERENCES

Abdollahi, M., Yu, J., Liu, P. K., Ciora, R., Sahimi, M. & Tsotsis, T. T. 2012. Ultra-pure hydrogen production from reformate mixtures using a palladium membrane reactor system. *Journal of Membrane Science*, 390, 32–42.

Abdulagatov, I. & Skripov, P. 2020. Thermodynamic and transport properties of supercritical fluids: review of thermodynamic properties (part 1). *Russian Journal of Physical Chemistry B*, 14, 1178–1216.

Abe, J. O., Popoola, A., Ajenifuja, E. & Popoola, O. M. 2019. Hydrogen energy, economy and storage: review and recommendation. *International Journal of Hydrogen Energy*, 44, 15072–15086.

Acar, C. & Dincer, I. 2020. The potential role of hydrogen as a sustainable transportation fuel to combat global warming. *International Journal of Hydrogen Energy*, 45, 3396–3406.

Agency, E. E. 2017. EN01 Energy related greenhouse gas emissions. *European Environment Agency.*

Allam, T., Guo, X., Lipińska-Chwałek, M., Hamada, A., Ahmed, E. & Bleck, W. 2020. Impact of precipitates on the hydrogen embrittlement behavior of a V-alloyed medium-manganese austenitic stainless steel. *Journal of Materials Research and Technology*, 9, 13524–13538.

Arafin, M. & Szpunar, J. 2009. A new understanding of intergranular stress corrosion cracking resistance of pipeline steel through grain boundary character and crystallographic texture studies. *Corrosion Science*, 51, 119–128.

Arafin, M. & Szpunar, J. 2011. Effect of bainitic microstructure on the susceptibility of pipeline steels to hydrogen induced cracking. *Materials Science and Engineering: A*, 528, 4927–4940.

Araújo, D., Vilar, E. & Carrasco, J. P. 2014a. A critical review of mathematical models used to determine the density of hydrogen trapping sites in steels and alloys. *International Journal of Hydrogen Energy*, 39, 12194–12200.

Araújo, D. F., Vilar, E. O. & Palma Carrasco, J. 2014b. A critical review of mathematical models used to determine the density of hydrogen trapping sites in steels and alloys. *International Journal of Hydrogen Energy*, 39, 12194–12200.

Atrens, A., Liu, Q., Tapia-Bastidas, C., Gray, E., Irwanto, B., Venezuela, J. & Liu, Q. 2018a. Influence of hydrogen on steel components for clean energy. *Corrosion and Materials Degradation*, 1, 3–26.

Atrens, A., Liu, Q., Zhou, Q., Venezuela, J. & Zhang, M. 2018b. Evaluation of automobile service performance using laboratory testing. *Materials Science and Technology*, 34, 1893–1909.

Baba, K., Mizuno, D., Yasuda, K., Nakamichi, H. & Ishikawa, N. 2016. Effect of Cu addition in pipeline steels on prevention of hydrogen permeation in mildly sour environments. *Corrosion*, 72, 1107–1115.

Bae, J.-H., Ro, Y.-J., Chon, S.-H., Sung, H. K., Lee, S. & Lee, C.-S. 2015. Development of X60 and X100 linepipe steels with high deformation capacity for strain-based design. *The Twenty-fifth International Ocean and Polar Engineering Conference, Kona, Hawaii, USA.*

Balat, M. 2008. Potential importance of hydrogen as a future solution to environmental and transportation problems. *International Journal of Hydrogen Energy*, 33, 4013–4029.

Barrera, O., Bombac, D., Chen, Y., Daff, T., Galindo-Nava, E., Gong, P., Haley, D., Horton, R., Katzarov, I. & kermode, J. 2018a. Understanding and mitigating hydrogen embrittlement of steels: a review of experimental, modelling and design progress from atomistic to continuum. *Journal of Materials Science*, 53, 6251–6290.

Barrera, O., Bombac, D., Chen, Y., Daff, T. D., Galindo-Nava, E., Gong, P., Haley, D., Horton, R., Katzarov, I., Kermode, J. R., Liverani, C., Stopher, M. & Sweeney, F. 2018b. Understanding and mitigating hydrogen embrittlement of steels: a review of experimental, modelling and design progress from atomistic to continuum. *Journal of Materials Science*, 53, 6251–6290.

Barthélémy, H. 2007. Compatibility of metallic materials with hydrogen review of the present knowledge. In *16th World Hydrogen Energy Conference*.

Beachem, C. D. 1972. A new model for hydrogen-assisted cracking (hydrogen "embrittlement"). *Metallurgical and Materials Transactions B*, 3, 441–455.

Boggess, T. B., Boggess, T. A. B., Ningileri, S. T. N. N., Stalheim, D. S. S., Ningileri, S. T. N. & Stalheim, D. S. 2013. *Materials Solutions for Hydrogen Delivery in Pipelines*. US Department of Energy, Lexington, KY.

Brilz, M., Hoche, H. & Oechsner, M. 2022. Hydrogen-assisted cracking (HAC) of high-strength steels as a function of the hydrogen pre-charging time. *Engineering Fracture Mechanics*, 261, 108246.

Briottet, L., Batisse, R., De Dinechin, G. D., Langlois, P. & Thiers, L. 2012. Recommendations on X80 steel for the design of hydrogen gas transmission pipelines. *International Journal of Hydrogen Energy*, 37, 9423–9430.

Cheng, Y. 2007. Fundamentals of hydrogen evolution reaction and its implications on near-neutral pH stress corrosion cracking of pipelines. *Electrochimica Acta*, 52, 2661–2667.

Choo, W. & Lee, J. Y. 1982. Thermal analysis of trapped hydrogen in pure iron. *Metallurgical Transactions A*, 13, 135–140.

Cremaschi, P., Yang, H. & Whitten, J. L. 1995. Ab initio chemisorption studies of H on Fe (110). *Surface Science*, 330, 255–264.

Crippa, M., Guizzardi, D., Solazzo, E., Muntean, M., Schaaf, E., Monforti-Ferrario, F., Banja, M., Olivier, J., Grassi, G. & Rossi, S. 2021. *GHG Emissions of All World Countries*. Publications Office of the European Union, Luxembourg. https://publications.jrc. ec.europa.eu/repository/handle/JRC126363.

Depover, T., Escobar, D. P., Wallaert, E., Zermout, Z. & Verbeken, K. 2014. Effect of hydrogen charging on the mechanical properties of advanced high strength steels. *International Journal of Hydrogen Energy*, 39, 4647–4656.

Depover, T., Monbaliu, O., Wallaert, E. & Verbeken, K. 2015. Effect of Ti, Mo and Cr based precipitates on the hydrogen trapping and embrittlement of Fe-C-X Q&T alloys. *International Journal of Hydrogen Energy*, 40, 16977–16984.

Djukic, M. B., Bakic, G. M., Zeravcic, V. S., Sedmak, A. & Rajicic, B. 2019. The synergistic action and interplay of hydrogen embrittlement mechanisms in steels and iron: localized plasticity and decohesion. *Engineering Fracture Mechanics*, 216, 106528.

Doe, U. 2020. Hydrogen strategy: enabling a low-carbon economy. *Office of Fossil Energy*, 24, 3–10.

Domizzi, G., Anteri, G. & Ovejero-Garcia, J. 2001. Influence of sulphur content and inclusion distribution on the hydrogen induced blister cracking in pressure vessel and pipeline steels. *Corrosion Science*, 43, 325–339.

Dong, C.-F., Xiao, K., Liu, Z.-Y., Yang, W.-J. & Li, X.-G. 2010. Hydrogen induced cracking of X80 pipeline steel. *International Journal of Minerals, Metallurgy, and Materials*, 17, 579–586.

Dong, C. F., Liu, Z. Y., Li, X. G. & Cheng, Y. F. 2009. Effects of hydrogen-charging on the susceptibility of X100 pipeline steel to hydrogen-induced cracking. *International Journal of Hydrogen Energy*, 34, 9879–9884.

Dunne, D. P., Hejazi, D., Saleh, A. A., Haq, A. J., Calka, A. & Pereloma, E. V. 2016. Investigation of the effect of electrolytic hydrogen charging of X70 steel: I. The effect of microstructure on hydrogen-induced cold cracking and blistering. *International Journal of Hydrogen Energy*, 41, 12411–12423.

Dwivedi, S. K. & Vishwakarma, M. 2018. Hydrogen embrittlement in different materials: a review. *International Journal of Hydrogen Energy*, 43, 21603–21616.

Eames, I., Austin, M. & Wojcik, A. 2022. Injection of gaseous hydrogen into a natural gas pipeline. *International Journal of Hydrogen Energy*, 47, 25745–25754.

Elhoud, A., Renton, N. C. & Deans, W. F. 2010. Hydrogen embrittlement of super duplex stainless steel in acid solution. *International Journal of Hydrogen Energy*, 35, 6455–6464.

Erdener, B. C., Sergi, B., Guerra, O. J., Chueca, A. L., Pambour, K., Brancucci, C. & Hodge, B.-M. 2023. A review of technical and regulatory limits for hydrogen blending in natural gas pipelines. *International Journal of Hydrogen Energy*, 48, 5595–5617.

Ganda, F. 2019. The impact of innovation and technology investments on carbon emissions in selected organisation for economic co-operation and development countries. *Journal of Cleaner Production*, 217, 469–483.

Gangloff, R. P. & Somerday, B. P. 2012. *Gaseous Hydrogen Embrittlement of Materials in Energy Technologies: Mechanisms, Modelling and Future Developments*. Elsevier, Amsterdam, Netherlands.

Gielen, D., Boshell, F., Saygin, D., Bazilian, M. D., Wagner, N. & Gorini, R. 2019. The role of renewable energy in the global energy transformation. *Energy Strategy Reviews*, 24, 38–50.

Gutierrez-Solana, F. & Elices, M. 1982. High-pressure hydrogen behavior of a pipeline steel. *Current Solutions to Hydrogen Problems in Steels*, 1, 181–185.

Han, Y., Wang, R., Wang, H. & Xu, L. 2019. Hydrogen embrittlement sensitivity of X100 pipeline steel under different pre-strain. *International Journal of Hydrogen Energy*, 44, 22380–22393.

Haq, A. J., Muzaka, K., Dunne, D. P., Calka, A. & Pereloma, E. V. 2013. Effect of microstructure and composition on hydrogen permeation in X70 pipeline steels. *International Journal of Hydrogen Energy*, 38, 2544–2556.

Hardie, D., Charles, E. & Lopez, A. 2006. Hydrogen embrittlement of high strength pipeline steels. *Corrosion Science*, 48, 4378–4385.

Haryanto, A., Fernando, S., Murali, N. & Adhikari, S. 2005. Current status of hydrogen production techniques by steam reforming of ethanol: a review. *Energy & Fuels*, 19, 2098–2106.

Hayden, L. E. & Stalheim, D. 2009. ASME B31. 12 hydrogen piping and pipeline code design rules and their interaction with pipeline materials concerns, issues and research. *ASME Pressure Vessels and Piping Conference*, 355–361.

Hickel, T., Nazarov, R., Mceniry, E., Leyson, G., Grabowski, B. & Neugebauer, J. 2014. Ab initio based understanding of the segregation and diffusion mechanisms of hydrogen in steels. *JOM*, 66, 1399–1405.

Hirth, J. P. 1980a. Effects of hydrogen on the properties of iron and steel. *Metallurgical Transactions A*, 11, 861–890.

Hirth, J. P. 1980b. Effects of hydrogen on the properties of iron and steel. *Metallurgical Transactions A*, 11, 861–890.

Hong, S. C., Lim, S. H., Hong, H. S., Lee, K. J., Shin, D. H. & Lee, K. S. 2003. Effects of Nb on strain induced ferrite transformation in C-Mn steel. *Materials Science and Engineering: A*, 355, 241–248.

Huang, S., Chen, D., Song, J., Mcdowell, D. L. & Zhu, T. 2017. Hydrogen embrittlement of grain boundaries in nickel: an atomistic study. *NPJ Computational Materials*, 3, 28.

Hwang, B., Kim, Y. M., Lee, S., Kim, N. J. & Yoo, J. Y. 2005. Correlation of rolling condition, microstructure, and low-temperature toughness of X70 pipeline steels. *Metallurgical and Materials Transactions A*, 36, 1793–1805.

Inés, M. N. & Mansilla, G. 2013. Effect of hydrogen concentration and MnS inclusions on the embrittlement of a high strength steel. *Acta Microscopica*, 22, 20–25.

Ishikawa, N., Sueyoshi, H. & Nagao, A. 2016. Hydrogen microprint analysis on the effect of dislocations on grain boundary hydrogen distribution in steels. *ISIJ International*, 56, 413–417.

Jack, T. A., Pourazizi, R., Ohaeri, E., Szpunar, J., Zhang, J. & Qu, J. 2020. Investigation of the hydrogen induced cracking behaviour of API 5L X65 pipeline steel. *International Journal of Hydrogen Energy*, 45, 17671–17684.

Januta, A. 2021. Global carbon emissions rebound to near pre-pandemic levels. Reuters. Retrieved on February 10th, 2024 from https://www.reuters.com/business/cop/global-carbon-emissions-rebound-near-pre-pandemic-levels-2021-11-04/.

Jayabal, S., Saranya, G., Wu, J., Liu, Y., Geng, D. & Meng, X. 2017. Understanding the high-electrocatalytic performance of two-dimensional MoS2 nanosheets and their composite materials. *Journal of Materials Chemistry A*, 5, 24540–24563.

Jiang, D. & Carter, E. A. 2003. Adsorption and diffusion energetics of hydrogen atoms on Fe (1 1 0) from first principles. *Surface Science*, 547, 85–98.

Jin, T., Liu, Z. & Cheng, Y. 2010a. Effect of non-metallic inclusions on hydrogen-induced cracking of API5L X100 steel. *International Journal of Hydrogen Energy*, 35, 8014–8021.

Jin, T. Y., Liu, Z. Y. & Cheng, Y. F. 2010b. Effect of non-metallic inclusions on hydrogen-induced cracking of API5L X100 steel. *International Journal of Hydrogen Energy*, 35, 8014–8021.

Jo'm, B., Beck, W., Genshaw, M., Subramanyan, P. & Williams, F. 1971. The effect of stress on the chemical potential of hydrogen in iron and steel. *Acta Metallurgica*, 19, 1209–1218.

Johnson, D. F. & Carter, E. A. 2010. First-principles assessment of hydrogen absorption into FeAl and Fe3Si: towards prevention of steel embrittlement. *Acta Materialia*, 58, 638–648.

Johnson, W. H. 1875. II. On some remarkable changes produced in iron and steel by the action of hydrogen and acids. *Proceedings of the Royal Society of London*, 23, 168–179.

Johnson, H. H. 1988. *Hydrogen in Iron*. Springer, New York.

Kappes, M., Iannuzzi, M. & Carranza, R. M. 2013. Hydrogen embrittlement of magnesium and magnesium alloys: a review. *Journal of The Electrochemical Society*, 160, C168.

Koyama, M., Tasan, C. C., Akiyama, E., Tsuzaki, K. & Raabe, D. 2014. Hydrogen-assisted decohesion and localized plasticity in dual-phase steel. *Acta Materialia*, 70, 174–187.

Kuramochi, T., Roelfsema, M., Hsu, A., Lui, S., Weinfurter, A., Chan, S., Hale, T., Clapper, A., Chang, A. & Höhne, N. 2020. Beyond national climate action: the impact of region, city, and business commitments on global greenhouse gas emissions. *Climate Policy*, 20, 275–291.

Laadel, N.-E., El Mansori, M., Kang, N., Marlin, S. & Boussant-Roux, Y. 2022. Permeation barriers for hydrogen embrittlement prevention in metals-a review on mechanisms, materials suitability and efficiency. *International Journal of Hydrogen Energy*, 47, 32707–32731.

Lan, L.-Y., Qiu, C.-L., Zhao, D.-W. & Gao, X.-H. 2011. Microstructural evolution and mechanical properties of Nb-Ti microalloyed pipeline steel. *Journal of Iron and Steel Research International*, 18, 57–63.

Lan, L., Kong, X., Qiu, C. & Du, L. 2021. A review of recent advance on hydrogen embrittlement phenomenon based on multiscale mechanical experiments. *Acta Metall Sin*, 57, 845–859.

Li, D., Gangloff, R. P. & Scully, J. R. 2004. Hydrogen trap states in ultrahigh-strength AERMET 100 steel. *Metallurgical and Materials Transactions A*, 35, 849–864.

Li, Y., Liu, J., Deng, Y., Han, X., Hu, W. & Zhong, C. 2016. Ex situ characterization of metallurgical inclusions in X100 pipeline steel before and after immersion in a neutral pH bicarbonate solution. *Journal of Alloys and Compounds*, 673, 28–37.

Li, L., Song, B., Cai, Z., Liu, Z. & Cui, X. 2019a. Effect of vanadium content on hydrogen diffusion behaviors and hydrogen induced ductility loss of X80 pipeline steel. *Materials Science and Engineering: A*, 742, 712–721.

Li, X., Zhang, J., Akiyama, E., Fu, Q. & Li, Q. 2019b. Hydrogen embrittlement behavior of Inconel 718 alloy at room temperature. *Journal of Materials Science & Technology*, 35, 499–502.

Li, H., Niu, R., Li, W., Lu, H., Cairney, J. & Chen, Y.-S. 2022a. Hydrogen in pipeline steels: recent advances in characterization and embrittlement mitigation. *Journal of Natural Gas Science and Engineering*, 105, 104709.

Li, M., Zhang, H., Zeng, Y. & Liu, J. 2022b. Adsorption and dissociation of high-pressure hydrogen on Fe (100) and Fe2O3 (001) surfaces: combining DFT calculation and statistical thermodynamics. *Acta Materialia*, 239, 118267.

Liang, P., Li, X., Du, C. & Chen, X. 2009. Stress corrosion cracking of X80 pipeline steel in simulated alkaline soil solution. *Materials & Design*, 30, 1712–1717.

Liu, Z., Li, X., Du, C., Lu, L., Zhang, Y. & Cheng, Y. 2009. Effect of inclusions on initiation of stress corrosion cracks in X70 pipeline steel in an acidic soil environment. *Corrosion Science*, 51, 895–900.

Liu, Q., Atrens, A. D., Shi, Z., Verbeken, K. & Atrens, A. 2014. Determination of the hydrogen fugacity during electrolytic charging of steel. *Corrosion Science*, 87, 239–258.

Liu, Q., Venezuela, J., Zhang, M., Zhou, Q. & Atrens, A. 2016. Hydrogen trapping in some advanced high strength steels. *Corrosion Science*, 111, 770–785.

Liu, Q., Zhou, Q., Venezuela, J., Zhang, M. & Atrens, A. 2018. Evaluation of the influence of hydrogen on some commercial DP, Q&P and TWIP advanced high-strength steels during automobile service. *Engineering Failure Analysis*, 94, 249–273.

Liu, J., Zhao, M. & Rong, L. 2023. Overview of hydrogen-resistant alloys for high-pressure hydrogen environment: on the hydrogen energy structural materials. *Clean Energy*, 7, 99–115.

Lovicu, G., Bottazzi, M., D'Aiuto, F., De Sanctis, M., Dimatteo, A., Santus, C. & Valentini, R. 2012. Hydrogen embrittlement of automotive advanced high-strength steels. *Metallurgical and Materials Transactions A*, 43, 4075–4087.

Luo, Y., Lu, H., Min, N., Li, W. & Jin, X. 2022. Effect of Mo and Nb on mechanical properties and hydrogen embrittlement of hot-rolled medium-Mn steels. *Materials Science and Engineering: A*, 844, 143108.

Lynch, S. 1979. Hydrogen embrittlement and liquid-metal embrittlement in nickel single crystals. *Scripta Metallurgica*, 13, 1051–1056.

Lynch, S. 2003. Mechanisms of hydrogen assisted cracking-a review. *Hydrogen Effects on Material Behaviour and Corrosion Deformation Interactions, Moran, WY*, 449–466.

Lynch, S. 2012a. Hydrogen embrittlement phenomena and mechanisms. *Corrosion Reviews*, 30, 105–123.

Lynch, S. 2012b. Metallographic and fractographic techniques for characterising and understanding hydrogen-assisted cracking of metals. In Richard P. Gangloff and Brian P. Somerday (Eds.), *Gaseous Hydrogen Embrittlement of Materials in Energy Technologies*. Elsevier, Amsterdam, Netherlands.

Mahajan, D., Tan, K., Venkatesh, T., Kileti, P. & Clayton, C. R. 2022. Hydrogen blending in gas pipeline networks-a review. *Energies*, 15, 3582.

Martin, M., Somerday, B., Ritchie, R., Sofronis, P. & Robertson, I. 2012. Hydrogen-induced intergranular failure in nickel revisited. *Acta Materialia*, 60, 2739–2745.

Masoumi, M., Silva, C. C., Béreš, M., Ladino, D. H. & De Abreu, H. F. G. 2017. Role of crystallographic texture on the improvement of hydrogen-induced crack resistance in API 5L X70 pipeline steel. *International Journal of Hydrogen Energy*, 42, 1318–1326.

Matsumoto, Y., Miyashita, T. & Takai, K. 2018. Hydrogen behavior in high strength steels during various stress applications corresponding to different hydrogen embrittlement testing methods. *Materials Science and Engineering: A*, 735, 61–72.

Mcqueen, S., Stanford, J., Satyapal, S., Miller, E., Stetson, N., Papageorgopoulos, D., Rustagi, N., Arjona, V., Adams, J. & Randolph, K. 2020. *Department of Energy Hydrogen Program Plan*. US Department of Energy (USDOE), Washington DC (United States).

Mehrer, H. 2007. *Diffusion in Solids: Fundamentals, Methods, Materials, Diffusion-Controlled Processes*. Springer Science & Business Media, New York.

Melaina, M. W., Antonia, O. & Penev, M. 2013. *Blending Hydrogen into Natural Gas Pipeline Networks: A Review of Key Issues*. National Renewable Energy Laboratory, Golden, CO.

Michler, T. & Naumann, J. 2010. Influence of high pressure hydrogen on the tensile and fatigue properties of a high strength Cu-Al-Ni-Fe alloy. *International Journal of Hydrogen Energy*, 35, 11373–11377.

Mohtadi-Bonab, M. & Eskandari, M. 2017a. A focus on different factors affecting hydrogen induced cracking in oil and natural gas pipeline steel. *Engineering Failure Analysis*, 79, 351–360.

Mohtadi-Bonab, M. A. & Eskandari, M. 2017b. A focus on different factors affecting hydrogen induced cracking in oil and natural gas pipeline steel. *Engineering Failure Analysis*, 79, 351–360.

Mohtadi-Bonab, M., Szpunar, J. & Razavi-Tousi, S. 2013a. A comparative study of hydrogen induced cracking behavior in API 5L X60 and X70 pipeline steels. *Engineering Failure Analysis*, 33, 163–175.

Mohtadi-Bonab, M., Szpunar, J. & Razavi-Tousi, S. 2013b. Hydrogen induced cracking susceptibility in different layers of a hot rolled X70 pipeline steel. *International Journal of Hydrogen Energy*, 38, 13831–13841.

Mohtadi-Bonab, M., Szpunar, J., Basu, R. & Eskandari, M. 2015. The mechanism of failure by hydrogen induced cracking in an acidic environment for API 5L X70 pipeline steel. *International Journal of Hydrogen Energy*, 40, 1096–1107.

Mohtadi-Bonab, M., Eskandari, M., Karimdadashi, R. & Szpunar, J. 2017. Effect of different microstructural parameters on hydrogen induced cracking in an API X70 pipeline steel. *Metals and Materials International*, 23, 726–735.

Moli-Sanchez, L., Martin, F., Leunis, E., Briottet, L., Lemoine, P. & Chêne, J. 2014. Comparison of the tensile behavior of a tempered 34CrMo4 steel exposed in situ to high pressure H2 gas or to cathodic H charging. *SteelyHydrogen2014 Conference Proceedings*, 448e61.

Mori, G. K., Feyerl, J. & Zitter, H. 2002. On the influence of hydrogen content and stress level on hydrogen embrittlement of bainitic carbon steel fasteners. Corrosion 2002, 51300-02430-SG.

Muritala, I. K., Guban, D., Roeb, M. & Sattler, C. 2020. High temperature production of hydrogen: assessment of non-renewable resources technologies and emerging trends. *International Journal of Hydrogen Energy*, 45, 26022–26035.

Nagumo, M. & Takai, K. 2019. The predominant role of strain-induced vacancies in hydrogen embrittlement of steels: overview. *Acta Materialia*, 165, 722–733.

Nanninga, N., Levy, Y., Drexler, E. S., Condon, R., Stevenson, A. & Slifka, A. J. 2012a. Comparison of hydrogen embrittlement in three pipeline steels in high pressure gaseous hydrogen environments. *Corrosion Science*, 59, 1–9.

Nanninga, N. E., Levy, Y. S., Drexler, E. S., Condon, R. T., Stevenson, A. E. & Slifka, A. J. 2012b. Comparison of hydrogen embrittlement in three pipeline steels in high pressure gaseous hydrogen environments. *Corrosion Science*, 59, 1–9.

Nayak, S., Misra, R., Hartmann, J., Siciliano, F. & Gray, J. 2008. Microstructure and properties of low manganese and niobium containing HIC pipeline steel. *Materials Science and Engineering: A*, 494, 456–463.

Neeraj, T., Srinivasan, R. & Li, J. 2012. Hydrogen embrittlement of ferritic steels: observations on deformation microstructure, nanoscale dimples and failure by nanovoiding. *Acta Materialia*, 60, 5160–5171.

Nibur, K., Bahr, D. & Somerday, B. 2006. Hydrogen effects on dislocation activity in austenitic stainless steel. *Acta materialia*, 54, 2677–2684.

Nikolaidis, P. & Poullikkas, A. 2017. A comparative overview of hydrogen production processes. *Renewable and Sustainable Energy Reviews*, 67, 597 611.

Novak, P., Yuan, R., Somerday, B., Sofronis, P. & Ritchie, R. 2010. A statistical, physical-based, micro-mechanical model of hydrogen-induced intergranular fracture in steel. *Journal of the Mechanics and Physics of Solids*, 58, 206–226.

Ohaeri, E., Omale, J., Eduok, U. & Szpunar, J. 2018. Effect of thermomechanical processing and crystallographic orientation on the corrosion behavior of API 5L X70 pipeline steel. *Metallurgical and Materials Transactions A*, 49, 2269–2280.

Olden, V., Thaulow, C. & Johnsen, R. 2008. Modelling of hydrogen diffusion and hydrogen induced cracking in supermartensitic and duplex stainless steels. *Materials & Design*, 29, 1934–1948.

Omale, J., Ohaeri, E., Tiamiyu, A., Eskandari, M., Mostafijur, K. & Szpunar, J. 2017. Microstructure, texture evolution and mechanical properties of X70 pipeline steel after different thermomechanical treatments. *Materials Science and Engineering: A*, 703, 477–485.

Onyebuchi, V. E., Kolios, A., Hanak, D. P., Biliyok, C. & Manovic, V. 2018. A systematic review of key challenges of CO_2 transport via pipelines. *Renewable and Sustainable Energy Reviews*, 81, 2563–2583.

Pfeil, L. B. 1926. The effect of occluded hydrogen on the tensile strength of iron. *Proceedings of the Royal Society of London. Series A, Containing Papers of a Mathematical and Physical Character*, 112, 182–195.

Porter, D. A., Easterling, K. E. & Sherif, M. Y. 2021. *Phase Transformations in Metals and Alloys*. CRC Press, Boca Raton, FL.

Pourazizi, R., Mohtadi-Bonab, M. & Szpunar, J. 2020. Investigation of different failure modes in oil and natural gas pipeline steels. *Engineering Failure Analysis*, 109, 104400.

Pradhan, P., Robi, P. & Roy, S. K. 2012. Micro void coalescence of ductile fracture in mild steel during tensile straining. *Frattura ed Integrità Strutturale*, 6, 51–60.

Qin, W. & Szpunar, J. 2017. A general model for hydrogen trapping at the inclusion-matrix interface and its relation to crack initiation. *Philosophical Magazine*, 97, 3296–3316.

Qin, W., Szpunar, J. & Kozinski, J. 2015. Hydride-induced degradation of zirconium alloys: a criterion for complete ductile-to-brittle transition and its dependence on microstructure. *Proceedings of the Royal Society A: Mathematical, Physical and Engineering Sciences*, 471, 20150192.

Ramamurthy, S. & Atrens, A. 2013. Stress corrosion cracking of high-strength steels. *Corrosion Reviews*, 31, 1–31.

Ramirez, E., González-Rodriguez, J., Torres-Islas, A., Serna, S., Campillo, B., Dominguez-Patiño, G. & Juárez-Islas, J. 2008. Effect of microstructure on the sulphide stress cracking susceptibility of a high strength pipeline steel. *Corrosion Science*, 50, 3534–3541.

Robertson, I. 2001. The effect of hydrogen on dislocation dynamics. *Engineering Fracture Mechanics*, 68, 671–692.

Robertson, I. M., Sofronis, P., Nagao, A., Martin, M. L., Wang, S., Gross, D. & Nygren, K. 2015. Hydrogen embrittlement understood. *Metallurgical and Materials Transactions A*, 46, 2323–2341.

Russo, T. N. 2020. Hydrogen: hype or a glide path to decarbonizing natural gas-Part 2. *Climate and Energy*, 37, 24–32.

San Marchi, C., Somerday, B. P. & Robinson, S. L. 2007. Permeability, solubility and diffusivity of hydrogen isotopes in stainless steels at high gas pressures. *International Journal of Hydrogen Energy*, 32, 100–116.

Shirband, Z., Shishesaz, M. R. & Ashrafi, A. 2011. Hydrogen degradation of steels and its related parameters, a review. *Phase Transitions*, 84, 924–943.

Singla, M. K., Nijhawan, P. & Oberoi, A. S. 2021. Hydrogen fuel and fuel cell technology for cleaner future: a review. *Environmental Science and Pollution Research*, 28, 15607–15626.

Smith, B., Frame, B., Eberle, C., Blencoe, J., Anovitz, L., Armstrong, T. & Mays, J. 2005. *New Materials for Hydrogen Pipelines*. Hydrogen Pipeline Working Group Meeting, Augusta.

Somerday, B. P. & San Marchi, C. W. 2006. *Effects of Hydrogen Gas on Steel Vessels and Pipelines*. Sandia National Lab. (SNL-CA), Livermore, CA.

Song, J. & Curtin, W. 2013. Atomic mechanism and prediction of hydrogen embrittlement in iron. *Nature Materials*, 12, 145–151.

Song, E. J., Bhadeshia, H. & Suh, D.-W. 2013. Effect of hydrogen on the surface energy of ferrite and austenite. *Corrosion Science*, 77, 379–384.

Sun, Y. & Cheng, Y. F. 2021. Thermodynamics of spontaneous dissociation and dissociative adsorption of hydrogen molecules and hydrogen atom adsorption and absorption on steel under pipelining conditions. *International Journal of Hydrogen Energy*, 46, 34469–34486.

Sun, Y., Ren, Y. & Cheng, Y. F. 2022. Dissociative adsorption of hydrogen and methane molecules at high-angle grain boundaries of pipeline steel studied by density functional theory modeling. *International Journal of Hydrogen Energy*, 47, 41069–41086.

Svoboda, J., Mori, G., Prethaler, A. & Fischer, F. D. 2014. Determination of trapping parameters and the chemical diffusion coefficient from hydrogen permeation experiments. *Corrosion Science*, 82, 93–100.

Tal-Gutelmacher, E. & Eliezer, D. 2005. The hydrogen embrittlement of titanium-based alloys. *JOM*, 57, 46–49.

Timmerberg, S. & Kaltschmitt, M. 2019. Hydrogen from renewables: supply from North Africa to Central Europe as blend in existing pipelines-potentials and costs. *Applied Energy*, 237, 795–809.

Trautmann, A., Mori, G., Oberndorfer, M., Bauer, S., Holzer, C. & Dittmann, C. 2020. Hydrogen uptake and embrittlement of carbon steels in various environments. *Materials (Basel)*, 13, 3604.

Turnbull, A. 2012. Hydrogen diffusion and trapping in metals. In Richard P. Gangloff and Brian P. Somerday (Eds.), *Gaseous Hydrogen Embrittlement of Materials in Energy Technologies*. Elsevier, Amsterdam, Netherlands.

Turnbull, A. 2015. Perspectives on hydrogen uptake, diffusion and trapping. *International Journal of Hydrogen Energy*, 40, 16961–16970.

Turnbull, A. & Hutchings, R. 1994. Analysis of hydrogen atom transport in a two-phase alloy. *Materials Science and Engineering: A*, 177, 161–171.

Vecchi, L., Pecko, D., Mamme, M. H., Van Laethem, D., Depover, T., Van Den Eeckhout, E., Van Den Steen, N., Ozdirik, B., Verbeken, K. & Van Ingelgem, Y. 2019. Numerical interpretation to differentiate hydrogen trapping effects in iron alloys in the Devanathan-Stachurski permeation cell. *Corrosion Science*, 154, 231–238.

Venezuela, J., Liu, Q., Zhang, M., Zhou, Q. & Atrens, A. 2015. The influence of hydrogen on the mechanical and fracture properties of some martensitic advanced high strength steels studied using the linearly increasing stress test. *Corrosion Science*, 99, 98–117.

Venezuela, J., Liu, Q., Zhang, M., Zhou, Q. & Atrens, A. 2016. A review of hydrogen embrittlement of martensitic advanced high-strength steels. *Corrosion Reviews*, 34, 153–186.

Venezuela, J., Zhou, Q., Liu, Q., Li, H., Zhang, M., Dargusch, M. S. & Atrens, A. 2018a. The influence of microstructure on the hydrogen embrittlement susceptibility of martensitic advanced high strength steels. *Materials Today Communications*, 17, 1–14.

Venezuela, J., Zhou, Q., Liu, Q., Zhang, M. & Atrens, A. 2018b. Hydrogen trapping in some automotive martensitic advanced high-strength steels. *Advanced Engineering Materials*, 20, 1700468.

Venezuela, J. J. D. G. 2017. The influence of hydrogen on MS980, MS1180, MS1300 and MS1500 martensitic advanced high strength steels used for automotive applications. PhD Thesis, School of Mechanical and Mining Engineering, The University of Queensland

Venkatasubramanian, T. V. & Baker, T. J. 1984. Role of MnS inclusions in hydrogen assisted cracking of steel exposed to H2S saturated salt solution. *Metal Science*, 18, 241–248.

Wang, B., Liu, X. & Wang, G. 2018. Inclusion characteristics and acicular ferrite nucleation in Ti-containing weld metals of X70 pipeline steel. *Steel Research International*, 89, 1700316.

Wang, W., Shan, Y. & Yang, K. 2009. Study of high strength pipeline steels with different microstructures. *Materials Science and Engineering: A*, 502, 38–44.

Wei, X., Dong, C., Chen, Z., Xiao, K. & Li, X. 2016. The effect of hydrogen on the evolution of intergranular cracking: a cross-scale study using first-principles and cohesive finite element methods. *RSC Advances*, 6, 27282–27292.

Wu, X., Zhang, H., Yang, M., Jia, W., Qiu, Y. & Lan, L. 2022. From the perspective of new technology of blending hydrogen into natural gas pipelines transmission: mechanism, experimental study, and suggestions for further work of hydrogen embrittlement in high-strength pipeline steels. *International Journal of Hydrogen Energy*, 47, 8071–8090.

Wulf, C., Reuß, M., Grube, T., Zapp, P., Robinius, M., Hake, J.-F. & Stolten, D. 2018. Life cycle assessment of hydrogen transport and distribution options. *Journal of Cleaner Production*, 199, 431–443.

Xie, D., Li, S., Li, M., Wang, Z., Gumbsch, P., Sun, J., Ma, E., Li, J. & Shan, Z. 2016. Hydrogenated vacancies lock dislocations in aluminium. *Nature Communications*, 7, 13341.

Xiong, X., Ma, H., Zhang, L., Song, K., Yan, Y., Qian, P. & Su, Y. 2023a. The hydrogen-resistant surface of steels designed by alloy elements doping: first-principles calculations. *Computational Materials Science*, 216, 111854.

Xiong, X. L., Ma, H. X., Zhang, L. N., Song, K. K., Yan, Y., Qian, P. & Su, Y. J. 2023b. The hydrogen-resistant surface of steels designed by alloy elements doping: first-principles calculations. *Computational Materials Science*, 216, 111854.

Xu, J., Misra, R., Guo, B., Jia, Z. & Zheng, L. 2013. Understanding variability in mechanical properties of hot rolled microalloyed pipeline steels: process-structure-property relationship. *Materials Science and Engineering: A*, 574, 94–103.

Xue, H. B. & Cheng, Y. F. 2011. Characterization of inclusions of X80 pipeline steel and its correlation with hydrogen-induced cracking. *Corrosion Science*, 53, 1201–1208.

Xue, H. & Cheng, Y. 2013. Hydrogen permeation and electrochemical corrosion behavior of the X80 pipeline steel weld. *Journal of Materials Engineering and Performance*, 22, 170–175.

Yagodzinskyy, Y., Ivanchenko, M. & Hänninen, H. 2012. Hydrogen-dislocation interaction in austenitic stainless steel studied with mechanical loss spectroscopy. *Solid State Phenomena*, 2012, 227–232.

Yamaguchi, M. 2011. First-principles study on the grain boundary embrittlement of metals by solute segregation: part I. iron (Fe)-solute (B, C, P, and S) systems. *Metallurgical and Materials Transactions A*, 42, 319–329.

Yigzawe, T. M. & Sadus, R. J. 2013. Intermolecular interactions and the thermodynamic properties of supercritical fluids. *The Journal of Chemical Physics*, 138, 194502.

Zaferani, S. H., Miresmaeili, R. & Pourcharmi, M. K. 2017. Mechanistic models for environmentally-assisted cracking in sour service. *Engineering Failure Analysis*, 79, 672–703.

Zhang, Q.-H., Li, X.-R., Liu, P., Meng, X.-Z., Wu, L.-K., Luo, Z.-Z. & Cao, F.-H. 2021. Quantitative study of the kinetics of hydrogen evolution reaction on aluminum surface and the influence of chloride ion. *International Journal of Hydrogen Energy*, 46, 39665–39674.

Zheng, Y., Jiao, Y., Jaroniec, M. & Qiao, S. Z. 2015. Advancing the electrochemistry of the hydrogen-evolution reaction through combining experiment and theory. *Angewandte Chemie International Edition*, 54, 52–65.

Zhou, X., Marchand, D., Mcdowell, D. L., Zhu, T. & Song, J. 2016. Chemomechanical origin of hydrogen trapping at grain boundaries in FCC metals. *Physical Review Letters*, 116, 075502.

Züttel, A. 2004. Hydrogen storage methods. *Naturwissenschaften*, 91, 157–172.

10 Cryogenic Liquid Tankers for Hydrogen Transportation

Fatemeh Salahi, Fatemeh Zarei-Jelyani, and Mohammad Reza Rahimpour

10.1 INTRODUCTION

LH is the first fuel on a large scale that is used for low-temperature heavy-duty rockets, such as the CZ-8, CZ-7 series, the operational CZ-5 series, and the upcoming generation of piloted rockets that rely on LH propulsion. The demand for this technology is experiencing significant growth due to advancements in deep space exploration, piloted landings on the moon, operations carried out on space stations, the BeiDou Navigation Satellite System, deployed internet satellite systems, and other missions. Furthermore, LH possesses an extensive array of industrial and civil applications, constituting a significant avenue for the advancement of new energy technologies. Referring to the draft of the energy law of the PRC published by the National Energy Administration on April 10, 2020, hydrogen energy is specifically classified as energy. This marks the initial instance in which China has formally acknowledged the energy classification of hydrogen energy. Civilians and the military alike recognize the significance of LH as a pure energy source (Qiu et al., 2021). Worldwide, the LH industry comprises dozens of facilities, each possessing a daily capacity of 470 tons of LH. North America possesses over 85% of the total capacity for LH on a global scale. The United States is home to over 15 LH facilities, combined with a daily production capacity of over 326 tons, which places it at the forefront of the global LH industry. Four LH facilities in Europe each produce 24 tons of LH per day. Asia is home to 16 LH facilities, collectively capable of producing 38.3 tons per day. Japan represents two-thirds of this capacity in Asia. The only active LH plants in China are located in Wenchang, at the 101 Institute in Beijing, and at the Xichang Base. These facilities are exclusively utilized for space rocket launches. Their expensiveness and limited production capacity impose significant obstacles on the utilization of LH in sectors such as energy, high-end manufacturing, metallurgy, and electronics. As a result, LH and the hydrogen energy sector have emerged as the primary drivers of energy advancement in China (Lu et al., 2013). The critical apparatus in the fields of space and energy is the cryogenic storage and transportation vessel of LH, which serves as a vital medium for the storage and transportation of hydrogen on a large scale. Scientific application of low-temperature materials facilitates the design and fabrication of vessels for transporting and storing LH.

DOI: 10.1201/9781003382553-12

A comprehensive understanding of the properties of the cryogenic materials used in the containers for storage and transit is a critical assurance for the safe, dependable, and extended and lasting operation of the cryogenic propellant storage for transportation. LH and other cryogenic propellant transportation and storage containers are widely used in chemical, civil energy, space propulsion systems, and other industries in the Japan, United States, Europe, and other locations. Experts in the fields of material performance research and the secure and efficient use of LH transportation and storage containers have gained significant knowledge in these fields. They attained a thorough comprehension of the test method for material performance at low temperatures and developed an optimal design, standardized selection process, and specification (Qiu et al., 2021).

The aerospace industry in China has conducted initial research on the approach and improvement of cryogenic propellants, specifically focusing on LH containers for storage and transportation, the mechanical characteristics, and selection criteria for associated materials. However, the civil sector is mostly undeveloped, with limited advancements in associated technology and an imperfect standard system. Previously, the performance standards for metal materials in China only specified requirements up to 77 K for pressure vessels and metal materials. However, there were no specific regulations regarding the design selection and efficiency parameters criteria for metal materials at temperatures of LH. Typically, the material characteristics at a temperature of 77 K are chosen and assessed, although there is insufficient evidence to support these selections. Research on the characteristics of low-temperature metal materials has gained significance in the field of materials science in recent years. Esteemed academics from several countries have made substantial contributions in this domain. Based on the published data, the Soviet Union conducted early research on this aspect, exploring various materials such as titanium alloy, aluminum alloy, nickel manganese steel, and copper alloy. In contrast, the United States conducted their studies at a later time. Subsequently, when US space technology advanced, they created a comprehensive material database specifically for low-temperature materials. However, the crucial facts inside this database were not made available to the general public. As an illustration, NIST provides information on the Young's modulus, coefficient of linear expansion, specific heat capacity, and thermal conductivity of various materials, such as aluminum alloy, stainless steel, titanium alloy, and others, at low temperatures. The primary characteristics of the materials are their physical parameters; however, there is a lack of pertinent data on the low-temperature mechanical properties of various material brands (Ren et al., 2020).

China's aerospace equipment has seen a steady improvement in its ability to be independent and controllable, leading to increased attention in this area. Nevertheless, there is currently a lack of established series and standards for low-temperature metal materials, and there is a dearth of detailed data support. Despite the fact that the standard specification has reduced the design temperature to the range of LH temperature, there have been enhancements in the material strength and pressure vessel strength in low-temperature conditions. The absence of fundamental mechanical property data, for example, impacts toughness and plasticity, significantly hinders the domestic self-sufficiency in the production of cryogenic propellant container for transportation and storage equipment (Kovač et al., 2021). This chapter presents a thorough examination of the existing research on cryogenic materials used in containers for storing and transporting LH.

10.2 STAINLESS STEEL

10.2.1 STAINLESS STEEL MATERIALS SUITABLE FOR CRYOGENIC APPLICATIONS

Austenitic stainless steel (austenitic SS) shows excellent performance at low temperatures, making it the preferred material for working in low-temperature circumstances. It is the most widely used material for storing and transporting LH. In 1955, the United States made an initial attempt to use LH as aviation fuel. As part of this effort, two LH tanks were built from stainless steel and placed on the wing tip (Maniaci, 2008). The austenite phase in stainless steel is characterized by a face-centered cubic crystal structure, which imparts exceptional plastic deformation capabilities. The addition of a greater amount of nickel (Ni) and chromium (Cr) to the alloy increases the stability of austenite at low temperatures. Consequently, the strength of austenitic SS may be enhanced with a drop in temperature, while still preserving its excellent flexibility and tolerance to low temperatures. As shown in Table 10.1 the composition constituents of various grades of stainless steel vary, which dictates their overall characteristics and uses. There is currently an absence of a theoretical or standard strategy regarding the appropriate stainless steel grade to use for various applications in the LH storage tank.

TABLE 10.1

Austenite Stainless Steel Chemical Composition Is Often Utilized at Cryogenic Temperatures

Alloy Grade	Chemical Composition %							
	C	Mn	Si	Cr	Ni	P	S	Other
201	0.15	5.5–7.5	1.0	16.0–18.0	3.5–5.5	0.06	0.03	0.25N
202	0.15	7.5–10.0	1.0	17.0–19.0	4.0–6.0	0.06	0.03	0.25N
205	0.12–0.25	14.0–15.5	1.0	16.5–18.0	1.0–1.75	0.06	0.03	0.32–0.40N
301	0.15	2.0	1.0	16.0–18.0	6.0–8.0	0.045	0.03	
302	0.15	2.0	1.0	17.0–19.0	8.0–10.0	0.045	0.03	
303	0.15	2.0	1.0	17.0–19.0	8.0–10.0	0.2	0.15	0.6Mo(b)
304	0.08	2.0	1.0	18.0–20.0	8.0–10.5	0.045	0.03	
304L	0.03	2.0	1.0	18.0–20.0	8.0–10.5	0.045	0.03	
304N	0.08	2.0	1.0	18.0–20.0	8.0–10.5	0.045	0.03	0.1–0.16N
316	0.08	2.0	1.0	16.0–18.0	10.0–14.0	0.045	0.03	2.0–3.0Mo
316L	0.03	2.0	1.0	16.0–18.0	10.0–14.0	0.045	0.03	2.0–3.0Mo
310	0.25	2.0	1.5	24.0–26.0	19.0–22.0	0.045	0.03	
321	0.08	2.0	1.0	17.0–19.0	9.0–12.0	0.045	0.03	5×% Cmin Ti

Source: Adapted from Qiu et al. (2021).

Additionally, there is a need for a comprehensive evaluation of the performance of stainless steel in the temperature zone of LH. Austenitic SS may be categorized based on its chemical composition into two groups: Cr-Ni (300 series) and Cr-Ni-Mn (200 series). The 300 series, including 304, 304L, 316, 316L, 321, 347, etc., is extensively used in low-temperature liquid containers for storage and transportation because of its exceptional overall performance (Park et al., 2010). The variation in alloy constituents across various grades of stainless steel significantly impacts the ultimate material's application and performance. 316L stainless steel is enhanced with the addition of a molybdenum (Mo) element, significantly enhancing its resistance to corrosion caused by chloride ions. This is appropriate for maritime conditions characterized by a high concentration of salt spray. 321 stainless steel, unlike other types, has a titanium element that enhances its ability to withstand intergranular corrosion and maintain high-temperature strength. As a result, it is well-suited for environments that need exceptional resistance to corrosion and heat. When selecting the most suitable material for a particular application, it is important to take into account the welding performance, corrosion resistance, cost of the materials, and mechanical qualities. In "50 project" hydrogen oxygen engine test run, a $100\,m^3$ LH tank was constructed using 304 stainless steel. Similarly, in the Hainan huge launch site, a $300\,m^3$ LH transport tank vehicle was built using 321 stainless steel. In order to meet the growing need for storing and transporting low-temperature materials using LH, it is crucial to conduct thorough and systematic research on stainless steel materials for these purposes. This research should aim to fully understand the performance of various stainless steel's brands and types. Based on this understanding, material selection standards can be developed for stainless steel that meets the requirements of low-temperature storage and transportation containers for LH (DeSisto and Carr, 1961).

10.2.2 HE of Austenitic SS

HE is the process by which the hydrogen molecules' diffusion into a material leads to the beginning and fracturing of cracks in the substance. The structure must be given extra consideration, particularly in a hydrogen environment, because of the significant harm caused by HE (Mansilla et al., 2018). During the smelting process or when the material is exposed to a hydrogen environment, hydrogen molecules will accumulate in areas of internal stress concentration within the material. This accumulation leads to the start, propagation, and eventual fracture of cracks (Lynch, 2012). Theoretical explanations for HE primarily include the hydrogen-induced local plastic deformation theory, hydrogen-induced weak bond theory, and hydride theory (McLellan and Harkins, 1975; Hwang et al., 2020). HE in austenitic SS mostly occurs due to the hypothesis of hydrogen-driven local plastic deformation, since the production of hydrides by Fe and H atoms is challenging. The strong HE resistance of austenitic SS is attributed to the stability of austenite at low temperatures. This characteristic enables its suitability for applications in the environment of LH. Nevertheless, the durability of the austenite phase significantly impacts the capacity to resist HE. Metastable 304 austenitic SS has inferior resistance to HE compared to 310 and 316 austenitic SS, which have superior phase stability. Furthermore, prolonged exposure to the environment of LH or the application of stress in such an environment will result in the austenitic SS's HE (Hwang et al., 2020).

The composition, processing method, and heat treatment procedure have a significant impact on the microstructure of stainless steel. Varying microstructures and surface conditions exhibit distinct levels of resistance to HE. Lee et al. (2021) conducted a comparison between the HE properties of austenitic SS produced through additive manufacturing and traditional austenitic SS. The study revealed that the HE resistance of stainless steel produced through additive manufacturing remained unaffected, while its strength was enhanced. Lucas (Queiroga et al., 2019) conducted a study on the surface roughness impact on the HE stainless steel properties. He discovered that the transition of surface martensite caused by cutting was the primary factor contributing to HE. 310 stainless steel exhibited superior stability and enhanced resistance to HE as compared to 304 stainless steel. Fan et al. (2019b) conducted a study on the impact of grain refinement on the HE of 304 stainless steel. The findings demonstrated that grain refinement has the ability to decrease stress concentration and enhance resistance to HE. Furthermore, the research demonstrated that austenitic SS with a nano scale twin structure exhibited significant resistance to HE (Fan et al., 2019a). Hence, to enhance the HE resistance of austenitic SS to the greatest extent, it is crucial to meticulously regulate all elements of material forming and preparation, optimize austenite stability, and minimize stress concentration. Furthermore, much study has been conducted to enhance the resistance of stainless steel to HE corrosion. This research has explored many approaches such as surface coating, cathodic protection, ion implantation, and laser shot peening (Huang et al., 2020, Zhou et al., 2016). As the use of stainless steel in LH transportation and storage expands, there is a need to further enhance our knowledge of HE. Hence, it is very practical to investigate the HE properties of several brands of austenitic SS materials and their diverse uses (Xiuqing et al., 2021).

10.2.3 Stainless Steel's Cryogenic Mechanical Properties

In general, when the temperature decreases, the material's yield strength, tensile strength, and elastic modulus rise, along with the endurance limit and fatigue strength. The impact toughness is intricately linked to the material's crystal structure. The face-centered cubic crystal has superior impact resistance at low temperatures, while the body-centered cubic crystal demonstrates worse impact resistance. The low-temperature material's plasticity is contingent upon the ductile brittle transition properties of these materials. Materials without a low-temperature brittle transition exhibit an increase in elongation as the temperature decreases. Materials exhibiting low-temperature brittle transition see a significant drop in flexibility at low temperatures, rendering them unsuitable for use in low-temperature environments. The evaluation of cryogenic materials primarily relies on cryogenic toughness, which serves as the most crucial assessment criterion. The test techniques directly associated with containers used for the storage and transportation of LH are as follows: The following tests are conducted to assess the impact toughness of materials at low temperatures: V-notch test according to the fracture mechanics test (involving crack tip opening displacement cod method and plane strain fracture toughness

KIC), full thickness test (including Esso test, double tensile test, and wide plate test), drop weight test, German Society for Materials and Testing (DVM) standards. The cryogenic Charpy (V-notch) impact toughness test is highly prevalent for evaluating the material's low-temperature toughness. It determines the temperature at which a specific absorption energy Akv or a certain percentage of fracture fiber (known as the brittle transition temperature) is achieved in the impact test. Austenitic SS is classified as having a face-centered cubic structure, which exhibits excellent stability at low temperatures. The flexibility and hardness of the material will not diminish greatly with the reduction in temperature (Duthil, 2015).

The assessment of cryogenic materials largely depends on their cryogenic toughness, which is the most critical factor for evaluation. The test methodologies specifically linked to containers used for the storage and conveyance of LH are as follows: The impact toughness of materials at low temperatures is evaluated through several tests, including the V-notch test following the standards of the DVM, test of drop weight, test of full thickness (which includes test of wide plate, Esso test, and double tensile test), and test of fracture mechanics (which involves the crack tip opening displacement cod method and assessment of plane strain fracture toughness KIC). The Charpy V-notch impact toughness test is commonly used to assess the toughness of materials at low temperatures. The test of impact is used to measure the temperature at which a specified absorption energy, known as Akv, or a given percentage of fracture fiber, referred to as the brittle transition temperature, is reached. Austenitic SS is categorized by its face-centered cubic structure, which demonstrates exceptional resilience at low temperatures. The material's flexibility and hardness will only see a little decrease when the temperature decreases. The low-temperature impact test may be waived for the base metal and heat-affected zone. However, if the carbon content exceeds 0.1%, the minimum service temperature without the low-temperature test of impact is −48°C or above. As per the Japanese Industrial Standards (JIS) for austenitic SS used in low-temperature conditions, a low-temperature impact test must be conducted when the service temperature falls below −196°C. Foreign standards impose more stringent criteria on the low-temperature qualities of stainless steel compared to domestic standards, indicating a greater emphasis on low-temperature impact toughness. To assess the impact toughness of materials used in practical LH storage tanks, it is important to conduct low-temperature testing on both the base metal and the weld. The microstructure and properties of the materials are influenced by factors such as heat treatment process, batch, material type, and forming process. Typically, the performance of welded parts is slightly inferior to that of the base metal. Therefore, it is crucial to evaluate the weld's low-temperature impact toughness (Qiu et al., 2021).

The tensile test may be used to determine the plasticity index of materials, elastic modulus, yield strength, and tensile strength. The electronic universal testing equipment is capable of assessing the low-temperature tensile characteristics of materials. The tensile property index, being a fundamental characteristic of materials, has significant importance in the design of low-temperature containers. Figure 10.1 demonstrates that the yield stainless steel strength and tensile strength rise as temperature decreases, but the stainless steel plasticity diminishes with decreasing temperature.

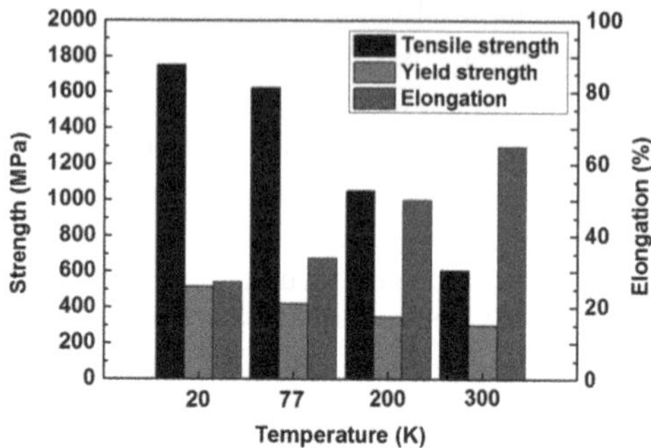

FIGURE 10.1 Temperature-dependent cryogenic characteristics of 18Cr-8Ni stainless steel. Adapted from Qiu et al. (2021).

10.3 ALUMINUM ALLOY

Aluminum alloy shares the face-centered cubic crystal structure with austenitic SS, resulting in the absence of a discernible ductile brittle transition temperature at low temperatures. Additionally, it exhibits exceptional impact toughness. Furthermore, aluminum alloy has the attributes of favorable processability, elevated specific strength, and reduced susceptibility to HE, thus making it extensively used in LH containers. An aluminum alloy is particularly well-suited for use as a low-temperature material, especially in aeronautical propellant tanks. The aluminum alloys used in low-temperature applications primarily consist of precipitation hardening and solution hardening. The alloys often used for solution hardening are typically 3000 series aluminum and 5000 series aluminum. The primary precipitation hardening alloys consist of 2000 series aluminum, 6000 series aluminum, and 7000 series aluminum. In the United States, the area of rocket launching has seen the employment of aluminum alloy LH storage tanks due to advancements in aeronautics and astronautics. Specifically, the tanks are made from 2195, 2029, and 2219 aluminum alloys. The collaborative effort between Boeing and NASA has implemented a LH rocket as the space launch mechanism in recent years, with the primary objective of facilitating human exploration of Mars. The fuel tank for LH is constructed using 2219 aluminum alloy and joined together by friction-stir-welding (FSW) (Verstraete et al., 2010). Chinese launch vehicle's propellant tank has transitioned from using 5A06 alloys to use 2219 aluminum copper alloys and 2A14 aluminum alloys. The 2A14 aluminum alloy has been chosen as the primary structural material for tanks up to this point and is the predominant material for established types now in use. In addition, in China, 2219 aluminum alloys are identified as the structural material for the launch vehicle tank's next generation. Furthermore, the LH storage tank for the Long March 5 is also constructed using 2219 aluminum alloy. The growing launch capabilities of rockets have led to a future trend in the use of LH and liquid oxygen rockets. Consequently, there will be an increased need for aluminum alloy LH storage tanks (Jha et al., 2003).

The yield strength and tensile strength of aluminum alloy exhibit an upward trend when the temperature decreases, although its ratio of yield strength in terms of notch sensitivity and plasticity remains relatively unaffected. Consequently, aluminum alloys serve as the primary materials for LH tanks used in rocket launches. The fabrication of LH tanks often involves processes such as milling, welding, sheet metal forming, etc. Among these, welding technique has consistently been a focal point in the study of LH tanks. As a result of changes in the microstructure of the heat-affected zone and weld, three distinct kinds of fractures may occur in the welding area: storage cracks, liquefaction cracks, and crystallization cracks. As per the production quality standard for propellant tanks in China, the presence of fractures in the welded seam is strictly prohibited. The welding procedure also results in residual stress creation. Eliminating residual stress in the tank after welding is crucial because it reduces fatigue strength and increases the likelihood of fracture (Viswanath et al., 2019; Peel et al., 2003). Furthermore, it is important to consider the stress corrosion and corrosion resistance properties of the welded components in aluminum alloy tanks. Several researchers have conducted studies on the corrosion resistance of aluminum alloy weldments (Jariyaboon et al., 2007, Entringer et al., 2019) and have suggested several surface treatment techniques to enhance their corrosion resistance (Wu et al., 2016). The material's welding performance is a crucial factor in selecting materials for the production process of low-temperature storage tanks, especially when considering the use of an LH environment. For instance, the welding junction of 2A14 aluminum alloy exhibits elongation segregation and a significant propensity for cracking (Jha et al., 2010). The fusion welding technique, characterized by high energy density and the absence of filler wire, is unsuitable for the material. Currently, the welding processes frequently used in engineering are appropriate for the 2219 aluminum alloy (Xian et al., 2015). Furthermore, the 7xxx aluminum alloy exhibits susceptibility to plastic deterioration or hydrogen-induced brittleness when exposed to an LH environment, rendering it unsuitable for the fabrication of LH tanks (Moshtaghi et al., 2021, Safyari et al., 2020).

10.4 TITANIUM ALLOY

Titanium alloys are extensively used in the aerospace industry because of their notable benefits, including high specific strength, excellent corrosion resistance, high-temperature resistance, low thermal conductivity, and small coefficient of expansion (Kuramoto et al., 2001). Furthermore, titanium alloys have exceptional performance at very low temperatures. As deep space exploration and aerospace continue to advance, the performance demands for low-temperature materials in aircraft structures are significantly enhanced. Titanium alloys are highly appropriate for aerospace applications as a novel material for low-temperature conditions. In 1981, NASA launched the Apollo space rocket and used titanium and titanium alloys for the purpose of containing liquid helium and LH, as well as for constructing structural pipelines. Titanium in its pure form has a compact hexagonal shape at ambient temperature, known as α phase. Isomeric transformation occurs when the temperature exceeds 855°C, resulting in the creation of a body-centered cubic structure known as the β phase, specifically the low-temperature β phase. The flexibility and hardness of phase α diminish with the reduction in temperature.

Consequently, the majority of titanium alloys used in low-temperature applications consist of pure titanium, α titanium alloy, or a β two-phase titanium alloy with a lower phase concentration (Boyer, 1987).

10.5 CRYOGENIC COMPOSITES

The need for lightweight fuel tanks is growing as the aerospace profession continues to advance. Composite materials have superior strength and reduced density, enabling a possible weight reduction of 25% as compared to aluminum alloy tanks (Verstraete et al., 2010). The cryogenic characteristics of composites, which serve as containers for storing and transporting cryogenic liquids, have garnered significant interest. Numerous studies have thoroughly examined this topic in recent years (Schutz, 1998, Sethi and Ray, 2013, Horiuchi and Ooi, 1995). The mechanical properties of composites at low temperatures are influenced by the characteristics of the resin, fiber, and the contact between them. Typically, when the temperature decreases, the polymer molecular chain's bonding strength rises, resulting in an increase in the tensile strength and Young's modulus of the resin matrix. Nevertheless, when the temperature is lowered, the resin matrix diminishes resilience. Carbon fiber has greater thermal stability compared to polymers, resulting in less impact on its strength with changes in temperature. As the temperature drops, the individual carbon fiber's strength diminishes. However, when considering the composite as a whole, the impact of this loss is negligible in comparison to the matrix resin or the fiber-resin interface. Furthermore, the mechanical properties of composite materials at low temperatures are intricately linked to factors such as structural configuration, composite manufacturing method, molding technique, and orientation of the fibers inside the composite materials. The fracture mechanism exhibits more complexity in low-temperature conditions. Currently, the majority of research has concentrated on examining the mechanical properties of the resin matrix at low temperatures. However, there has been limited investigation into composite systems that incorporate fiber reinforcement. Furthermore, there is a lack of a dependable fracture criterion and evaluation method for assessing the low-temperature fiber-reinforced composites (FRC) fracture. Further study is required in the future for composite materials in this approach (Hohe et al., 2021).

The mechanical characteristics of the composite are mostly influenced by the thermal expansion during temperature changes. The disparity in the expansion coefficient between the fiber and resin results in distinct modes and levels of deformation throughout the cooling phase (Slifka and Smith, 1997). The deformation modes vary across various kinds of microscopes and also alter depending on the orientation of the fiber being seen. This also dictates the intricacy of the micromechanical coordination mechanism during the composite temperature alteration. The deformation characteristics of composite materials, fiber, and resin during the cooling process can be described as follows: as the temperature decreases, the resin undergoes shrinkage, leading to compressive stress on the surface. Conversely, the majority of carbon fibers experience both horizontal and longitudinal expansion as the temperature decreases, causing tensile stress on the surface. As the temperature drops, both basalt fibers and glass fibers experience contraction in both horizontal and longitudinal directions, causing compressive stress on the surface. In contrast, aramid fibers undergo horizontal shrinkage and longitudinal expansion (Sápi and Butler, 2020). Hence, it serves as the fundamental basis for investigating the alteration of FRC and developing the composites fracture criterion in order to gain a comprehensive understanding of the expansion properties of resin and fiber. This understanding will enable the establishment of a correlation mechanism between resin and fiber and the initiation of micro cracks in composites. The Cycom 5320-1/IM7 composite material as an option for LH storage tanks, serving as a material for both storage and transit containers, has been developed by NASA. This material has the ability to entirely prevent the formation of tiny fractures resulting from hydrogen infiltration. Additionally, it may achieve a 30% decrease in weight and a 20% decrease in cost when compared to conventional aluminum alloy metal storage tank (Gomez and Smith, 2019). Composite materials for LH storage tanks are typically constructed using carbon fibers and cured with resin. The primary issue is in the infiltration of hydrogen, which results in the formation of minute cracks, fractures, leaks, and several other significant complications. NASA has created an inner wall barrier membrane that efficiently prevents hydrogen penetration. The study on composite structure in China started around the 1970s. Composite materials have been increasingly used in the load-bearing construction of launch vehicles in recent years. Nevertheless, the utilization of composite materials in LH storage tanks necessitates thorough and methodical investigation. There remain numerous technical advancements to be made in resin materials, hydrogen permeation, low-temperature performance of materials, the molding process, and other related areas (Heydenreich, 1998).

10.6 CONCLUSION

The need for LH transportation and storage containers is expanding due to the anticipated growth of the LH sector. Hence, conducting systematic research on cryogenic materials and their cryogenic characteristics is very important to develop a comprehensive database of relevant materials. Stainless steel is highly resistant to hydrogen brittleness, exhibits good performance at low temperatures, is easily weldable, and has great corrosion resistance. As a result, it is extensively used in the storage and transportation of LH, particularly in ground-based facilities. Nevertheless, there is a need for further enhancement of the mechanical characteristics of materials at low temperatures, while considering unique environmental conditions. For instance, it is necessary to take into account the low-temperature qualities and corrosion resistance of storage tanks, as well as conduct a thorough assessment of the low-temperature mechanical capabilities and weldment stress corrosion properties in a salt spray environment. Aluminum alloys are extensively used in LH storage tanks for space launches, both internationally and domestically. Aluminum alloys have significant benefits over stainless steel in terms of their lightweight nature, exceptional formability, welding capabilities, and commendable resistance to corrosion. Currently, there is a greater emphasis on researching the enhancement of an aluminum alloy welding technique, as well as investigating the mechanical characteristics at low temperatures, stress corrosion, and corrosion resistance. Titanium alloys provide notable advantages in terms of low-temperature performance, lightweight properties, and high strength. However, their benefits in welding and forming capabilities are not prominent, and they are very expensive, resulting in limited use. In the realm of space launch and on-orbit operations, the future direction of growth is in the use of lightweight storage and transportation containers. Among the several options, composite materials show the most promise as an alternative material for LH transportation and storage containers. Nevertheless, the fundamental theory, procedure, and mechanism governing the composite materials' low-temperature performance are not sufficiently developed. This area also presents a research focus that requires further enhancement in the future. In broad terms, the future development and research of materials for LH transportation and storage containers can be categorized into the following areas: (i) creating a database of mechanical properties for conventional low-temperature materials (such as stainless steel, titanium alloy, and aluminum alloy) within the temperature range of LH; (ii) designing new materials with superior performance and affordability for low-temperature applications; and (iii) investigating the fundamental theory and technology behind FRC.

REFERENCES

Boyer, H. E. 1987. *Atlas of Stress-Strain Curves*. ASM, Metals Park, OH.

Desisto, T. S. & Carr, L. 1961. Low temperature mechanical properties of 300 series stainless steel and titanium. Advances in Cryogenic Engineering: Proceedings of the 1960 Cryogenic Engineering Conference University of Colorado and National Bureau of Standards Boulder, August 23–25, 1960, 1961. Springer, Colorado, 577–586.

Duthil, P. 2015. Material properties at low temperature. *arXiv preprint arXiv:1501.07100*.

Entringer, J., Meisnar, M., Reimann, M., Blawert, C., Zheludkevich, M. & Dos Santos, J. F. 2019. The effect of grain boundary precipitates on stress corrosion cracking in a bobbin tool friction stir welded Al-Cu-Li alloy. *Materials Letters: X*, 2, 100014.

Fan, Y., Cui, F., Lu, L. & Zhang, B. 2019a. A nanotwinned austenite stainless steel with high hydrogen embrittlement resistance. *Journal of Alloys and Compounds*, 788, 1066–1075.

Fan, Y., Zhang, B., Wang, J., Han, E.-H. & Ke, W. 2019b. Effect of grain refinement on the hydrogen embrittlement of 304 austenitic stainless steel. *Journal of Materials Science & Technology*, 35, 2213–2219.

Gomez, A. & Smith, H. 2019. Liquid hydrogen fuel tanks for commercial aviation: Structural sizing and stress analysis. *Aerospace Science and Technology*, 95, 105438.

Heydenreich, R. 1998. Cryotanks in future vehicles. *Cryogenics*, 38, 125–130.

Hohe, J., Neubrand, A., Fliegener, S., Beckmann, C., Schober, M., Weiss, K.-P. & Appel, S. 2021. Performance of fiber reinforced materials under cryogenic conditions-a review. *Composites Part A: Applied Science and Manufacturing*, 141, 106226.

Horiuchi, T. & Ooi, T. 1995. Cryogenic properties of composite materials. *Cryogenics*, 35, 677–679.

Huang, S., Ma, D., Sheng, J., Agyenim-Boateng, E., Zhao, J. & Zhou, J. 2020. Effects of laser peening on tensile properties and martensitic transformation of AISI 316L stainless steel in a hydrogen-rich environment. *Materials Science and Engineering: A*, 788, 139543.

Hwang, J.-S., Kim, J.-H., Kim, S.-K. & Lee, J.-M. 2020. Effect of PTFE coating on enhancing hydrogen embrittlement resistance of stainless steel 304 for liquefied hydrogen storage system application. *International Journal of Hydrogen Energy*, 45, 9149–9161.

Jariyaboon, M., Davenport, A., Ambat, R., Connolly, B., Williams, S. & Price, D. 2007. The effect of welding parameters on the corrosion behaviour of friction stir welded AA2024-T351. *Corrosion Science*, 49, 877–909.

Jha, A. K., Murty, S. N., Diwakar, V. & Kumar, K. S. 2003. Metallurgical analysis of cracking in weldment of propellant tank. *Engineering Failure Analysis*, 10, 265–273.

Jha, A. K., Narayanan, P. R., Sreekumar, K. & Sinha, P. 2010. Cracking of Al-4.5 Zn-1.5 Mg aluminium alloy propellant tank-A metallurgical investigation. *Engineering Failure Analysis*, 17, 562–570.

Kovač, A., Paranos, M. & Marciuš, D. 2021. Hydrogen in energy transition: a review. *International Journal of Hydrogen Energy*, 46, 10016–10035.

Kuramoto, S., Okahana, J. & Kanno, M. 2001. Hydrogen assisted intergranular crack propagation during environmental embrittlement in an Al-Zn-Mg-Cu Alloy. *Materials Transactions*, 42, 2140–2143.

Lee, D.-H., Sun, B., Lee, S., Ponge, D., Jägle, E. A. & Raabe, D. 2021. Comparative study of hydrogen embrittlement resistance between additively and conventionally manufactured 304L austenitic stainless steels. *Materials Science and Engineering: A*, 803, 140499.

Lu, J., Zahedi, A., Yang, C., Wang, M. & Peng, B. 2013. Building the hydrogen economy in China: drivers, resources and technologies. *Renewable and Sustainable Energy Reviews*, 23, 543–556.

Lynch, S. 2012. Hydrogen embrittlement phenomena and mechanisms. *Corrosion Reviews*, 30, 105–123.

Maniaci, D. 2008. Relative performance of a liquid hydrogen-fueled commercial transport. 46th AIAA Aerospace Sciences Meeting and Exhibit, 152.

Mansilla, C., Bourasseau, C., Cany, C., Guinot, B., Le Duigou, A. & Lucchese, P. 2018. Hydrogen applications: overview of the key economic issues and perspectives. *Hydrogen Supply Chains*, 2018, 271–292.

Mclellan, R. & Harkins, C. 1975. Hydrogen interactions with metals. *Materials Science and Engineering*, 18, 5–35.

Moshtaghi, M., Safyari, M. & Hojo, T. 2021. Effect of solution treatment temperature on grain boundary composition and environmental hydrogen embrittlement of an Al-Zn-Mg-Cu alloy. *Vacuum*, 184, 109937.

Park, W. S., Yoo, S. W., Kim, M. H. & Lee, J. M. 2010. Strain-rate effects on the mechanical behavior of the AISI 300 series of austenitic stainless steel under cryogenic environments. *Materials & Design*, 31, 3630–3640.

Peel, M., Steuwer, A., Preuss, M. & Withers, P. 2003. Microstructure, mechanical properties and residual stresses as a function of welding speed in aluminium AA5083 friction stir welds. *Acta Materialia*, 51, 4791–4801.

Qiu, Y., Yang, H., Tong, L. & Wang, L. 2021. Research progress of cryogenic materials for storage and transportation of liquid hydrogen. *Metals*, 11, 1101.

Queiroga, L. R., Marcolino, G. F., Santos, M., Rodrigues, G., Dos Santos, C. E. & Brito, P. 2019. Influence of machining parameters on surface roughness and susceptibility to hydrogen embrittlement of austenitic stainless steels. *International Journal of Hydrogen Energy*, 44, 29027–29033.

Ren, X., Dong, L., Xu, D. & Hu, B. 2020. Challenges towards hydrogen economy in China. *International Journal of Hydrogen Energy*, 45, 34326–34345.

Safyari, M., Moshtaghi, M. & Kuramoto, S. 2020. Effect of strain rate on environmental hydrogen embrittlement susceptibility of a severely cold-rolled Al-Cu alloy. *Vacuum*, 172, 109057.

Sápi, Z. & Butler, R. 2020. Properties of cryogenic and low temperature composite materials-a review. *Cryogenics*, 111, 103190.

Schutz, J. 1998. Properties of composite materials for cryogenic applications. *Cryogenics*, 38, 3–12.

Sethi, S. & Ray, B. C. 2013. Mechanical behavior of polymer composites at cryogenic temperatures. In: Kalia, S. & Fu, S. Y. (eds) *Polymers at Cryogenic Temperatures*. Springer, Berlin, Heidelberg, 59–113.

Slifka, A. & Smith, D. 1997. Thermal expansion of an E-glass/vinyl ester composite from 4 to 293 K. *International Journal of Thermophysics*, 18, 1249–1256.

Verstraete, D., Hendrick, P., Pilidis, P. & Ramsden, K. 2010. Hydrogen fuel tanks for subsonic transport aircraft. *International Journal of Hydrogen Energy*, 35, 11085–11098.

Viswanath, V., Asraff, A., Jayesh, P., Thomas, S. M. & Krishnakumar, R. 2019. Structural integrity assessment of a propellant tank in presence of welding residual stresses. *Procedia Structural Integrity*, 14, 442–448.

Wu, S.-D., Cheng, W.-Y., Yang, J. H., Yang, C.-H. & Liao, S.-Y. 2016. The corrosion protection study on inner surface from welding of aluminum alloy 7075-T6 hydrogen storage bottle. *International Journal of Hydrogen Energy*, 41, 570–596.

Xian, L., Yonglun, S., Zhenyang, L. & Yujuan, S. 2015. High frequency energy coupling pulsed TIG welding process on 2219 aluminum alloy. *Transactions of the China Welding Institution*, 36, 17–20.

Xiuqing, X., Junwei, A., Chen, W. & Jing, N. 2021. Study on the hydrogen embrittlement susceptibility of AISI 321 stainless steel. *Engineering Failure Analysis*, 122, 105212.

Zhou, P., Li, W., Li, Y. & Jin, X. 2016. The effect of MoS2 content on the protective performance of Ni-MoS2 composite coatings against hydrogen embrittlement in high strength steel. *Journal of the Electrochemical Society*, 164, D23.

Section III

Hydrogen Storage and Transportation Safety Considerations

11 Prevention of Hydrogen Pipeline Cracking and Leakage

Xinmeng Jiang, Hongfang Lu, Shaohua Dong, Zhao-Dong Xu, and Bohong Wang

11.1 INTRODUCTION

With the advancement of human science and technology and the increasing awareness of the deteriorating ecological environment, the utilization of environmentally friendly energy sources has become a prominent issue. Alongside wind energy, hydrogen energy has garnered significant attention in recent years (Karaca et al., 2020). Hydrogen boasts a high calorific value, with a combustion value three times that of gasoline, 3.9 times that of alcohol, and 4.5 times that of coke. The combustion product of hydrogen is water, which is recognized as the cleanest energy source globally due to its zero-waste production and its potential to supplement electricity (Pivovar et al., 2018; Holechek et al., 2022).

Nevertheless, multiple factors currently impede the development of hydrogen energy. Firstly, there is the issue of hydrogen production, which is often performed through the electrolysis of water. As water resources are unevenly distributed across different regions, further research is needed to address this disparity. Secondly, the transportation of hydrogen presents a challenge due to its flammability and high reactivity with pipeline materials, ultimately impacting pipeline longevity. Of all the obstacles, the most concerning is the potential failure of high-pressure hydrogen pipelines resulting from hydrogen embrittlement. Thus, the study of metal hydrogen embrittlement in high-pressure gas environments is crucial for energy utilization and development (Abdalla et al., 2018).

This chapter aims to provide a comprehensive understanding of the hydrogen embrittlement phenomenon in metallic materials within the context of hydrogen pipeline transportation. By reviewing the research progress and achievements in this field, the chapter aims to elucidate the three major types of metal hydrogen embrittlement phenomena, namely hydrogen-induced cracking, hydrogen blistering, and degradation in the mechanical properties of metals. The primary mechanisms underlying these phenomena will be discussed to enhance knowledge regarding hydrogen embrittlement and enable effective solutions for preventing the cracking and leakage of hydrogen pipelines.

DOI: 10.1201/9781003382553-14

11.2 MECHANISM OF CRACKING OF HYDROGEN PIPELINE

11.2.1 Classification of Metal Hydrogen Embrittlement

Hydrogen embrittlement refers to the phenomenon where the degradation or deterioration of metals occurs after interactions between hydrogen atoms and metals. It is important to note that hydrogen embrittlement only occurs between hydrogen atoms and metals, and not between hydrogen molecules and metals. Moreover, there are no restrictions on the types of metals that undergo the reaction of hydrogen embrittlement. Any metal that interacts with hydrogen atoms and results in property degradation is classified as experiencing hydrogen embrittlement (Barrera et al., 2018; Djukic et al., 2016).

Currently, there is a dearth of research on hydrogen embrittlement of metals both domestically and internationally, and there are numerous classifications of hydrogen embrittlement. Most publications divide hydrogen embrittlement into three categories: "hydrogen-induced cracking", "hydrogen bubbling", and "degradation of metal mechanical properties".

11.2.1.1 Hydrogen-Induced Cracking

11.2.1.1.1 *Definition of Hydrogen-Induced Cracking*

The most prevalent failure mode associated with metallic hydrogen embrittlement is hydrogen-induced cracking, which is considered one of the most hazardous forms of hydrogen embrittlement. Hydrogen-induced cracking occurs due to the initiation and propagation of cracks in the metal's interior or surface caused by various mechanisms of hydrogen effect (Nagao et al., 2018).

11.2.1.1.2 *Steps for Hydrogen-Induced Cracking*

The process of hydrogen-induced cracking can be divided into the following steps (Laureys et al., 2015):

> Step 1: Hydrogen molecules are dissociated and adsorbed within the metal surface to produce hydrogen atoms.
> Step 2: Some of the hydrogen atoms penetrate the metal and become attached to the metal's inner surface.
> Step 3: Hydrogen atoms diffuse within the metal, accumulate in localized areas, and become trapped by hydrogen traps.
> Step 4: Cracks occur inside the metal under applied or residual stress when the pipeline steel's strength is high but its toughness is low.

11.2.1.2 Mechanism of Hydrogen-Induced Cracking

Lan et al. (2021) summarized the latest progress in hydrogen embrittlement in 2021 and reported that countries worldwide have employed various experimental methods to elucidate the macroscopic and microscopic hydrogen embrittlement phenomena. Despite this, the scientific community has not yet widely accepted the mechanism of hydrogen embrittlement.

The hydrogen-enhanced localized plasticity (HELP) mechanism has garnered substantial empirical support, as evidenced by various experimental investigations. The current body of supporting evidence for the HELP mechanism encompasses the following key aspects:

1. Elasticity theory (Lan et al., 2021; Birnbaum, 1990; Birnbaum et al., 1996)
2. Atomistic calculations (Lu et al., 2001)
3. In situ transmission electron microscopy observations of dislocations in thin foils exposed to hydrogen gas (Robertson, 1999)
4. Softening and strain localization in bulk specimens under certain conditions (Birnbaum, 1990, 1994; Birnbaum et al., 1996, 2000; Birnbaum and Sofronis, 1994)
5. Comparisons of slip-line characteristics in hydrogen-charged specimens versus hydrogen-free specimens (Birnbaum, 1990, 1994; Birnbaum et al., 1996)
6. Nano-indentation tests (Barnoush and Behoff, 2006)

In the HELP theory, dislocations in metals play a crucial role. The concept of dislocation was first introduced by Italian mathematician and physicist Vito Volterra in 1905. In materials science, dislocation refers to an internal microscopic defect in a crystalline material, which is a local irregular arrangement of atoms or a crystal defect. From a geometrical standpoint, dislocation is a type of linear defect that can be viewed as the boundary between the slipped and unslipped portions of the crystal. The presence of dislocations has a significant impact on the material's physical properties, particularly its mechanical properties (Sun et al., 2022). Metallic defects such as dislocations typically possess higher hydrogen binding energy than the lattice. Therefore, when hydrogen atoms penetrate the metal, they can be easily captured by metal defects such as dislocations, leading to their aggregation and a reduction in interfacial elastic energy between moving dislocations. Consequently, when subjected to stress, the metal is more susceptible to plastic deformation due to the weakening of its elastic properties, ultimately leading to crack formation (Birnbaum and Sofronis, 1994; Wang et al., 2013).

Hydrogen-enhanced decohesion (HEDE) is another essential mechanism that explains hydrogen-induced cracking. This was first proposed by Pfeil (1926), who suggested that "hydrogen reduces the cohesion of the cubic cleavage plane". Unlike the HELP mechanism, the HEDE mechanism does not emphasize the influence of metal defects such as dislocations on the capture of hydrogen atoms. Instead, it emphasizes that the aggregation of hydrogen atoms within the metal leads to a reduction in bonding strength within the metal and a decrease in interatomic bonding force. Critical cracks are more likely to occur when tensile stress is applied from the outside. At the tip of the region where the critical crack occurs, hydrogen atoms are more likely to accumulate, and their aggregation further promotes the crack's development, ultimately leading to hydrogen cracking and metal failure (Troiano, 1959). There is little supporting evidence for this theory as there is currently no technique to directly observe changes at the atomic scale of crack tips in bulk materials. This is the most direct and valid evidence required to support this theory (Wada et al., 1987). Some scholars propose that adsorption-induced dislocation emission (AIDE) is the mechanism that explains hydrogen-induced cracking.

In practical scenarios, it is important to recognize that the AIDE, HELP, and HEDE mechanisms do not operate independently. Djukic et al. (2019) have proposed that the interaction between these mechanisms offers a more comprehensive and accurate explanation for hydrogen-induced cracking. However, the significance of these three factors in hydrogen-induced cracking primarily depends on the fracture mode of the metal, which is influenced by its material properties. For example, the AIDE and HEDE mechanisms occur sequentially, with AIDE taking place initially until the back-stress in the dislocation increases, leading to the subsequent occurrence of HEDE. As the crack tip moves away from the front and releases the misplaced stress field, AIDE is reactivated (Balitskii and Ivaskevich, 2019; Lynch, 2003).

11.2.1.3 Hydrogen Blistering

11.2.1.3.1 Definition of Hydrogen Blistering

Hydrogen bubbling is a phenomenon in which hydrogen atoms combine to form hydrogen molecules at a localized location within a metal, causing a buildup of local pressure inside the metal and leading to the formation of bubbles on the metal surface (Zhang et al., 2019).

11.2.1.3.2 Steps for Hydrogen Blistering

The process of hydrogen bubbling can be divided into the following steps:

Step 1: Hydrogen molecules are dissociated and adsorbed within the metal surface to produce hydrogen atoms.

Step 2: Some of the hydrogen atoms penetrate the metal and become attached to the metal's inner surface.

Step 3: Hydrogen atoms diffuse within the metal, accumulate in localized areas, and become trapped by hydrogen traps.

Step 4: Hydrogen atoms trapped by irreversible hydrogen traps gather at a particular location and recombine to form gaseous hydrogen molecules, resulting in a local pressure increase.

Step 5: Bubbling occurs near the metal surface.

Step 6: The bubble triggers a crack, and hydrogen continues to accumulate, causing the crack to gradually spread to the metal surface.

Step 7: After the crack reaches the metal surface, the hydrogen gas inside the crack is released, and the crack stops growing and no longer produces bubbles.

11.2.1.3.3 Mechanism of Hydrogen Bubbling

The mechanism underlying hydrogen failure caused by hydrogen bubbles is generally explained by the Hydrogen Internal Pressure (HIP) mechanism. This mechanism suggests that after hydrogen atoms are captured by irreversible hydrogen traps when the concentration of atomic hydrogen exceeds a critical level (the solubility limit of the material), atomic hydrogen recombines to produce gaseous hydrogen within the metal lattice (Liu, 2011). It is important to note that the value of this critical level is determined by the metal material. When the resulting pressure is not balanced with the pressure of the gas inside and outside the pipe to the metal, cracks occur due to the action of internal pressure. These cracks often bend toward the surface due to the strong shear facilitated by hydrogen (Ju and Rigsbee, 1985). As the gaseous hydrogen builds up, the pressure increases and the crack propagates toward the metal surface (Garofalo et al., 1960). The bubbles burst, and hydrogen gas is released, causing the cracks to stop growing. Neutron tomography has shown that some cracks under blisters did not contain hydrogen, while other cracks and deeper cracks unrelated to the blisters did contain hydrogen (Griesche et al., 2014). The initial site of the hydrogen bubble phenomenon is generally the place with the highest concentration of hydrogen (Schastlivtsev et al., 2011), and in most cases, it occurs near the surface of steel. Hydrogen bubbles produce round bubbles that deform metal surfaces (Isakov et al., 2000).

11.2.1.4 Degradation in Mechanical Properties of Metals

11.2.1.4.1 Understanding of Mechanical Degradation of Metals

Hydrogen embrittlement can result in the degradation of the mechanical properties of metals, primarily characterized by a decrease in toughness and plasticity, as well as changes in fracture mode. For instance, in the absence of hydrogen, metal fracture occurs as a dimple plastic fracture, whereas in the presence of hydrogen, it transforms into a brittle quasi-cleavage fracture or intergranular fracture (Chen et al., 2022). This degradation of metal mechanical properties can cause embrittlement of pipes under load conditions.

11.2.1.4.2 Mechanism of Mechanical Degradation of Metals

There are several mechanisms through which hydrogen embrittlement can lead to the degradation of mechanical properties of metals:

Hydrogen atoms can interact with strain-induced dislocations (Cheng, 2022; Zheng et al., 2020). At higher strain rates, X80 pipeline steel did not show significant hydrogen embrittlement after hydrogen charging, while pre-strained steel samples showed a significant decrease in elongation after hydrogen charging. High concentrations of hydrogen atoms can limit the movement of metal atoms and dislocations inside the crystal, thus increasing the micro-elastic modulus and microhardness of the metal, leading to embrittlement (Zhang et al., 2018).

Hydrogen-induced cracking and bubbling phenomena have been observed to result in a reduction in the binding energy of metal matrix atoms, leading to a localized decrease in the microscopic density of the metal within a specific range (Wasim and Djukic, 2020). The experimental study conducted by Wasim in 2019 revealed irreversible hydrogen-induced damage to the material's microstructure, characterized by the presence of a significant number of hydrogenated micropores, microcracks, and blisters. In line with the relationship between the elastic modulus of a metal and the second derivative of its cohesive energy, the experiment included measurements of the metal's elastic modulus, which demonstrated a reduction in the cohesive energy of the matrix. Furthermore, an increase in hydrogen concentration was found to correspond with a decrease in both the bulk elastic modulus and the nano elastic modulus of the material.

11.2.2 The Difference and Relation of Three Kinds of Hydrogen Embrittlement in Metals

It is important to note that hydrogen-induced cracking and hydrogen bubbling can occur at different concentrations of hydrogen atoms, and both can lead to a decrease in the toughness and plasticity of the metal. In addition, the critical value of hydrogen concentration for hydrogen-induced cracking and hydrogen bubbling is dependent on the material being used (Ohaeri et al., 2018).

Although hydrogen-induced cracking and hydrogen bubbling may appear similar in the early stages, hydrogen bubbling often results in visible blisters or surface bulges, while hydrogen-induced cracking only appears as a crack.

The occurrence of hydrogen bubbling is often associated with metals that have strong toughness and lower strength.

Regardless of the mechanism involved, hydrogen embrittlement will lead to the formation of cracks, and the presence of hydrogen may not always be detected in cracks that form beneath blisters.

11.2.3 The Relationship between Hydrogen Embrittlement and Local Stress

The critical concentration of hydrogen atoms required to induce cracking is strongly influenced by the local stress state. Specifically, under conditions of high local stress concentration, the critical hydrogen concentration necessary for crack initiation decreases, while a decrease in stress results in an increase in the critical hydrogen concentration required for crack initiation (Cheng, 2022).

11.3 THE BEHAVIOR OF HYDROGEN IN THE PIPELINE

11.3.1 The Current Mode of Transportation of Hydrogen Gas

At present, there are three main modes of hydrogen transportation: high-pressure hydrogen tank truck transportation, long pipeline trailer transportation, and pipeline transportation (Martin and Sofronis, 2022; Griesche et al., 2014; Demir and Dincer, 2018). Table 11.1 illustrates the transport capacity, cost, and efficiency of these three modes of transportation.

TABLE 11.1

Capacity, Cost, and Efficiency of Three Modes of Hydrogen Transport (Martin and Sofronis, 2022)

Transport Means	Pipeline	Tanker Trucks	Tube Trailers
Physical state	Gas	Liquid	Gas
Capability	100,000 kg/h	4,000 kg/truck	400 kg/truck
Cost ($/kg 100 km)	0.1–1	0.3–0.5	0.5–2
Efficiency (%)	99.2/100 km	99/100 km	94/100 km

The current transportation of hydrogen is mainly accomplished through two commonly used modes, namely high-pressure hydrogen tank truck transportation and long pipeline trailer transportation. However, as shown in Table 11.1, hydrogen pipeline transportation is the most efficient and has the largest capacity.

11.3.2 About Pure Hydrogen Pipeline and Mixed Hydrogen Pipeline

Despite the advantages of large-scale and high efficiency, the use of hydrogen pipeline transportation requires a high initial investment and time cost for pipeline construction. According to estimates, the construction cost of a 6.9 MPa hydrogen pipeline with a diameter of 2–4 in. is between $329 and $590 per meter, which is 10%–20% higher than that of a natural gas pipeline of the same specification (Van der Zwaan et al., 2011). Additionally, due to hydrogen's low density, low molecular weight, and flammability, it is prone to leakage and explosion during transportation. Therefore, the most practical method currently used is to transport hydrogen at medium and low pressure through existing natural gas pipelines.

The utilization of natural gas pipelines for hydrogen transport offers several advantages. First, it can overcome the high costs associated with constructing dedicated hydrogen pipelines. Second, natural gas pipelines possess existing infrastructure, technical feasibility, and a skilled workforce, which facilitates the rapid expansion of hydrogen transmission capacity. Third, this approach can accelerate the comprehensive development of the hydrogen economy. Finally, given that the construction of pure hydrogen pipelines is still in its nascent stages, adding hydrogen to the natural gas pipeline system can enhance public recognition and acceptance of hydrogen, laying the groundwork for future pure hydrogen pipeline transportation. The addition of 5%–15% hydrogen into the existing natural gas pipeline system does not pose risks to domestic applications, public safety, or pipelines. Therefore, mixed gas and hydrogen transport is generally considered to be a feasible hydrogen transport scheme.

11.4 THE PRODUCTION OF HYDROGEN ATOMS

The sources of hydrogen atoms can be classified into two main categories: those produced in a solution environment and those produced in a gaseous environment.

Hydrogen atoms in a gaseous environment are the main source of hydrogen inside high-pressure hydrogen pipelines. These hydrogen atoms can come from various sources, including the production and transportation process of hydrogen, as well as leaks and purges during storage and transportation (Sun and Cheng, 2021; Benziger, 1980). During hydrogen production, hydrogen atoms can be generated through various methods such as steam methane reforming, water electrolysis, and coal gasification. During transportation, hydrogen atoms can be produced due to the reaction of hydrogen with air, moisture, or impurities in the pipeline or storage tanks (Burke and Madix, 1990; Baro and Erley, 1981). Therefore, it is crucial to implement proper measures to prevent or minimize the generation and accumulation of hydrogen atoms during the production, storage, and transportation processes of hydrogen.

In a high-pressure hydrogen pipeline, hydrogen atoms are primarily produced in a gaseous environment through two processes. The first process is the dissociative adsorption of hydrogen atoms. This involves a hydrogen molecule dissociating into two hydrogen atoms on the surface of the pipeline and being adsorbed onto the metal surface. The second process is the spontaneous decomposition of hydrogen molecules inside the pipeline, where one hydrogen molecule is converted into two hydrogen atoms, which are then adsorbed onto the metal surface (Sun and Cheng, 2021).

11.4.1　Hydrogen Atoms Adsorb on the Outer Surface of the Metal

It has been observed that hydrogen chemisorption can occur on various metal surfaces, and there are significant differences in the adsorption energy of hydrogen atoms on different crystal planes of pipeline steel (Benziger, 1980; Burke and Madix, 1990). High-resolution electron energy loss spectroscopy, thermal atom scattering, and angular analytical thermal desorption spectroscopy are common techniques used to investigate hydrogen dissociation adsorption and generate adsorption configuration information (Baro and Erley, 1981). It has been found that, excluding metal defects, low-index crystal surfaces have relatively low surface energy and hydrogen adsorption energy, making them preferable for hydrogen adsorption (Cheng, 2022). However, for the same low-index crystal surface, different stacking sites are assigned different adsorption priorities (Moritz et al., 1985). For example, in the Fe(100) crystal plane, Hartree-Fock calculation shows that its four-fold site (4-Fold, 4F) is the most stable configuration, and hydrogen atoms will preferentially adsorb at the 4F site (Walch, 1984). For the Fe(110) crystal surface, hydrogen preferentially adsorbs on quasi-three-fold sites (QT) (Moritz et al., 1985). It has been observed that different metals have different preferential adsorption sites (Zhai et al., 2021). Recent studies have introduced the Fe-H dynamic correction term and applied stress. Results showed that stress can make the adsorption energy of hydrogen positive, which is not conducive to the formation of a stable adsorption configuration. However, stress fluctuations of the pipeline have no significant influence on the adsorption energy (Sun and Cheng, 2021). The results show that the Gibbs free energy is negative under typical pipeline transport conditions, indicating that it is thermodynamically feasible to dissociate and adsorb hydrogen molecules on the pipeline steel surface. Additionally, the change in Gibbs free energy of H_2 molecules adsorbed and dissociated at specific sites of pipeline steel is negative, indicating that H_2 molecules can be adsorbed on the outer surface of metals through adsorption and dissociation under conditions such as temperature, pressure, and hydrogen-mixing ratio during pipeline operation (Sun and Cheng, 2021).

As there are currently few pure hydrogen transport pipelines, hydrogen is often mixed with natural gas for transportation. The gas in the pipeline can also be accompanied by impurities such as oxygen and carbon monoxide, making it important to study their influence on the dissociation and adsorption of hydrogen molecules in the pipeline. Staykov et al. (2014) used the DFT method to find that oxygen has greater electronegativity than hydrogen, leading to a charge deflection process where oxygen molecules are more attractive on the surface of Fe and more likely to undergo dissociation adsorption. However, when oxygen atoms are attached to the surface of Fe, their ability to provide electrical charge is impaired, inhibiting their catalytic effect on hydrogen dissociation adsorption. Both oxygen and carbon monoxide molecules can inhibit the dissociation and adsorption of hydrogen on the surface of Fe, with oxygen having a stronger inhibitory effect. There is no competitive adsorption relationship between the adsorption of natural gas and hydrogen on the surface of Fe.

11.4.2 HYDROGEN ATOMS PASS FROM THE OUTER SURFACE TO THE INNER SURFACE OF THE METAL

The diffusion process of hydrogen atoms in metals is often studied using the Devanathan-Stachurski cell model (Devanathan and Stachurski, 1964). When hydrogen is adsorbed on the outer surface of the metal, three types of chemical reactions may occur simultaneously, including chemical recombination, electrochemical recombination, and absorption into the metal.

$$H_{ads} + H_{ads} \rightarrow H_{2(gas)} \text{ Tafel reaction} \tag{11.1}$$

$$H_{ads} + H_2O + e^- \rightarrow H_{2(gas)} + OH^- \text{ Heyrovsky reaction} \tag{11.2}$$

$$H_{ads} \rightarrow H_{ads} \text{ Absorption reaction} \tag{11.3}$$

The competition between these reactions is influenced by various factors such as pH, overpotential, applied current, and the surface state, and the dynamics of the process are dependent on these factors. Among the three reactions, reaction (11.3) is the most important despite being the least favorable as it is responsible for providing hydrogen atoms into the metal sample. These hydrogen atoms serve as the source of diffusion within the metal and ultimately lead to hydrogen embrittlement. The amount of hydrogen that enters the metal is not solely determined by the chemical reactions mentioned above, but also by the balance between the amount of hydrogen absorbed on the surface of the metal and the amount of hydrogen that enters the surface of the metal, reaching equilibrium between the sub-surface layer and the bulk of the metal (Vecchi et al., 2018).

11.4.3 ABSORPTION OF HYDROGEN ATOMS ON THE INNER SURFACE OF A METAL

The hydrogen atoms that enter the inner surface of pipeline steel typically occupy interstitial sites within the crystal lattice, and their presence is largely independent of metal defects. The tetrahedral site has a lower absorption energy barrier for hydrogen atoms compared to the octahedral site, making it a preferred location for hydrogen accumulation and stability (Sun and Cheng, 2021). In metals, including pipeline steel, hydrogen atoms possess a certain solubility within each tetrahedral site. Therefore, hydrogen embrittlement does not occur significantly until the concentration of hydrogen atoms in the tetrahedral sites reaches a certain solubility limit. To prevent hydrogen embrittlement, it is important to ensure that hydrogen atoms are uniformly distributed within the tetrahedral sites of the crystal lattice, maintaining low hydrogen atom concentration throughout the metal (Chen et al., 2022).

11.4.4 Diffusion of Hydrogen Atoms in Pipeline Steel

Once the metal is penetrated by hydrogen atoms, they start to diffuse throughout the metal lattice. Due to the small size of the hydrogen atom compared to the Fe atom and its low binding energy with the tetrahedral gap, the hydrogen atom can easily diffuse through the lattice. The rate of hydrogen diffusion is dependent on the rate at which hydrogen atoms jump from one interstitial lattice to another. However, the presence of traps in bulk metals strongly influences the diffusivity of hydrogen. The hydrogen residence time in the trap is determined by the binding energy. During the diffusion process, some hydrogen atoms move toward the opposite side of the metal due to the concentration gradient and leave the metal to reach the outer surface, completing the process of hydrogen penetration (Vecchi et al., 2018).

11.4.5 Capture of Hydrogen Atoms in Pipeline Steel

Pipeline steel and other metals often contain various types of defects, such as dislocations, inclusions, and phase boundaries, which can act as traps for hydrogen atoms due to their higher binding energy compared to the lattice sites. These defects are commonly referred to as "hydrogen traps". Two types of hydrogen traps are typically dominant within the spectrum of trap energies: reversible and irreversible traps (Vecchi et al., 2018) (Table 11.2). Reversible hydrogen traps have low hydrogen binding energy, and the trapped hydrogen atoms can escape the trap and enter the lattice at room temperature. Examples of reversible traps include dislocations and small-angle grain boundaries. Irreversible hydrogen traps, on the other hand, have high hydrogen binding energy, and hydrogen atoms cannot escape the trap at room temperature. Examples of irreversible traps include non-metallic inclusions and phase boundaries (Chen et al., 2022). These traps can significantly slow down the hydrogen diffusion process.

Localized accumulation of hydrogen atoms in irreversible hydrogen traps can increase the hydrogen solubility of pipeline steel, restrict the diffusion of hydrogen atoms to high-stress areas, and decrease the susceptibility of steel to hydrogen embrittlement. However, if the concentration of hydrogen atoms in irreversible hydrogen traps reaches a critical level, hydrogen-induced cracking may also occur.

TABLE 11.2

Main Hydrogen Traps Contained in Steel and Their Hydrogen Binding Energies (Chen et al., 2022)

Hydrogen Trap	Hydrogen Binding Energy	The Type of Hydrogen Trap
Dislocation	0.25–0.31	Reversible type
Interstitial solute atoms	0.13–0.27	
Grain boundary (small angle)	0.27	
Grain boundary (large angle)	0.55–0.61	Irreversible type
Secondary particle	0.8–0.98	
Alumina inclusion	0.82	
Manganese sulfide inclusion	0.75	

Hydrogen atoms that diffuse freely in the lattice and can escape the reversible hydrogen trap can move toward high-stress areas and cause cracks in the pipeline steel (Cheng, 2022).

11.5 THE EFFECT OF HYDROGEN ON PIPE MATERIAL

11.5.1 The Current Mode of Transportation of Hydrogen

Based on survey data presented in Figure 11.1, it has been observed that steel holds the dominant position as the primary material employed in diverse pipeline applications. In the United States, steel accounts for approximately 99.5% of transmission lines and 89.2% of gathering lines (Zhang and Tian, 2022). This trend is also

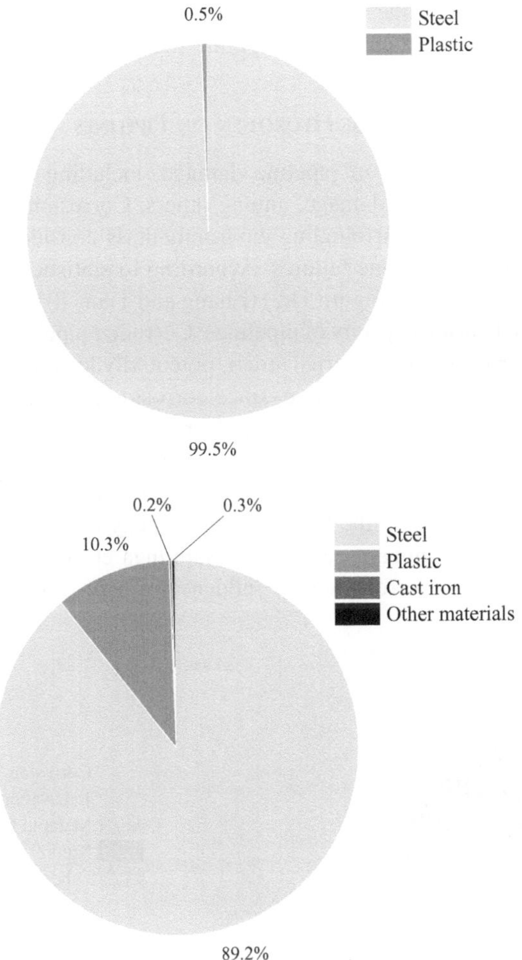

FIGURE 11.1 Proportions of pipe materials in gathering and transmission lines in the United States (Zhang and Tian, 2022).

prevalent in other nations, establishing a strong link between pipeline protection and material research, with particular emphasis on steel. Consequently, the investigation of the detrimental effects of hydrogen on pipeline steel and ensuring the structural integrity and operational dependability of pipelines in hydrogen environments bear immense significance in advancing the renewable energy sector.

As of 2017, Europe had approximately 1,598 km of hydrogen pipelines with a hydrogen transport pressure of 2–10 MPa. The majority of these pipelines utilize seamless steel pipes with diameters ranging from 0.3 to 1.0 m, primarily composed of low-strength pipeline steel, such as X42, X52, and X56. In the United States, the total length of hydrogen pipelines is around 2,575 km, with the majority of pipelines buried, and the hydrogen transport pressure generally being less than 7 MPa. Pipeline materials primarily consist of pipeline steel in the range of X52 to X80, with an expected service life of 15–30 years (Liu et al., 2020). As a result, our research focuses primarily on pipeline steel materials, and we must also investigate the differences in hydrogen's impact on various pipeline steel material specifications.

11.5.2 Corrosion Effect of Hydrogen on Pipelines

There are numerous causes of pipeline damage, including corrosion, excavation damage, and external force damage, among others. Corrosion, which is a material loss defect caused by the surrounding environment, is considered one of the most significant reasons for pipeline failures. According to statistics, corrosion accounts for 23% of pipeline failures (Figure 11.2) (Zhang and Tian, 2022). Corrosion can partially weaken the bearing capacity of pipelines. Corroded pipelines are more prone to crack or leak under the same external stress, potentially leading to serious accidents such as explosion or fire. Furthermore, since corrosion is an inherent feature of the entire material life cycle, there is practically no way to avoid it, highlighting the importance of studying the effect of hydrogen on pipeline corrosion.

Hydrogen-induced damage induces a substantial decline in the ductility of pipeline steel, thereby impacting both the load-bearing capacity and the damage behavior of corroded pipelines (Nanninga et al., 2012). Nanninga et al. (2012) conducted comparative experiments to corroborate the influence of hydrogen on pipeline ductility.

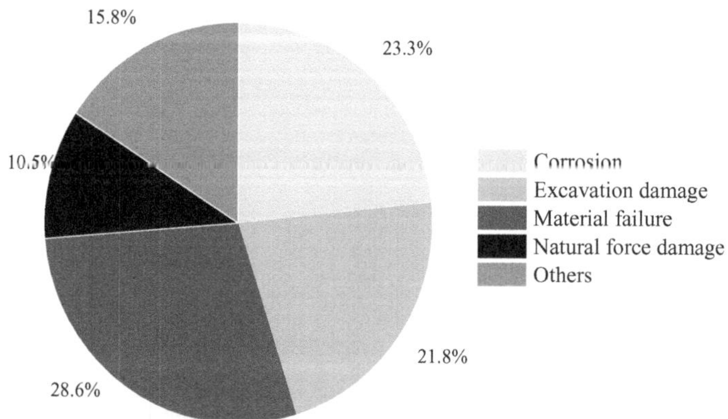

FIGURE 11.2   The causes of pipeline damage in the United States (Zhang and Tian, 2022).

They assessed the tensile properties of X52, X65, and X100 pipeline steels under high-pressure (13.8 MPa) hydrogen gas, with stretching experiments under helium gas serving as the control group. The findings reveal that, in comparison to the control group, the hydrogen environment significantly reduces elongation and cross-sectional area reduction of the failed pipeline steel. Furthermore, they investigated the effects of hydrogen on metal failure at varying pressures and determined that hydrogen embrittlement tends to occur more prominently as hydrogen pressure gradually increases. Additionally, the phenomenon of hydrogen embrittlement becomes more prevalent in alloys with higher strength (Moro et al., 2010). Moro et al. (2010) conducted tensile tests on X80 pipeline steel under different pressures and strain rates, deducing that the primary cause of hydrogen embrittlement lies in the presence of diffused hydrogen near the material's surface, as observed microscopically. Briotet et al. (2012) systematically performed slow strain rate tensile tests, fracture toughness tests, wafer tests, fatigue crack growth tests, and WOL tests on X80 pipeline steel within high-pressure hydrogen environments. The results indicate that the elastic modulus, yield strength, and tensile strength of the materials exhibit negligible changes in hydrogen environments. However, plasticity and fracture toughness of the materials decrease significantly, while the rate of fatigue crack growth increases substantially. Hardie et al. (2006) examined the hydrogen embrittlement sensitivity of X60, X80, and X100 pipeline steels using electrochemical hydrogen charging. The findings demonstrate that as the hydrogen charging current density reaches a certain threshold, the susceptibility of materials to hydrogen embrittlement escalates significantly with increasing material strength. Consequently, when utilizing cathodic electric protection for buried pipelines, careful attention should be given to the current density.

11.6 METHODS FOR CRACKING AND LEAKAGE PREVENTION OF HYDROGEN PIPELINES (CAZENAVE ET AL., 2021)

11.6.1 CONTROL OF IMPURITIES

In addition to oxygen molecules, the presence of impurities in the pipeline environment can have a significant impact on hydrogen embrittlement. Carbon monoxide (CO) and hydrogen sulfide (H_2S) are two common impurities that can effectively inhibit hydrogen atom infiltration and reduce hydrogen embrittlement. These impurities compete with hydrogen for adsorption sites on the material's surface, limiting the availability of sites for hydrogen absorption. By occupying these sites, CO and H_2S prevent hydrogen atoms from penetrating the material and mitigating the embrittlement process. Therefore, it is crucial to monitor and control the levels of impurities in the pipeline to ensure their concentration remains within acceptable limits (Paul and Snyder, 2012).

11.6.2 ALLOY SELECTION

The choice of alloy is an important factor in preventing hydrogen embrittlement in pipeline steel. It is recommended to select lower strength alloys when designing pipelines that transport hydrogen. Higher strength alloys tend to exhibit increased sensitivity to hydrogen, making them more prone to embrittlement under the same pressure and environmental conditions. Lower strength alloys have a reduced suscep-tibility to hydrogen embrittlement, as they possess a more favorable microstructure and dislocation density, which can effectively resist the detrimental effects of hydro-gen (Hardie et al., 2006).

11.6.3 Non-steel Pipes

While steel is commonly used for hydrogen pipelines due to its strength and dura-bility, alternative materials like polyethylene pipes have gained attention for their potential in mitigating hydrogen embrittlement. Polyethylene pipes have been found to exhibit limited susceptibility to hydrogen-induced embrittlement. The unique molecular structure and chemical properties of polyethylene make it less prone to hydrogen absorption and subsequent embrittlement compared to steel. Therefore, considering non-steel pipe options, such as polyethylene, can provide a viable solu-tion for preventing hydrogen-related issues in pipeline systems (Byrne et al., 2023).

11.6.4 Heat Treatment Techniques

Heat treatment techniques play a crucial role in influencing the sensitivity of pipe-line materials to hydrogen embrittlement. Annealing, normalizing, quenching, and tempering are common heat treatment methods employed to modify the microstruc-ture and mechanical properties of steel. By carefully adjusting the temperature and duration of these heat treatment processes, it is possible to reduce the susceptibility of pipeline steel to hydrogen embrittlement. Techniques like subtemperature quench-ing and secondary quenching have shown promise in increasing the critical stress required for hydrogen embrittlement fracture, thereby enhancing the resistance of the material to hydrogen-induced cracking and leakage (Torres-Islas et al., 2005).

11.6.5 Alloy Modification

The addition of specific elements during the smelting process of pipeline steel can significantly influence its resistance to hydrogen embrittlement. For instance, the addition of vanadium or copper has been found to improve the dispersion of hydrogen atoms within the tetrahedral space of the material. This improved dispersion helps maintain the concentration of hydrogen atoms at lower levels, minimizing the likeli-hood of local hydrogen embrittlement. The presence of these alloying elements alters the material's microstructure, resulting in enhanced resistance to hydrogen-induced cracking and leakage (Li et al., 2018).

11.6.6 Surface Treatment

Surface treatment techniques offer an additional layer of protection against hydrogen embrittlement. Pellet bombardment, also known as shot peening, is a common sur-face treatment method that involves bombarding the steel surface with small pellets. This process creates a compressive stress layer on the material's surface, which effec-tively inhibits the diffusion of hydrogen into the steel. By introducing compressive stresses, the pellet bombardment technique reduces the effective stress acting on the material and lowers the hydrogen diffusion coefficient, thus minimizing the likelihood of hydrogen-induced cracking and leakage (Brass et al., 1991).

11.6.7 MAINTENANCE PRACTICES

Regular maintenance practices are essential for preventing hydrogen infiltration in metals and maintaining the integrity of hydrogen pipelines. Several measures can be implemented to reduce the amount of hydrogen ingress. Firstly, using sandblasting instead of pickling for surface preparation helps remove surface contaminants without introducing hydrogen. Secondly, the use of corrosion inhibitors can be employed to mitigate hydrogen ingress by forming a protective layer on the surface of the pipeline, reducing the chance of hydrogen absorption. Additionally, avoiding the use of electrochemical oil removal methods, which can introduce hydrogen into the pipeline material, is recommended (Li et al., 2020).

11.6.8 COATINGS WITH LOW HYDROGEN DIFFUSIVITY AND SOLUBILITY

The application of coatings with low hydrogen diffusivity and solubility can effectively prevent hydrogen infiltration in steel pipelines. Certain coatings, such as copper (Cu), molybdenum (Mo), aluminum (Al), silver (Ag), gold (Au), and tungsten (W), have demonstrated lower hydrogen solubility and diffusivity compared to bare steel. These coatings act as barriers, limiting the interaction between hydrogen and the pipeline material, thereby reducing the risk of hydrogen embrittlement. By selecting coatings with improved resistance to hydrogen, the overall integrity and safety of the hydrogen pipeline can be enhanced (Samanta et al., 2020).

11.6.9 STRESS TREATMENT AND DEHYDROGENATION

Stress treatment and dehydrogenation processes are vital steps in mitigating hydrogen embrittlement, particularly in electroplating applications. Prior to electroplating, stress treatment is carried out to relieve any residual stresses in the material. This treatment helps minimize stress concentrations that can accelerate hydrogen-induced cracking. After electroplating, dehydrogenation treatment is performed to eliminate any hydrogen absorbed during the plating process. Dehydrogenation is typically conducted in controlled ovens, with temperature and duration tailored to the specific requirements of the parts, including their size, material strength, coating performance, and plating time. By implementing stress treatment and dehydrogenation processes, the risk of hydrogen embrittlement in electroplated components can be effectively managed (Dwivedi and Vishwakarma, 2018).

By incorporating these preventive measures and techniques, the chapter aims to provide a comprehensive understanding of the strategies available for preventing hydrogen pipeline cracking and leakage. Through the utilization of appropriate alloy selection, heat treatment methods, surface treatments, coatings, and maintenance practices, the integrity and safety of hydrogen pipelines can be significantly enhanced, ensuring the reliable and sustainable transportation of hydrogen (Djukic et al., 2016).

11.7 CONCLUSION

To enable the utilization of new energy and promote environmental protection, the key challenge is to advance the development of hydrogen pipeline transportation. However, the main limitation to this technology is hydrogen embrittlement. It is crucial to fully comprehend the behavior of hydrogen in pipelines and its effects on pipeline materials. Furthermore, the absence of a uniform standard for the reconstruction of hydrogen and natural gas pipelines, along with a comprehensive evaluation system for pipelines, is also a critical development direction for hydrogen pipeline transportation. Therefore, the development of hydrogen pipelines requires multidisciplinary and interdisciplinary collaboration. Scholars worldwide should focus on this development trend and conduct more extensive research.

ACKNOWLEDGMENT

This work is funded by the Natural Science Foundation of Jiangsu Province (Grant No. BK20220848).

REFERENCES

Abdalla, A. M., Hossain, S., Nisfindy, O. B., Azad, A. T., Dawood, M., & Azad, A. K., 2018. Hydrogen production, storage, transportation and key challenges with applications: a review. *Energy Conversion and Management*, 165, 602–627.

Balitskii, A. & Ivaskevich, L., 2019. Hydrogen effect on cumulation of failure, mechanical properties, and fracture toughness of Ni-Cr alloys. *Advances in Materials Science and Engineering*, 2019, 1–8.

Barnoush, A. & Vehoff, H., 2006. Electrochemical nanoindentation: a new approach to probe hydrogen/deformation interaction. *Scripta Materialia*, 55(2), 195–198.

Baro, A. M. & Erley, W., 1981. The chemisorption of hydrogen on a (110) iron crystal studied by vibrational spectroscopy (EELS). *Surface Science Letters*, 112(1–3), L759–L764.

Barrera, O., Bombac, D., Chen, Y., Daff, T. D., Galindo-Nava, E., Gong, P., & Sweeney, F., 2018. Understanding and mitigating hydrogen embrittlement of steels: a review of experimental, modelling and design progress from atomistic to continuum. *Journal of Materials Science*, 53(9), 6251–6290.

Benziger, J. B., 1980. Thermodynamics of adsorption of diatomic molecules on transition metal surfaces. *Applications of Surface Science*, 6(2), 105–121.

Birnbaum, H. K. & Sofronis, P., 1994. Hydrogen-enhanced localized plasticity-a mechanism for hydrogen-related fracture. *Materials Science and Engineering: A*, 176(1–2), 191–202.

Birnbaum, H. K., 1990. Mechanisms of hydrogen related fracture of metals. *Hydrogen Effects on Materials Behavior*, Technical Report. University of Illinois Materials Research Laboratory, Urbana, IL, 639–658.

Birnbaum, H. K., 1994. Hydrogen effects on deformation--relation between dislocation behavior and the macroscopic stress-strain behavior. *Scripta Metallurgica et Materialia*, 31(2), 149–153.

Birnbaum, H. K., Robertson, I. M., & Sofronis, P., 2000. Hydrogen effects on plasticity. In: Lépinoux, J., Mazière, D., Pontikis, V., Saada, G. (eds) *Multiscale Phenomena in Plasticity: From Experiments to Phenomenology, Modelling and Materials Engineering*, Springer, Dordrecht, 367–381.

Birnbaum, H. K., Robertson, I. M., Sofronis, P., & Teter, D., 1996. Mechanisms of hydrogen related fracture – a review. In *Second International Conference on Corrosion-Deformation Interactions. CDI'96,* Moran, WY, 172–195.

Brass, A. M., Chene, J., Anteri, G., Ovejero-Garcia, J., & Castex, L., 1991. Role of shot-peening on hydrogen embrittlement of a low-carbon steel and a 304 stainless steel. *Journal of Materials Science,* 26, 4517–4526.

Briottet, L., Moro, I., & Lemoine, P., 2012. Quantifying the hydrogen embrittlement of pipeline steels for safety considerations. *International Journal of Hydrogen Energy,* 37(22), 17616–17623.

Burke, M. L. & Madix, R. J., 1990. Hydrogen on Pd (100)-S: the effect of sulfur on precursor mediated adsorption and desorption. *Surface Science,* 237(1–3), 1–19.

Byrne, N., Ghanei, S., Espinosa, S. M., & Neave, M., 2023. Influence of hydrogen on vintage polyethylene pipes: slow crack growth performance and material properties. *International Journal of Energy Research,* 2023(24), 1–8.

Cazenave, P., Jimenez, K., Gao, M., Moneta, A., & Hryciuk, P., 2021. Hydrogen assisted cracking driven by cathodic protection operated at near-1200 mV CSE-an onshore natural gas pipeline failure. *Journal of Pipeline Science and Engineering,* 1(1), 100–121.

Chen, Y. F., Shun, Y. H., & Zheng, Y. D., 2022. Hydrogen pipeline development and pipeline steel hydrogen embrittlement challenge. *Journal of Yangtze University (from Science Edition),* 19(1), 54–69.

Cheng, Y. F., 2022. Essence and gap analysis for hydrogen embrittlement of pipelines in high-pressure hydrogen environments. *Oil & Gas Storage and Transportation,* 42(1), 1–8.

Demir, M. E. & Dincer, I., 2018. Cost assessment and evaluation of various hydrogen delivery scenarios. *International Journal of Hydrogen Energy,* 43(22), 10420–10430.

Devanathan, M. A. V. & Stachurski, Z., 1964. The mechanism of hydrogen evolution on iron in acid solutions by determination of permeation rates. *Journal of the Electrochemical Society,* 111(5), 619.

Djukic, M. B., Bakic, G. M., Zeravcic, V. S., Sedmak, A., & Rajicic, B., 2016. Hydrogen embrittlement of industrial components: prediction, prevention, and models. *Corrosion,* 72(7), 943–961.

Djukic, M. B., Bakic, G. M., Zeravcic, V. S., Sedmak, A., & Rajicic, B., 2019. The synergistic action and interplay of hydrogen embrittlement mechanisms in steels and iron: Localized plasticity and decohesion. *Engineering Fracture Mechanics,* 216, 106528.

Dwivedi, S. K. & Vishwakarma, M., 2018. Hydrogen embrittlement in different materials: a review. *International Journal of Hydrogen Energy,* 43(46), 21603–21616.

Garofalo, F., Chou, Y. T., & Ambegaokar, V., 1960. Effect of hydrogen on stability of micro cracks in iron and steel. *Acta Metallurgica,* 8(8), 504–512.

Griesche, A., Dabah, E., Kannengiesser, T., Kardjilov, N., Hilger, A., & Manke, I., 2014. Three-dimensional imaging of hydrogen blister in iron with neutron tomography. *Acta Materialia,* 78, 14–22.

Griesche, A., Dabah, E., Kannengiesser, T., Kardjilov, N., Hilger, A., & Manke, I., 2014. Three-dimensional imaging of hydrogen blister in iron with neutron tomography. *Acta Materialia,* 78, 14–22.

Hardie, D., Charles, E. A., & Lopez, A. H., 2006. Hydrogen embrittlement of high strength pipeline steels. *Corrosion Science,* 48(12), 4378–4385.

Holechek, J. L., Geli, H. M., Sawalhah, M. N., & Valdez, R., 2022. A global assessment: can renewable energy replace fossil fuels by 2050?. *Sustainability,* 14(8), 4792.

Isakov, M. G., Izotov, V. I., Karpel'ev, V. A., & Filippov, G. A., 2000. Kinetics of the damage formation in a low-carbon low-alloy steel upon hydrogenation. *The Physics of Metals and Metallography,* 90(3), 302–308.

Ju, C. P. & Rigsbee, J. M., 1985. The role of microstructure for hydrogen-induced blistering and stepwise cracking in a plain medium carbon steel. *Materials Science and Engineering,* 74(1), 47–53.

Karaca, A. E., Dincer, I., & Gu, J., 2020. Life cycle assessment study on nuclear based sustainable hydrogen production options. *International Journal of Hydrogen Energy*, 45(41), 22148–22159.

Lan, L. K., Kong, X. W., Qiu, C. L., & Du, L. X., 2021. A review of recent advance on hydrogen embrittlement phenomenon based on multiscale mechanical experiments. *Acta Metall Sin*, 57(7), 845–859.

Laureys, A., Depover, T., Petrov, R., & Verbeken, K., 2015. Characterization of hydrogen induced cracking in TRIP-assisted steels. *International Journal of Hydrogen Energy*, 40(47), 16901–16912.

Li, L., Song, B., Cheng, J., Yang, Z., & Cai, Z., 2018. Effects of vanadium precipitates on hydrogen trapping efficiency and hydrogen induced cracking resistance in X80 pipeline steel. *International Journal of Hydrogen Energy*, 43(36), 17353–17363.

Li, X., Ma, X., Zhang, J., Akiyama, E., Wang, Y., & Song, X., 2020. Review of hydrogen embrittlement in metals: hydrogen diffusion, hydrogen characterization, hydrogen embrittlement mechanism and prevention. *Acta Metallurgica Sinica (English Letters)*, 33, 759–773.

Liu, W. J., 2011. Modeling nucleation of hydrogen induced cracking in steels during sour service. *Materials Science Forum*, 675, 983–986.

Liu, Z., Xiong, S., Zheng, J., Zhang, Y., Hua, Z., Gu, C., 2020. Comparative analysis of hydrogen pipeline and natural gas pipeline. *Pressure Vessel Technology*, 37(2), 56–63.

Lu, G., Zhang, Q., Kioussis, N., & Kaxiras, E., 2001. Hydrogen-enhanced local plasticity in aluminum: an ab initio study. *Physical Review Letters*, 87(9), 095501.

Lynch, S. P., 2003. Mechanisms of hydrogen assisted cracking – a review. In: Neville R. Moody, Anthony W. Thompson, Richard E. Ricker, Gray S. Was and Russell H. Jones (eds) *Hydrogen effects on material behaviour and corrosion deformation interactions*, TMS, Moran, WY, 449–466.

Martin, M. L. & Sofronis, P., 2022. Hydrogen-induced cracking and blistering in steels: a review. *Journal of Natural Gas Science and Engineering*, 101, 104547.

Moritz, W., Imbihl, R., Behm, R. J., Ertl, G., & Matsushima, T., 1985. Adsorption geometry of hydrogen on Fe (110). *The Journal of Chemical Physics*, 83(4), 1959–1968.

Moro, I., Briottet, L., Lemoine, P., Andrieu, E., Blanc, C., & Odemer, G., 2010. Hydrogen embrittlement susceptibility of a high strength steel X80. *Materials Science and Engineering: A*, 527(27–28), 7252–7260.

Nagao, A., Dadfarnia, M., Somerday, B. P., Sofronis, P., & Ritchie, R. O. 2018. Hydrogen-enhanced-plasticity mediated decohesion for hydrogen-induced intergranular and "quasi-cleavage" fracture of lath martensitic steels. *Journal of the Mechanics and Physics of Solids*, 112, 403–430.

Nanninga, N. E., Levy, Y. S., Drexler, E. S., Condon, R. T., Stevenson, A. E., & Slifka, A. J., 2012. Comparison of hydrogen embrittlement in three pipeline steels in high pressure gaseous hydrogen environments. *Corrosion Science*, 59, 1–9.

Ohaeri, E., Eduok, U., & Szpunar, J., 2018. Hydrogen related degradation in pipeline steel: a review. *International Journal of Hydrogen Energy*, 43(31), 14584–14617.

Paul, B. D. & Snyder, S. H. 2012. H2S signalling through protein sulfhydration and beyond. *Nature Reviews Molecular Cell Biology*, 13(8), 499–507.

Pfeil, L. B., 1926. The effect of occluded hydrogen on the tensile strength of iron. *Proceedings of the Royal Society of London. Series A, Containing Papers of a Mathematical and Physical Character*, 112(760), 182–195.

Pivovar, B., Rustagi, N., & Satyapal, S. 2018. Hydrogen at scale (H2@ Scale): key to a clean, economic, and sustainable energy system. *The Electrochemical Society Interface*, 27(1), 47.

Robertson, I. M., 1999. The effect of hydrogen on dislocation dynamics. *Engineering Fracture Mechanics*, 64(5), 649–673.

Samanta, S., Singh, C., Banerjee, A., Mondal, K., Dutta, M., & Singh, S. B., 2020. Development of amorphous Ni-P coating over API X70 steel for hydrogen barrier application. *Surface and Coatings Technology*, 403, 126356.

Schastlivtsev, V. M., Tabatchnikova, T. I., Tereshchenko, N. A., & Yakovleva, I. L., 2011. Degradation of the pipe-steel structure upon long-term operation in contact with a hydrogen sulfide-containing medium. *The Physics of Metals and Metallography*, 111, 281–293.

Staykov, A., Yamabe, J., & Somerday, B. P., 2014. Effect of hydrogen gas impurities on the hydrogen dissociation on iron surface. *International Journal of Quantum Chemistry*, 114(10), 626–635.

Sun, X., Wu, D., Zou, L., House, S. D., Chen, X., Li, M., Zakharov, D. N., Yang, J. C., & Zhou, G., 2022. Dislocation-induced stop-and-go kinetics of interfacial transformations. *Nature*, 607(7920), 708–713.

Sun, Y. & Cheng, Y. F., 2021. Thermodynamics of spontaneous dissociation and dissociative adsorption of hydrogen molecules and hydrogen atom adsorption and absorption on steel under pipelining conditions. *International Journal of Hydrogen Energy*, 46(69), 34469–34486.

Torres-Islas, A., Salinas-Bravo, V. M., Albarran, J. L., & Gonzalez-Rodriguez, J. G., 2005. Effect of hydrogen on the mechanical properties of X-70 pipeline steel in diluted NaHCO3 solutions at different heat treatments. *International Journal of Hydrogen Energy*, 30(12), 1317–1322.

Troiano, A. R., 1959. Delayed failure of high strength steels. *Corrosion*, 15(4), 57–62.

Van der Zwaan, B. C. C., Schoots, K., Rivera-Tinoco, R., & Verbong, G. P. J., 2011. The cost of pipelining climate change mitigation: an overview of the economics of CH4, CO2 and H2 transportation. *Applied Energy*, 88(11), 3821–3831.

Vecchi, L., Simillion, H., Montoya, R., Van Laethem, D., Van den Eeckhout, E., Verbeken, K., Terryn, H., Deconinck, J., & Van Ingelgem, Y., 2018. Modelling of hydrogen permeation experiments in iron alloys: characterization of the accessible parameters-Part I-The entry side. *Electrochimica Acta*, 262, 57–65.

Wada, M., Akaiwa, N., & Mori, T., 1987. Field evaporation of iron in neon and in hydrogen and its rate-controlling processes. *Philosophical Magazine A*, 55(3), 389–403.

Walch, S. P., 1984. Model studies of the interaction of H atoms with BCC iron. *Surface Science*, 143(1), 188–203.

Wang, S., Hashimoto, N., Wang, Y., & Ohnuki, S., 2013. Activation volume and density of mobile dislocations in hydrogen-charged iron. *Acta Materialia*, 61(13), 4734–4742.

Wasim, M. & Djukic, M. B., 2020. Hydrogen embrittlement of low carbon structural steel at macro-, micro-and nano-levels. *International Journal of Hydrogen Energy*, 45(3), 2145–2156.

Zhai, F., Tian, Y., Song, D., Li, Y., Liu, X., Li, T., Zhang, Z., & Shen, X., 2021. A thermo-dynamics study of hydrogen interaction with (110) transition metal surfaces. *Applied Surface Science*, 545, 148961.

Zhang, H. & Tian, Z., 2022. Failure analysis of corroded high-strength pipeline subject to hydrogen damage based on FEM and GA-BP neural network. *International Journal of Hydrogen Energy*, 47(7), 4741–4758.

Zhang, S., Zhao, Q., Liu, J., Huang, F., Huang, Y., & Li, X., 2019. Understanding the effect of niobium on hydrogen-induced blistering in pipeline steel: a combined experimental and theoretical study. *Corrosion Science*, 159, 108142.

Zhang, T., Zhao, W., Li, T., Zhao, Y., Deng, Q., Wang, Y., & Jiang, W., 2018. Comparison of hydrogen embrittlement susceptibility of three cathodic protected subsea pipeline steels from a point of view of hydrogen permeation. *Corrosion Science*, 131, 104–115.

Zheng, Y., Zhang, L., Shi, Q., Zhou, C., & Zheng, J., 2020. Effects of hydrogen on the mechanical response of X80 pipeline steel subject to high strain rate tensile tests. *Fatigue & Fracture of Engineering Materials & Structures*, 43(4), 684–697.

12 Delayed Hydrogen Ignition and Explosion

*Yihuan Wang, Ruiling Li, Ailin Xia, Siming Liu,
Zhenwei Zhang, and Guojin Qin*

12.1 INTRODUCTION

Hydrogen is a quite promising clean energy and has a wide range of applications (Wang et al., 2021; Cheng et al., 2022; Yang et al., 2021). Compared with traditional energy sources, hydrogen energy has the characteristics of high caloric, pollution-free, good thermal conductivity, and combustion performance (Zhang et al., 2017). The combustion calorific value of hydrogen is much higher than that of other fuels, and the calorific value of conventional fuels is shown in Figure 12.1. The development of hydrogen energy is conducive to solving the issues of fossil energy depletion and the greenhouse effect (Zhuo et al., 2023). The storage and transportation of hydrogen are dangerous due to its flammability and low ignition energy. Especially, the spontaneous combustion of leaked high-pressure hydrogen gas can cause fire and explosion disasters (Li et al., 2023). Hydrogen safety has become increasingly essential for integrity management (Chang et al., 2022). Recently, explosions and related damage accidents have demonstrated the importance of hydrogen safety (Hu et al., 2023). Hydrogen has dangerous characteristics such as flammability, explosiveness, and diffusibility. Once hydrogen leakage occurs, an explosion accident can happen. A hydrogen explosion requires a mixture of hydrogen and air (or oxygen) to reach the explosion limit in the presence of an ignition source. Combustible gases and ignition sources can easily lead to explosions and fires, which may result in incalculable life and economic losses. Therefore, the research on the explosion prevention and suppression of hydrogen is increasingly receiving social attention. Some measures have been taken to continuously reduce the risk of explosion (Li et al., 2019).

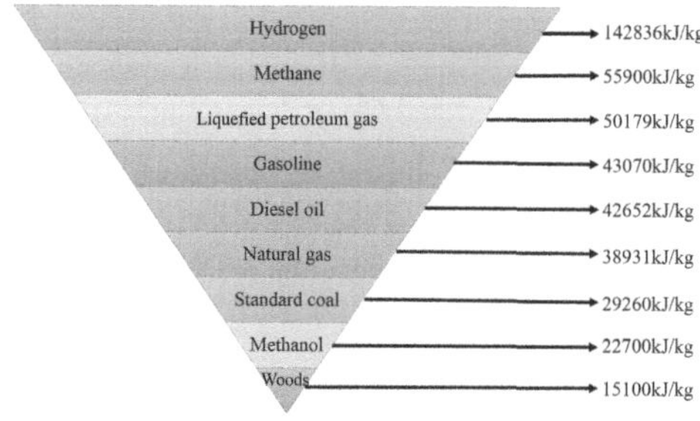

FIGURE 12.1 Calorific values of conventional fuels (Yuan et al., 2023).

DOI: 10.1201/9781003382553-15

The security of hydrogen production hinders the development of the hydrogen energy industry. From the perspective of manufacturing sources, the possibility of hydrogen leakage and accumulation can be reduced. Methods of suppressing hydrogen explosions and reliable hydrogen explosion suppressors also need to be investigated (Rasul et al., 2022). This chapter introduces the principles of delayed hydrogen ignition and explosion and summarizes the challenges and solutions for suppressing hydrogen ignition and explosion. Finally, the shortcomings of current research are presented.

12.2 PRINCIPLES

12.2.1 HYDROGEN IGNITION PRINCIPLE

Gas leaks can release a large mixture of combustible gases including hydrogen gas. The gas mixture can burn violently if ignited. Igniting a gas mixture can cause intense combustion. The ignited gas can form a spherical or hemispherical fireball in the air. The leaked gas mixture containing hydrogen gas mixes with air to form a gas cloud. Specifically, a small number of gas clouds or ignition sources with low energy can lead to flash fires.

Figure 12.2 is a schematic diagram of delayed hydrogen ignition. A hydrogen leakage accident occurs, which may cause ignition after a period of leakage. Generally, the scale of the leak, environment, and obstacles can cause the hydrogen mixture without being ignited immediately. The gas cloud is ignited in a relatively short leakage time and the affected spatial range is relatively small. On the contrary, a long leakage time leads to an extension of the ignition time. The diffusion degree of combustible gases is influenced by the spatial range, in which leakage disasters have serious consequences on a large scale (Kai et al., 2017).

The self-ignition mechanism of hydrogen fuel can be attributed to the categories of premixed flame ignition theory and diffusion flame ignition theory. The process is divided into five stages:

a. Periodic oscillatory flow field formation for specific high-temperature and high-pressure condition processes.
b. The process of mixing fuel and air to achieve the appropriate equivalence ratio.
c. Ignition occurs in a localized process with the right ratio of equivalents, where the heat absorption is greater than the heat release.

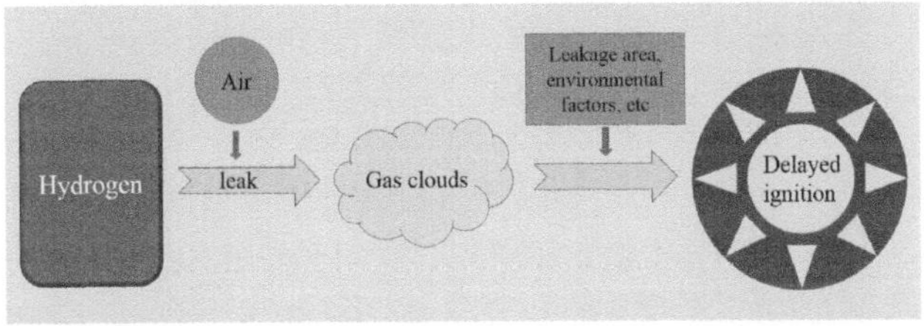

FIGURE 12.2 Principle of delayed hydrogen ignition (Zhang et al., 2023).

d. Combustion chemistry continues to occur and the OH activation process increases rapidly.

e. The expansion of combustion leads to a widening of the low-velocity zone and an increasing temperature and pressure, which sustains the subsequent reactions until the intense combustion reaction takes place.

12.2.2 Hydrogen Explosion Principle

Combustion is a prerequisite for the explosion. The combustion of hydrogen can be divided into two categories: one is the combustion of hydrogen and air under non-premixed conditions, and the other is the combustion of premixed gases. The flame range generated by non-premixed gas combustion is very wide. For different flow rates, the flame can be a micro flame or a high-pressure jet flame with a long length. The combustion of premixed gases may produce deflagration or detonation.

The explosion generated by hydrogen combustion is a chemical explosion including deflagration and detonation. Deflagration indicates a combustion process in which the flame or chemical reaction advances to the unburned combustible gas mixture at a rate below the speed of sound. Also, detonation refers to a combustion phenomenon in which shock waves generated by combustion propagate at supersonic speeds. For deflagration and detonation, the main difference is that the supersonic shock wave generated by detonation produces a sharp explosion sound. The detonation wave relies on the heat of the flame and the diffusion of chemical free radicals, igniting the combustible mixture ahead to implement propagation. Detonation is pushed outward by the adiabatic shock wave. Compared to deflagration, the formation of detonation requires more initial energy. Specifically, the pressure generated by detonation can even be eight times higher than the original mixture of gases. The high pressure generated by detonation could be 20 times the pressure of the original mixed gas, and could even reach 50 times if the pressure is reflection overlap (ISO/TR, 2015).

The explosion can be regarded as a process of rapidly releasing energy from deflagration to detonation, including a special condition detonation. Rapid deflagration is also an explosion, while low-speed deflagration is not considered an explosion as it does not generate shock waves. The coupling effect of fluid mechanics instability, thermal diffusion instability, and boundary layer effect may lead to the gradual evolution of laminar combustion to turbulent combustion until the transition from deflagration to detonation occurs (Figure 12.3). The shock wave generated by the continuous

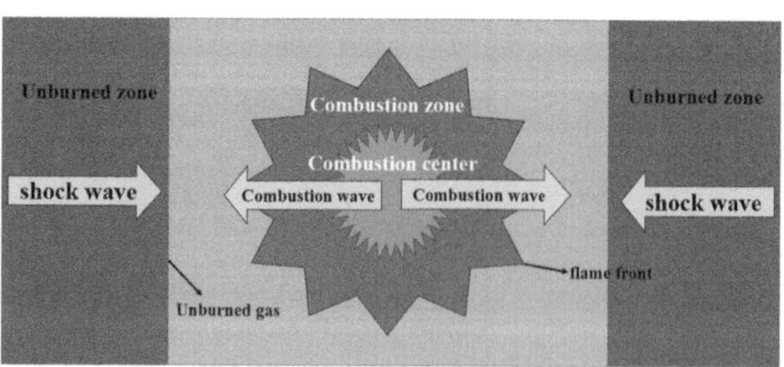

FIGURE 12.3   Development process of turbulent flame (Zhang et al., 2023).

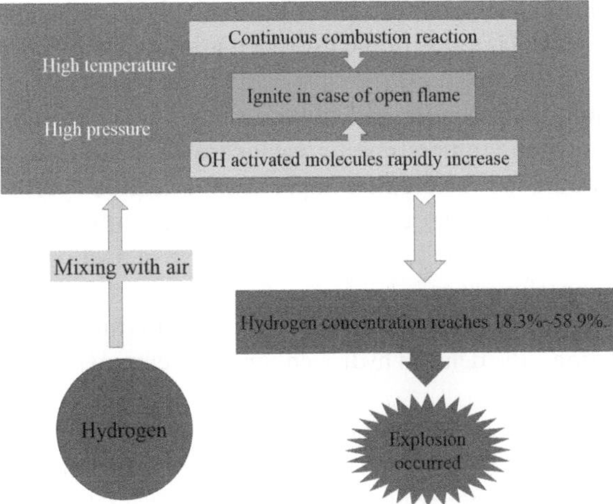

FIGURE 12.4 Schematic diagram of a hydrogen explosion (Zhang et al., 2023).

increase in pressure is coupled with the combustion reaction to form detonation. A gas explosion is a process that involves changes in physical and chemical states. The entire process releases a large amount of energy while generating impacts of light, heat, or mechanical impacts. The enhancement and impact of pressure are the main characteristics of explosions, as well as causing mechanical damage effects (Jai et al., 2023).

Hydrogen (H_2) is a colorless and odorless gas at ambient temperature and pressure. H_2 reacts with oxygen (O_2) to form water, a process that can explode violently. Research has found that hydrogen may explode when the concentration exceeds 10%, especially when the concentration range of hydrogen is between 18.3% and 58.9%. The schematic diagram of the hydrogen explosion is shown in Figure 12.4. Also, mixing hydrogen and oxygen with a 2:1 ratio may cause explosions (Yusuke et al., 2023).

Severe unrestricted hydrogen release explosions are usually associated with the release of generous hydrogen into crowded spaces. Dense spaces can bring about relatively large hydrogen-induced explosion accidents (Hoi et al., 2008). Turbulence is the main factor affecting explosive behavior. The change of external turbulence intensity is accompanied by occurring flame and an increasing overpressure. Under turbulent conditions, flowing flame clusters can be formed. The accumulated energy of combustion is sufficient to ignite the surrounding mixture, and the increased contact area can accelerate chemical reactions. The accumulated energy of combustion is sufficient to ignite the surrounding mixture (Chang et al., 2022; Jiang et al., 2022).

12.3 APPLICATIONS AND CASES

12.3.1 Delayed Hydrogen Ignition

This subsection first summarizes the effect of initial temperature, pressure, equivalence ratio, and hydrogen mixing ratio on ignition delay time in a hydrogen mixture. Then some of the technical challenges and solutions faced by the industry in engineering practice are enumerated.

The effects of initial temperature, pressure, equivalence ratio, and hydrogen mixing ratio on the delayed ignition of the hydrogen mixture are given as follows:

a. As the initial temperature gradually decreases, the predicted value of the delayed hydrogen ignition time for each case gradually increases. The initial temperature has the greatest impact on the delayed hydrogen ignition.
b. The increase in pressure can alter the delayed ignition time. Thus, high pressure can accelerate the chemical reaction of hydrogen and improve combustion performance.
c. The equivalence ratio has an impact on the fuel mixture. For the same pressure conditions, the delayed hydrogen ignition decreases as the equivalence ratio increases.
d. In the early stages of combustion, the addition of hydrogen can provide a concentration of H radicals and promote the production of OH radicals. The consumption rate of ammonia (NH_3) has been accelerated. Thus, the hydrogen mixing ratio can significantly affect the ignition performance.

Harman-Thomas et al. (2023) proposed the supercritical carbon dioxide mechanism ($UoSCO_2$-Mech) (Harman-Thomas et al., 2022). The ignition delay time (IDT) data for methane, hydrogen (H_2), and syngas were developed at a given pressure and carbon dioxide (CO_2) dilution. All measured IDT data were used to validate the $UoSCO_2$ 2.0 chemical kinetic mechanism. The mechanism was developed to simulate the combustion of methane, H_2, and syngas in a CO_2 bath gas. Quantitative evaluation was performed by comparing the average percentage difference between the experiment and simulated IDT. $UoSCO_2$ 2.0 outperformed AramcoMech 2.0 for the simulated IDT dataset.

Snehasish et al. (2022) proposed that H_2 and mixed H_2/methane (CH_4) mixtures were less reactive compared to pure CH_4 mixtures. The new experimental IDT results were compared with an updated mechanism NUIGMech 1.3. NUIGMech 1.3 accurately reproduced the IDT of H_2, CH_4, and 50/50 blends. For lower temperatures, the IDTs of all blends (H_2, CH_4, and 50/50 blends) became the same. In addition, the IDTs of H_2 and 50/50 mixtures were longer compared to the measured values of pure CH_4 at TC$\leq$930 K. At TC$\geq$890–925 K, H_2 became the slowest igniting fuel. The IDT of H_2 was 2.5 times that measured for pure CH_4 at 40 bar and TC at 890 K.

Tropin et al. (2020) revealed the inert particle effect of the diameter and volume concentration on the delayed ignition of silylene-hydrogen-air mixtures. The results indicated that decreased particle diameter and increased volume concentration led to an increase in delayed ignition. For special conditions, a temperature-dependent rupture could have happened at the ultimate ignition of temperature. An approximate equation related to the final ignition temperature of the mixture was obtained in the presence of inert particles, including a function of mixture composition, volume concentration, and particle diameter. Besides, the particle type depended on the final ignition temperature of the mixture. The results also indicated that the particle type has a dependent effect on the final ignition temperature of the mixture.

Liang et al. (2022) investigated the effect of actual gas effect on physicochemical properties under different pressure, temperature, and equivalence ratio conditions.

The results showed that the actual gas effect had a significant effect on the delayed ignition of hydrocarbon fuels. The actual gas effects-caused non-ideal deviation of delayed ignition was strongly correlated with temperature and pressure. Thus, there was a positive correlation between stress and non-ideal deviation. The non-ideal deviation of ignition delay time caused by the actual gas effect had a strong correlation with temperature and pressure. As the temperature increases, the non-ideal deviation increased with fluctuance to reach a peak value at the NTC region. The non-ideal deviation was also correlated with the contrasting pressure, the carbon chain length of hydrocarbon molecules, and the critical pressure. The non-ideal deviation was generally significant for long carbon chains and low-critical pressure fuels.

Wang et al. (2021) used six cases to research the effects of initial temperature, pressure, equivalence ratio, and hydrogen doping ratio on IDT with different hydrogen doping ratios. The different mechanisms for pure hydrogen under the fixed condition (equivalence ratio of 1, pressure of 40 bar, temperature range of 1,063–1,560 K) were studied, and the predicted values of IDT for pure hydrogen were verified.

The Otomo mechanism was selected as the target machine for implementing the ignition process of ammonia/hydrogen mixtures. The boundary conditions were initial temperature 900–1,200 K, initial pressure 100–200 bar, equivalence ratio 0.4–4, and hydrogen mixture ratio (mol) 0%–50%.

The results indicated that IDT is related to initial temperature and pressure. The value of IDT increases proportionally with pressure and temperature. The influence of the hydrogen ratio on IDT decreases with a decreased temperature. However, the combustion performance of hydrogen is superior to that of pure ammonia.

The IDT of the NH_3/H_2 mixture is affected by multiple variables. The value of IDT gradually decreases as the temperature increases. In addition, the equivalence ratio of 10% hydrogen doping concentration has little effect on IDT. The increased hydrogen mixing ratio does not enhance the impact of the equivalence ratio on IDT. As the hydrogen mixing ratio increases, the value of IDT shows an inverse relationship with the pressure value. Zhao et al. (2020) used CHEMKIN-PRO simulation to calculate the IDT and laminar flame velocity of natural gas engines, in which hydrogen doping ratios were 0%, 5%, 10%, 15%, 20%, and 25%, respectively. The operating conditions are reaction time (0.05 s), initial reaction temperatures (800 K, 900 K, 1,000 K, and 1,100 K), initial reaction pressure (4 MPa), and equivalence ratio (1). Then the IDT was calculated for operating conditions. For the same initial temperature conditions, a high proportion of mixed hydrogen gas leads to a shortened IDT. For example, the ignition delay time of a fuel mixture with a 25% hydrogen doping rate decreased by 71.99% at a temperature of 800 K. Chen et al. (2022) used FLACS software to study hydrogen-doped natural gas explosion accidents in confined spaces. Factors such as hydrogen incorporation ratio and ignition position have impacts on the characteristics of hydrogen-doped natural gas explosion accidents. The effects of hydrogen doping ratios of 0%, 5%, 10%, 15%, and 20% on hydrogen ignition and explosion were studied. The initial conditions include temperature (293 K), pressure (0.1 MPa), and equivalence ratio (1). The monitoring points with different hydrogen doping ratios were recorded, and the increased hydrogen doping rate shortened the IDT. The results confirmed that the minimum ignition energy of hydrogen in the air (0.017 MJ) was lower than that of methane (0.274 MJ).

Michał et al. (2023) studied the energy fraction effect of hydrogen and ammonia co-combustion on the performance, stability, and emissions for engines with compression ratios of 8 (CR8) and 10 (CR10). The energy fraction of hydrogen is increased to 12% to eliminate the instability of the combustible mixture during the ignition process. In addition to the ignition feature, the hydrogen component of the mixture cuts down IDT and combustion duration. Specifically, the IDT has been reduced by more than twice, and the combustion duration of CR8 has been reduced by about twice. Also, the combustion duration of CR10 has been reduced by nearly three times.

Eldakamawy et al. (2023) proposed the hydrogen-air combustion characteristics of a low-pressure ratio split flow cycle engine. Two typical situations of delayed hydrogen ignition were investigated. One case is a precondition of a fixed equivalence ratio of 0.15 to investigate the effects of different temperatures and pressures. The other case supposes the equivalence ratio of 0.15 and 0.3, which can implement the consequence of completely delayed ignition. In summary, the equivalence ratio is less affected under high-pressure conditions, and the ignition delay decreases with the increase in temperature.

To further research the IDT, Fordoei et al. (2021) discussed the combustion characteristics of a mixture of hydrogen and methane. The ignition delay time of the mixture was significantly reduced. A mixed hydrogen ratio of 10% reduced the ignition delay time by about 35%–40%, while the reduced rate expressed a decreasing trend. The results show that the presence of H_2 in the fuel results in a significant reduction in ignition delay time. After a certain mass fraction of hydrogen, the mixture ratio does not change the IDT.

12.3.2 Delayed Explosion

Although the principles of explosion are relatively mature, delay explosion still faces many challenges due to the complexity of real projects. This section provides an overview of the methods and techniques used to delay the explosion.

For the effect of porous materials on explosion suppression, researchers conducted the following studies. Cui et al. (2017) studied the barrier effect of different parameters of wire mesh structure and filling the position on a combustible gas explosion. It was found that the number of layers and mesh of wire had a significant effect on the explosion suppression effect. Sun et al. (2011) discussed the flame retardant and explosion suppression effects of metal mesh and foam ceramic materials. The study showed that metal mesh has good explosion-proof performance and poor flame resistance. On the contrary, the foam ceramic material had poor explosion-proof performance and good flame resistance performance. At the highest flame temperature, the attenuation rate of metal mesh can reach 60%, and the flame retardant effect is better than that of foam ceramic material. The explosion suppression effect of Fe-Ni and Cu foam with different pore sizes (20 and 40 ppi) was experimented with by Wang et al. (2022). A comparative analysis of the opening blockage rates (0.36, 0.64, and 0.84) was also carried out. Thus, the Fe-Ni-foam with 20 ppi can extinguish the flame with a fixed opening blockage rate (0.84).

Zhang et al. (2023) presented the multistage suppression of hydrogen-air explosions with non-premixed nitrogen (N_2) and CO_2. The multi-level suppression of hydrogen-air explosions by explosion parameters such as explosion pressure, pressure rise rate, combustion duration, and flame propagation has been discussed. The results confirmed that inert gases play an important role in suppressing explosions. Wu et al. (2022) proposed the minimum explosive concentrations of three agricultural dust points and two coal dust clouds, concluding that low N_2/CO_2 ratios can suppress the explosions of carbon dust clouds. Wang et al. (2022). found that the addition of N_2 and CO_2 reduced the maximum explosion pressure and maximum pressurization rate of the H_2-LPG air mixture. Also, the minimum suppression concentrations of inert gases were 35% N_2 and 25% CO_2, respectively. Compared to adding N_2, adding CO_2 with the same volume fraction provides a good inhibitory effect. Luo et al. (2023) discussed the suppression of 9.5% methane explosion pressure and flame propagation characteristics by CO_2 gas. The maximum explosion pressure and maximum explosion pressure growth rate of methane decrease with the increase of gas suppression volume fraction. The final stable state is delayed by the maximum IDT.

Researchers had conducted the following studies on the effects of fine water mist on explosion suppression. Tan et al. (2019) investigated the effect of halide additives on explosion overpressure. It was shown that 30% potassium iodide mist reduced the suppression of explosion overpressure by 27.46% compared to pure water mist under the same conditions. The synergistic effect of CO_2 solution and effective additives had a significant suppression effect on the explosion. Cheikhravat et al. (2015) studied the role of fine water mist particle size in the explosion of hydrogen/air mixtures. The study showed the coupling effect between fine water mist penetration velocity and flame propagation velocity. Boeck et al. (2015) investigated the explosion suppression performance of fine water mist with a hydrogen/air mixture and revealed the mechanism of fine water mist explosion suppression. Ingram et al. (2013) added sodium hydroxide (NaOH) to the ultra-fine fine water mist. It was found that the chemical suppression effect increased based on the original physical heat absorption and the explosion suppression effect was significantly improved.

On the suppression of explosions by inert gases, researchers had conducted studies. Oikawa et al. (2017) provided an electrical device to control an inert gas refinery unit. The electrical device is an electrical device with an explosion-proof structure (hydrogen does not ignite inside the device in case of hydrogen leakage). Experimental and numerical evaluation of hydrogen cloud explosion in an inert gas environment was carried out by Li et al. (2018). It was found that the average flame propagation velocity, maximum explosion pressure, maximum pressure rise rate, positive pressure pulse, and explosion pressure decay all decreased monotonically in the order of He, Ar, and N_2. For the additives He, Ar, N_2, and CO_2, CO_2 was the most effective additive for reducing the risk of hydrogen cloud explosion. Zhang et al. (2012) found that the minimum ignition energy increased exponentially when the inert gas dilution rate exceeded a certain value. Qiao (2007) compared the effects of He, Ar, N_2, and CO_2 on the laminar combustion velocity of air flames. The results showed that the decreasing order of laminar combustion velocity was He, Ar, N_2, and CO_2.

The explosion control of hydrogen/methane/air mixtures had been studied by a number of researchers. Lv et al. (2016) investigated the effect of aluminum alloy metal mesh materials on the suppression of the explosion of hydrogen/methane-air mixtures with different hydrogen contents. It was shown that the explosion suppression efficiency of the metal mesh material gradually decreased as the hydrogen content of the gas mixture increased. When the hydrogen content exceeded 68.95%, the aluminum alloy metal mesh material failed. Jin et al. (2017) investigated the effects of parameters such as the number of layers and mesh size of metal mesh materials on the suppression of hydrogen/methane/air mixtures in closed pipelines. The results showed that the grid size of the grid material plays a dominant role in heat absorption and cutting explosion flames. Thus, the multi-layer metal mesh material can effectively reduce the explosion pressure by up to 78.6%.

12.4 CONCLUSION

This chapter systematically developed an introduction to the principle of delayed hydrogen ignition and explosion. Some application examples of delayed ignition and explosion were reviewed, including methods and technology. However, some technical challenges are faced due to the complexity and uncertainty of actual projects, including the unknown impact of initial temperature, pressure, equivalence ratio, and hydrogen mixing ratio on the IDT of mixed hydrogen. At the same time, opportunities have also brought opportunities to industries. For example, the commonly used explosion suppression method is to add various types of inhibitors, including porous materials, fine water mist, and inert gases. The addition of inert gases (such as N_2, He, CO_2, etc.) is currently the most environmentally friendly and effective method. As a limited inhibitor, inert gases mainly rely on the physical dilution of inert additives. Therefore, high concentrations of inert gases are required, while high-pressure storage of inert gases would become difficult and lead to poor fluidity.

DECLARATION OF COMPETING INTEREST

The authors declare that they have no known competing financial interests or personal relationships that could have appeared to influence the work reported in this chapter.

ACKNOWLEDGMENT

This work was supported by the Young Scholars Development Fund of SWPU (grant no. 202299010027), China.

REFERENCES

Boeck, L. R., Kink, A., Oezdin, D., et al. (2015). Influence of water mist on flame acceleration, DDT, and detonation in H2-air mixtures. *International Journal of Hydrogen Energy*, 40(21), 6995–7004.

Chang, X., Chunhua, B., & Bai, Z. (2022). The effect of gas jets on the explosion dynamics of hydrogen-air mixtures. *Process Safety and Environmental Protection*, 162, 384–394.

Cheikhravat, H., Goulier, J., Bentaib, A., et al. (2015). Effects of water sprays on flame propagation in hydrogen/air/steam mixtures. *Proceedings of the Combustion Institute*, 35(3), 2715–2722.

Chen, Z., Li, J., & Yu, B. (2022). Simulation of hydrogen-doped natural gas explosions in indoor confined spaces. *Science and Engineering*, 22(14), 5608–5614.

Cheng, Y., Zhao, J., Liao, S., & Shu, C. (2022). Explosion mechanism of aluminum powder mixed with low-concentration hydrogen. *International Journal of Hydrogen Energy*, 47, 27293–27302.

Cui, Y., Wang, Z., Zhou, K., et al. (2017). Effect of wire mesh on double-suppression of CH4/air mixture explosions in a spherical vessel connected to pipelines. *Journal of Loss Prevention in the Process Industries*, 45, 69–77.

Eldakamawy, M. H. & Picard, M. (2023). Combustion characteristics for a hydrogen-fuelled low-pressure-ratio split-cycle engine. *International Journal of Hydrogen Energy*, 48(20), 7476–7487.

Fordoei, E. E., Mazaheri, K., & Amirreza, M. (2021). Effects of hydrogen addition to methane on the thermal and ignition delay characteristics of fuel-air, oxygen-enriched, and oxy-fuel MILD combustion. *International Journal of Hydrogen Energy*, 46(68), 34002–34017.

Harman-Thomas, J.M., Hughes, K. J., & Pourkashanian, M. (2022). The development of a chemical kinetic mechanism for combustion in supercritical carbon dioxide. *Energy*, 255, 124490.

Harman-Thomas, J. M., Kashif, T. A., Hughes, K. J., Pourkashanian, M., & Farooq, A. (2023). Experimental and modeling study of hydrogen ignition in CO_2 bath gas. *Fuel*, 334, 126664.

Hoi, D. N. & John, H. S. L. (2008). Comments on explosion problems for hydrogen safety. *Journal of Loss Prevention in the Process Industries*, 21, 136–146.

Hu, Q., Zhang, X., & Hao, H. (2023). A review of hydrogen-air cloud explosions: the fundamentals, overpressure prediction methods, and influencing factors. *International Journal of Hydrogen Energy*, 48, 13705–13730.

Ingram, J. M., Averill, A. F., Battersby, P., et al. (2013). Suppression of hydrogen/oxygen/nitrogen explosions by fine water mist containing sodium hydroxide additive. *International Journal of Hydrogen Energy*, 38(19), 8002–8010.

Jai, M. M. & Kenneth, B. (2023). Chapter 3- natural gas for combustion systems. In K. Brezinsky (Ed.), *Developments in Physical & Theoretical Chemistry, Combustion Chemistry and the Carbon Neutral Future* (pp. 63–90). Elsevier, Amsterdam, Netherlands.

Jiang, Y., Li, Y., Zhou, U., Jiang, H., Zhang, K., Gao, Z., & Gao, W. (2022). Investigation on unconfined hydrogen cloud explosion with external turbulence. *International Journal of Hydrogen Energy*, 47, 8658–8670.

Jin, K. Q., Duan, Q. L., Chen, J., et al. (2017). Experimental study on the influence of multi-layer wire mesh on dynamics of premixed hydrogen-air flame propagation in a closed duct. *International Journal of Hydrogen Energy*, 42(21), 14809–14820.

Kai, Y., Shuran, L., Gao, J., & Pang, L. (2017). Research on the coupling degree measurement model of urban gas pipeline leakage disaster system. *International Journal of Disaster Risk Reduction*, 22, 238–245.

Li, P., Zeng, Q., Duan, Q., Chen, X., & Jinhua, S. (2023). Effects of obstacles inside the tube on initial self-ignition of high-pressure hydrogen release through a tube. *Fuel*, 339, 127354.

Li, Y., Bi, M., Huang, L., Liu, Q., Li, B., & Ma, D. (2018). Wei Gao hydrogen cloud explosion evaluation under inert gas atmosphere. *Fuel Processing Technology*, 180, 96–104.

Li, Y., Bi, M., Yan, C., Liu, Q., Zhou, Y., & Gao, W. (2019). Inerting effect of carbon dioxide on confined hydrogen explosion. *International Journal of Hydrogen Energy*, 44, 22620–22631.

Liang, X., Liao, J., Zhu, S., Shen, W., & Wang, K. (2022). The impact of actual gas effects on the ignition delay time of hydrocarbon fuels. *Journal of Xi'an Jiaotong University*, 56(01), 1–11.

Luo, Z., Nan, F., Cheng, F., Xiao, Y., Wang, T., Shao, J., & Wang, C. (2023). Experimental study on CO2/CF3I suppression of methane-air explosion and flame propagation. *Journal of Loss Prevention in the Process Industries*, 83, 0950–4230.

Lv, P. F., Pang, L., Jin, J. H., et al. (2016). Effects of hydrogen addition on the deflagration characteristics of hydrocarbon fuel/air mixture under a mesh aluminum alloy. *International Journal of Hydrogen Energy*, 41(18), 7511–7517.

Michał, P., Michał, G., Wojciech, T., & Arkadiusz, J. (2023). Assessment of the co-combustion process of ammonia with hydrogen in a research VCR piston engine. *International Journal of Hydrogen Energy*, 48(7), 2821–2834.

Oikawa, Y., Akiyama, T., Kanamori, S., & Akiyama, Y. (2017). Electrical unit. Chinese Patent JP2017094293(A).

Qiao, L., Gu, Y., Dahm, W. J. A., Oran, E. S., & Faeth, G. M. (2007). A study of the effects of diluents on near-limit H2-air flames in microgravity at normal and reduced pressures. *Combustion & Flame*, 151, 196–208.

Rasul, M. G., Hazrat, M. A., Sattar, M. A., Jahirul, M. I., & Shearer, M. J. (2022). The future of hydrogen: Challenges on production, storage and applications. *Energy Conversion and Management*, 272, 116326.

SIS-ISO/TR 15916:2015. (2015). *Basic Considerations for the Safety of Hydrogen Systems*. MATGAS building, Campus UAB, Bellaterra, Barcelona, Spain.

Snehasish, P., Mohamed, A. A. E., Wang, P., Gilles, B., & Henry, J. C. (2022). When hydrogen is slower than methane to ignite. Proceedings of the Combustion Institute, 39(1), 253-263.

Sun, J., Yi, Z., Wei, C., Xie, S., & Huang, D. (2011). The comparative experimental study of the porous materials suppressing the gas explosion. *Procedia Engineering*, 26, 954–960.

Tan, W., Lü, D., Liu, L., Zhu, G., & Jiang, N. (2019). Suppression of methane/air explosion by water mist with potassium halide additives driven by CO_2. *Chinese Journal of Chemical Engineering*, 27, 1004–9541.

Tropin, D. A. & Bochenkov, E. S. (2020). Influence of inert particles on the ignition processes of hydrogen-silane-air mixtures. *International Journal of Hydrogen Energy*, 45(35), 17953–17960.

Wang, J., Liang, Y., & Zhao, Z. (2022). Effect of N_2 and CO_2 on explosion behavior of H2-Liquefied petroleum gas-air mixtures in a confined space. *International Journal of Hydrogen Energy*, 47, 0360–3199.

Wang, J., Liu, G., Zheng, L., Pan, R., Lu, C., Wang, Y, Fan, Z., & Zhao, Y. (2022). Effect of opening blockage ratio on the characteristics of methane/air explosion suppressed by porous media. *Process Safety and Environmental Protection*, 164, 129–141.

Wang, L., Ma, H., & Shen, Z. (2021). Effects of vessel height and ignition position upon explosion dynamics of hydrogen-air mixtures in vessels with low asymmetry ratios. *Fuel*, 289, 119926.

Wang, Y., Zhou, X., & Liu, L. (2021). Study on the mechanism of the ignition process of ammonia/hydrogen mixture under high-pressure direct-injection engine conditions. *International Journal of Hydrogen Energy*, 46(78), 38871–38886.

Wu, W., Huang, W., Wei, A., Martin, S., Ulrich, K., & Wu, D. (2022). Inhibition effect of N2/CO2 blends on the minimum explosion concentration of agriculture and coal dust. *Powder Technology*, 399, 117195.

Yang, W., Zheng, L., Wang, C., Wang, X., Jin, H., & Fu, Y. (2021). Effect of ignition position and inert gas on hydrogen/air explosions. *International Journal of Hydrogen Energy*, 46, 8820–8833.

Yuan, J., Mao, W., Hu, C., Zheng, J., Zheng, D., & Yang, Y. (2023). Leak detection and localization techniques in oil and gas pipeline: a bibliometric and systematic review. *Engineering Failure Analysis*, 146, 107060.

Yusuke, I., Shin-Ichi, H., Bunpei, S., Haru, Y., Yoshiyasu, T., & Fumitake, S. (2023). Guidelines for the selection of hydrogen gas inhalers based on hydrogen explosion accidents. *Medical Gas Research*, 13(2), 43–48.

Zhang, S., Wen, X., Guo, Z., Zhang, S., & Ji, W. (2023). Experimental study on the multi-level suppression of N2 and CO2 on hydrogen-air explosion. *Process Safety & Environmental Protection*, 169, 970–981.

Zhang, K., Wang, Z., Gong, J., Liu, M., Dou, Z., & Jiang, J. (2017). Experimental study of effects of ignition position, initial pressure and pipe length on H2-air explosion in linked vessels. *Journal of Loss Prevention in the Process Industries*, 50, 295–300.

Zhang, S., Wen, X., Guo, Z., Zhang, S., & Ji, W. (2023). Experimental study on the multi-level suppression of N2 and CO2 on a hydrogen-air explosion. *Process Safety and Environmental Protection*, 169, 970–981.

Zhang, W. K., Chen, Z., & Kong, W. J. (2012). Effects of diluents on the ignition of premixed H2/air mixtures. *Combustion & Flame*, 159(1), 151–160.

Zhao, R., Xu, L., Feng, S., Li, C., Yao, L., Hu, S., & Wang, Z. (2020). The impact of hydrogen blending on the combustion and emissions of marine natural gas engines. *Propulsion Technology*, 41(11), 2549–2557.

Zhuo, J., Zhang, C., Zhang, S., Chen, L., Yang, H., Liu, F., Du, Y., Lu, L., & Zheng, S. (2023). Influence of hydrogen environment on fatigue fracture morphology of X80 pipeline steel. *Journal of Materials Research and Technology*, 22, 1039–1047.

13 Liquid Hydrogen Release from Pressurized and Non-pressurized Tanks

Hisham Khaled Ben Mahmud
and Mian Umer Shafiq

13.1 INTRODUCTION

Transforming the EU into a modern, prosperous, and competitive economy is the aim of the European Green Deal. In May 2022, the REPowerEU proposal was released as part of the Green Deal, indicating the determination of several countries to achieve energy independence and address the global crisis. The presentation of the European Green Deal took place in December 2019 (European Commission, 2022), to address various issues such as climate change and environmental deterioration. Given its potential to be combined with renewable energy sources and contribute to the achievement of the Green Deal's objectives, hydrogen is set to play a crucial role in the energy transition phase. As part of the REPowerEU initiative, one of the key targets is to produce 10 million tons of domestic renewable hydrogen and import an additional 10 million tons by 2030 (European Commission, 2022).

Over the past 20 years, hydrogen technologies, such as fuel cells for propulsion, have been utilized on ships, largely in the form of trial projects. Vessels were equipped with pressurized tanks that contained gaseous hydrogen. Nonetheless, utilizing such storage methods on large, high-energy-demand ships that travel long distances would require vast amounts of space. Therefore, storing this fuel in its cryogenic liquid state appears to be one of the most feasible options. Hydrogen has a low density ($0.0883\,\mathrm{kg/m^3}$ at atmospheric temperature and pressure) (NIST, 2022), which makes large-scale use of it difficult and requires enormous storage quantities. Therefore, after production, hydrogen is frequently compressed or liquefied. A potential remedy for this problem is liquid hydrogen (LH_2), which has a density 1,000 times higher than that of atmospheric circumstances ($70.9\,\mathrm{kg/m^3}$) (NIST, 2022). LH_2 is one of the coldest cryogenic fluids, and its exceptionally low temperature ($-253°C$ at atmospheric pressure) (NIST, 2022) makes handling and storage of the fuel extremely difficult. For the past few decades, LH_2 has been mostly employed in the aerospace sector to power engines in rockets and spaceships that need massive amounts of high-energy fuels. Thus, the technology hasn't become widely accessible yet because of the high prices and scarcity of certain storage equipment needed for LH_2. However, despite the difficulties in

DOI: 10.1201/9781003382553-16

handling and storing LH_2, it is still one of the few practical ways to move substantial amounts of hydrogen over long distances and power long-range vehicles like ships and airplanes.

13.2 HOW IS HYDROGEN STORED?

Hydrogen will not be created on-site or at the moment of use for a variety of reasons. Instead, hydrogen can be created centrally and stored for later distribution. As a result, storage and/or transportation systems are required along the renewable hydrogen chain. The optimal storage option is influenced by several factors, including the required storage volume, the length of time the storage is needed, the discharge rate required, the geographical availability of storage options, and whether the storage is intended for small or large-scale use. When it comes to stationary applications in large-scale storage, energy density and filling time are not limiting factors (Grasemann & Kalidindi, 2015). However, for mobility applications in small-scale storage, energy density and charging durations are critical considerations. The purity of the hydrogen is an extra crucial issue, as high-quality hydrogen is critical for its use in fuel cells, whereas purity will not be a critical point for burning.

Hydrogen has a high energy content per unit mass, with 120.1 MJ/kg. However, its energy density is very low at room temperature, measuring just 0.01 MJ/L. This means that larger amounts of hydrogen need to be stored to provide the same energy as other fuels. To store 1 kg of hydrogen, which is enough to drive around 100 km, a space of 11.2 Nm^3 is required. This is equivalent to the volume of a big utility or commercial vehicle's boot (Laura, 2022). Indeed, increasing the storage density of hydrogen is crucial for making hydrogen storage economically viable. Hydrogen storage can be categorized into two types based on whether it involves physical or material storage as shown in Figure 13.1.

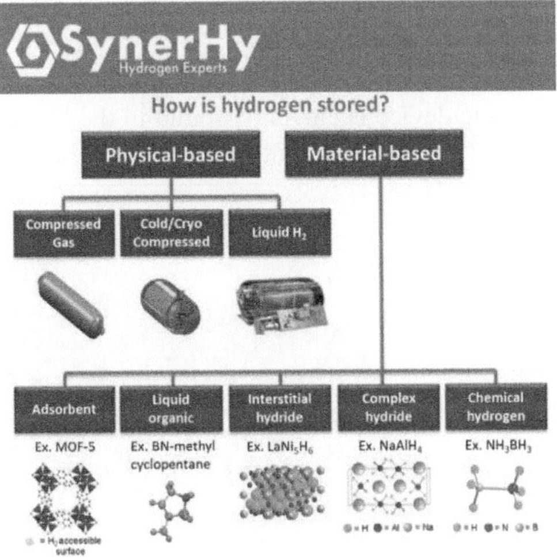

FIGURE 13.1 Categories of hydrogen storage methods (Laura, 2022).

In physical storage, hydrogen is stored as a gas or liquid in its pure molecular compound form, without any significant physical or chemical bonding to other materials. First, we examine the storage technologies that are currently in use on a large basis. To compress hydrogen in its gaseous state, it can be stored either in tanks or vessels for modest storage volumes or in geological storage for larger volumes. For tank storage, there are four different types of high-pressure cylinders (Grasemann & Kalidindi, 2015).

- **Type I** pressure vessels are constructed using metal materials. These tanks are designed to store hydrogen at pressures ranging from 150 to 300 bar and are typically used for industrial gas storage purposes. These high-pressure tanks are currently the most popular and least expensive, but their weight prevents them from being used in vehicles.
- **Type II** pressure vessels are constructed from a thick metal ring liner that is encased with carbon or glass fiber. Because they can resist up to 1,000 bar, they are employed as high-pressure tanks in hydro generators.
- **Type III** tanks are pressure containers with a full wrap of composite materials that can endure mechanical stress and an inside metal liner to prevent hydrogen leaking by diffusion. These containers are lighter than Types I and II because thick metal walls have been removed and composite materials have been added.
- **Type IV** tanks are shown in Figure 13.2, which are pressure vessels consisting of a high-density polymer liner that serves as a gas diffusion barrier and is completely encased in a carbon fiber substance. Metal valves are installed in these tanks to allow for the supply and refilling of hydrogen. They can stand up to 700 bars of pressure. Type III and Type IV vessels are made for portable applications where weight reduction is crucial and pressures ranging from 350 to 700 bar are needed. However, the usage of carbon fiber makes these vessels significantly more expensive.

With a storage capacity of between 10,000 and 20,000 tons, the largest operational salt cave hydrogen storage device is currently found in the United States. It can hold roughly 30 days' worth of hydrogen output. Salt caves are often smaller than oil and

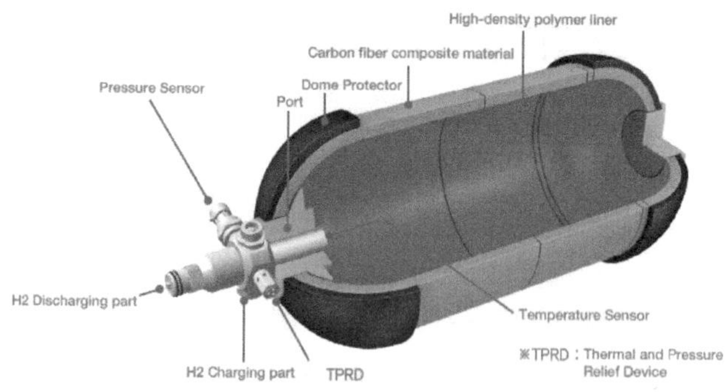

FIGURE 13.2 High-pressure tank type IV (Laura, 2022).

gas fields, but oil and gas fields are larger and more porous, and they also contain contaminants that need to be removed before using hydrogen in fuel cells. Water aquifers are the least developed alternative and the debate over their suitability is still ongoing (Winters, 2009).

13.3 LIQUID HYDROGEN STORAGE

Liquid hydrogen is often stored in cylindrical, double-walled cryogenic storage tanks. According to Verfondern (2008), passenger car tanks may hold up to 10 kg of LH_2 or about 170 L of volume. Up to 70 layers of aluminum foils or aluminized polymers, spaced by glass fibers or polymer spacers, with a total thickness of about 30 mm, can make up the insulation system utilized in these tanks (Verfondern, 2008). This design allows for a daily boil-off loss rate of less than 1.5%. Therefore, these tanks are very important for the safe transportation and storage of hydrogen, especially when used in fuel cell vehicles.

The insulation system of cryogenic storage tanks has a safety vent, pressure release valve, and double-walled filling tube (for passage). Such tanks usually run at about 4 bars although they can go up to 7 bars before having to be vented (Verfondern, 2008). The electric heating mechanism manages the internal pressure (Tzimas et al., 2003; Krainz et al., 2004). These safety measures ensure no potential overpressure hazards exist and that the tank's internal pressure is kept within allowable operating limits. Michel et al. (2006) considered an inventive method of managing pressure involving returning part of petrol into the receiver to transfer heat effectively on liquid fuel depending on sensed exit line pressure. This design eliminates power consumption and recovers exhaust heat from the engine (Michel et al., 2006). An important breakthrough was made in hydrogen fuel cell vehicles in 2006 when Magna Steyr manufactured LH_2 tanks with two walls that were then utilized by BMW Hydrogen 7. The inside wall measures around 2 mm in thickness. It is an interior wall and it has a thickness of approximately 2 mm whereby it is surrounded by one more external wall of about 2 mm and encased in a 30 mm high-vacuum super-insulation system. As per Müller et al. (2007), the pressure for the vacuum jacket is maintained at 0.1 bar for all cryogenic service conditions down to the temperature of 20 K. The hibernation process takes not more than 1 month for such cryogenic tanks to become operational again. Figure 13.3 shows an LH_2 tank schematic meant for automotive use. This design of the tank ensures safety while storing hydrogen efficiently, which becomes a key factor in fuel cell vehicles.

Alternatively, double-walled spherical tanks are used to store large amounts of LH_2 aiming at raising the ratio between its volume and surface area as much as possible. For example, NASA Kennedy Space Center has had the largest vacuum insulated LH_2 tank since the 1960s with a total capacity of about 3,800 m^3 (Peschka, 1992). Instead of multilayer insulation (MLI), the vacuum jacket in these tanks is filled with perlite powder. More recently, NASA has commenced building a larger fixed-site tank with a volume approximately equivalent to that of 5,683 m^3 or roughly equal to 1.25 million gallons that will be insulated by glass bubbles instead of perlite. This vessel will be used in support of future undertakings to the Moon and

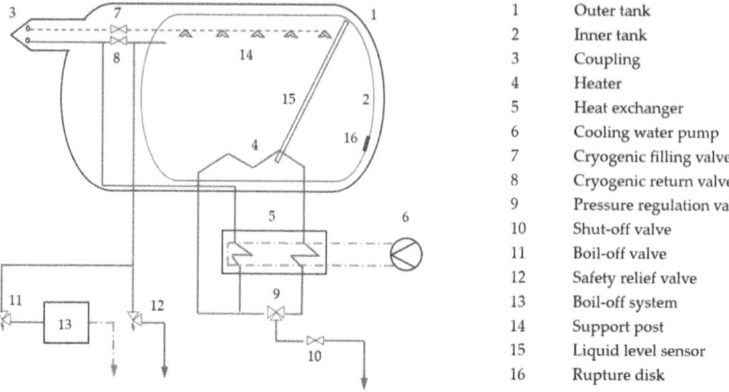

1	Outer tank
2	Inner tank
3	Coupling
4	Heater
5	Heat exchanger
6	Cooling water pump
7	Cryogenic filling valve
8	Cryogenic return valve
9	Pressure regulation valve
10	Shut-off valve
11	Boil-off valve
12	Safety relief valve
13	Boil-off system
14	Support post
15	Liquid level sensor
16	Rupture disk

FIGURE 13.3 Cryogenic tank for onboard storage of liquid hydrogen (Ustolin et al., 2020).

Mars (Sempsrott, 2022). Moreover, a revolutionary heat exchanger system known as integrated refrigeration and storage (IRAS) will also be incorporated into the tank to reduce boil-off rates and periodic venting through the pressure relief valve (PRV) (Granath, 2022; Notardonato et al., 2017).

13.3.1 Materials for LH$_2$ Tanks

The materials must be able to work in a cold, liquid hydrogen environment, be resistant to hydrogen embrittlement, and have good mechanical and thermophysical properties at very low temperatures (Ustolin et al., 2020). However, the specific requirements for container materials depend on whether they are being used in maritime, aerospace, automotive, or other applications. Liquid hydrogen vessels commonly use stainless steel as cryogenic material, especially in applications where the weight of the containment system is not a limiting factor. The preferred choice for transporting liquid hydrogen is austenitic stainless steel, as it demonstrates dependable performance at cryogenic temperatures (Qiu et al., 2021). This is in part due to the fact that austenitic steel's crystal structure is face-centered cubic, providing it with a superior ability to undergo plastic deformation.

Park et al. (2010) state that though materials keep the proper degrees of flexibility and impact resistance when the temperature drops, they also get stronger. Each grade of stainless steel's acceptability for a given application is ultimately determined by the alloy composition, which influences its macro properties. At low temperatures, alloys with higher concentrations of Ni and Cr tend to be more stable (Edeskuty & Stewart, 1996). At cryogenic temperatures, metals frequently show improved endurance limits and fatigue performance in addition to increases in elastic modulus, tensile strength, and yield strength (Anoop et al., 2021). The low-temperature plasticity of a material is determined by its ductile-brittle transition low-temperature behavior (Ustolin et al., 2020).

Unlike this, bcc (body-centered cubic) type metallic materials usually display a sharp decrease in pliability below the temperature at which they undergo ductile-to-brittle transformation. The danger of brittle fractures and cracks that result from reduced flexibility makes these materials unsuitable for cryogenic applications (Duthil, 2014). On the other hand, those materials where no brittle transition occurs normally have an increase in elongation with decreasing temperature.

Tests are often used to determine if a material suitable for low-temperature service includes fracture mechanics tests, drop weight tests, full thickness tests, and low-temperature impact toughness (LIT) tests. This is particularly important because low-temperature impact toughness testing is commonly required on base metals, welds, and heat-affected zones (HAZs) whenever assessing materials for use in liquid hydrogen storage tanks. These tests help in determining whether the material will fit into its intended purpose and more about how it behaves at low temperatures (Duthil, 2014).

13.4   RELEASE OF HYDROGEN

However, the method varies depending on the type of tank whether it is pressurized or not. In general high-pressure tanks are utilized for hydrogen storage while a gas release regulator is attached to them. Hydrogen tanks which are non-pressurized are commonly liquid or solid based. One way to control the gas release is to heat the tank or manually fracture the seal (Jarvi & Jarvi, 2018).

In pressurized tanks, hydrogen is typically stored at high pressures of up to 10,000 psi. A valve, which controls the release of hydrogen, is often located near the top of the tank. When the valve is opened, hydrogen is released from the tank and put to various uses. Generally, hydrogen is stored in either a liquid or solid state in non-pressurized tanks. The temperature of hydrogen in liquid hydrogen tanks is kept at −253°C to retain its liquid state. A pressurized tank releases liquid hydrogen, which can be vaporized and used in many ways when hydrogen is needed. In solid hydrogen storage, the hydrogen is maintained solid within a porous matrix. When hydrogen is required, the solid material is heated to liberate the hydrogen gas. It's important to keep in mind that, to mitigate any potential risks, handling hydrogen should always be done with the proper safety protocols in place (Jarvi & Jarvi, 2018).

13.4.1   RELEASE OF HYDROGEN FROM PRESSURIZED TANKS

Usually, pressurized hydrogen-releasing tanks are equipped with a valve that controls the hydrogen flow out of the tank. The following are standard procedures for managing the release of hydrogen from pressurized tanks (Jarvi & Jarvi, 2018):

- Ensure that the tank and any associated equipment are in excellent working order and that all safety precautions, such as putting in the appropriate personal protective equipment, are being followed.
- Find the valve that controls the discharge of the hydrogen tank. This valve, which is often located at the top of the tank, may be labeled for convenience. Carefully open the valve to begin the discharge of hydrogen. The release rate will depend on the specific application; however, it's important to keep in mind that releasing hydrogen too quickly could endanger public safety.

- Check hydrogen flow rate as well as tank pressure to make sure they are functioning properly.
- Dispose of waste or empty tanks of hydrogen after releasing the desired amount and store any remaining excess hydrogen in the right way.
- To stop the hydrogen from flowing, close the valve.

13.4.1.1 Procedures

The system, in which the hydrogen storage is released from pressurized tanks, will usually go through several stages to ensure it is functioning properly and in a safe manner (Grasemann & Kalidindi, 2015).

- Make sure that the hydrogen storage tank assembly, repair, maintenance, inspection, and other such works are carried out according to the safety requirements given in relevant regulations.
- Use the mechanism to which you can connect the tank and ensure that it will regulate both the pressure and flow of hydrogen during the release phase.
- Open the tank valve to allow hydrogen to depart the tank and enter the distribution system.
- Monitor the hydrogen pressure and flow as it is released from the tank to prevent the system from being over- or under-pressurized.
- Be sure to use all appropriate safety equipment, such as gas detectors and protective clothing, to prevent any incidents during the hydrogen release procedure.
- The hydrogen storage tank should be ventilated correctly so that there is no pressure build-up or accumulation of hydrogen gas.
- After releasing the required amount of hydrogen, you can stop the process by closing the valve.
- Remember, though, that the specific procedures for decompressing hydrogen tanks may differ according to the system and the safety rules defined by an application.

Remember that depending on the kind of tank and application, several protocols can be used to control the release of hydrogen from pressurized tanks. Always follow the manufacturer's instructions and take all necessary safety precautions when handling hydrogen. The pros and cons of releasing hydrogen under pressure from tanks are as follows:

13.4.1.2 Applications

In diverse fields, there are numerous applications of pressurized tank hydrogen discharge. The significant uses of pressurized tank hydrogen storage are as follows:

- ***Fuel cell vehicles***: Electric vehicles, which include battery electric vehicles and plug-in hybrid electric vehicles, are a fuel-less option for reducing carbon dioxide emissions as they are powered by electricity (Ogden & Lewis, 2019).

- ***Industrial applications:*** Hydrogen, which is contained in pressurized tanks, is utilized in many industrial processes, such as the manufacturing of glass and metal. Hydrogen can be used as a reducing agent to make metals in addition to generating heat for industrial purposes (Song, 2003).
- ***Energy storage:*** Hydrogen-filled pressurized tanks can be utilized to store energy for renewable energy sources such as wind and solar power. If renewable energy sources are not available, this allows excess energy to be stored as hydrogen for use at a later time (Hussain & Islam, 2016).

13.4.1.3   Advantages

- This makes it possible to use it in some applications that have space constraints since more hydrogen can be contained in one vessel. Pressurized tanks enable the filling and emptying of hydrogen quickly. This is important for applications that require fast refueling, such as the automotive industry, where short refueling times are key for user convenience.
- Pressurized tanks can cost less to make and maintain than cryogenic storage systems or high-technology solid-state storage. This economic advantage could be particularly substantial in the wide application of hydrogen as a source of energy.
- Hydrogen storage that is pressurized typically does not undergo any energy loss during the storage and retrieval process compared to some other methods that may require energy-demanding procedures. Hydrogen-based power systems need to maintain the overall efficiency of the entire system (Song, 2003).

13.4.1.4   Disadvantages

- Although hydrogen has high-pressure storage, it is still considered to have a lower energy density per mass or volume than other fuels. This will require the need for larger and heavier storage tanks that may not be practical under some circumstances, especially in cars where weight is a significant factor.
- It could be complicated to get the appropriate materials in terms of their price and feasibility at once, for high-pressure hydrogen storage causes stress on the materials. Moreover, the materials should not be susceptible to hydrogen embrittlement as this might change the mechanical properties of some metals.
- Hydrogen leakage can be one of the safety issues because pressurized hydrogen storage systems are leakable. Any amount of hydrogen released can generate explosive mixtures due to the highly combustible nature of hydrogen gas (Song, 2003).

13.4.2   RELEASE OF HYDROGEN FROM NON-PRESSURIZED TANKS

The specific type of storage system for hydrogen will determine how the release of hydrogen from non-pressurized tanks occurs. Different non-pressurized tank types, such as liquid hydrogen tanks, solid-state hydrogen storage, and metal hydride storage, can be used to store hydrogen. The general steps listed below can be taken to prevent hydrogen leaks from non-pressurized tanks (Jarvi & Jarvi, 2018):

- Confirm that the tank and related equipment function well, and ensure all safety measures, including the use of necessary protective gear, are in place.
- Determine the method of hydrogen removal from the tank, which will depend on the type of storage system used; for instance, liquid hydrogen tanks may require heating or pressurizing before releasing hydrogen gas.
- To extract hydrogen from the tank, you are supposed to follow the specific steps given by the manufacturer, and these may involve using a valve or another form of control that can regulate the flow of hydrogen.
- Once the necessary quantity of hydrogen has been released, cease the flow and properly store any excess hydrogen.
- When handling non-pressurized tanks, it's important to follow all applicable safety regulations and consult the manufacturer's instructions.
- When you are done with your waste, dispose of it safely and legally.
- You should constantly check the hydrogen tank as well as the flow rate to ensure that everything is going well and there are no leaks or other safety concerns.

13.4.2.1 Applications

Hydrogen storage, particularly in non-pressurized tanks, has a variety of uses, often in stationary energy storage systems. Here are some significant purposes for these types of tanks:

- ***Renewable energy storage:*** Renewable energy storage stores excessive renewable solar or wind energy in hydrogen form, contained in low-pressure tanks when stored, without the need for cooling. Subsequent release of hydrogen through a fuel cell or other device enables the hydrogen to power energy production. The grid can then 'store' excess energy in the form of renewable energy, thereby providing the same benefits as the grid and making the supply of energy not intermittent (Zhang et al., 2020).
- ***Portable power systems:*** Non-pressurized hydrogen storage is also useful in small portable power systems, such as backup power generators for outdoor events, camping equipment, and other alternatives to fossil fuel-based conventional generators. It's much greener and cheaper than petrol (Meyers, 2014).
- ***Industrial applications:*** Non-pressurized hydrogen tanks are used for industrial uses such as electronics, chemicals, and other materials. Hydrogen can be stored and used for heating systems in both buildings and homes. The hydrogen is released when needed for heating. This is a greener, fuel-efficient, and fossil fuel-based alternative to the heating systems that are available today (Pharoah, 2014).

13.4.2.2 Advantages

- Non-pressurized hydrogen storage systems could potentially cost less and need less upkeep than those that are pressurized.
- These tanks can sometimes hold more hydrogen than their pressurized counterparts, giving them the edge in certain situations.
- Some non-pressurized systems, like those using metal hydride storage, are easier to manage and transport. They safely keep hydrogen at room temperature and regular pressure (Pharoah, 2014).

13.4.2.3 Disadvantages

- It may be more challenging to release hydrogen from non-pressurized tanks than from pressurized ones. To release the hydrogen in some non-pressurized storage systems, specific tools or procedures are needed.
- Regardless of the type of storage system being utilized, handling or releasing hydrogen poses a danger of fire or explosion due to its combustible nature.
- Additionally, not all applications are suited for non-pressurized hydrogen storage systems. For instance, non-pressurized storage systems might not be able to release hydrogen fast enough or under high enough pressure for some applications (Pharoah, 2014).

13.4.3 Selection of Pressurized and Non-pressurized Tanks

The choice of whether to use pressurized or non-pressurized hydrogen release will depend on several factors explained by Winters et al. (2009, 2011, 2013), including:

13.4.3.1 Application Requirements

The type of release of hydrogen required will depend on the application. Pressurized tanks would be necessary in certain applications, such as those that require the quick and effective release of hydrogen under high pressure. Pressurized tanks are not as appropriate for other applications that require a slower or more regulated hydrogen discharge.

13.4.3.2 Safety Considerations

When handling and storing hydrogen, safety is always the primary concern. High-pressure hydrogen gas is contained in pressurized tanks, which poses a risk to public safety (if not handled carefully). Conversely, to discharge the hydrogen from non-pressurized tanks, specific tools or procedures could be needed. The safety requirements of the application will determine which storage solution is the optimum.

13.4.3.3 Space Constraints

Large volumes of hydrogen can be stored in comparatively small spaces by pressurized tanks; however, non-pressurized tanks need more room to hold the same volume of hydrogen. Based on the amount of space available for the project, a storage solution will be chosen.

In conclusion, the application-specific needs, safety concerns, financial limits, and available space will all play a role in the decision of whether to employ pressurized or non-pressurized hydrogen release. It's crucial to give each of these aspects serious thought when choosing a hydrogen storage system.

13.4.4 Hydrogen Release Modeling

Comparing bulk liquid hydrogen storage to comparably sized gaseous systems reveals a larger storage potential that permits higher station throughput. However, there is a dearth of data for developing and validating models for liquid hydrogen releases due to the absence of sufficient science-based test platforms that offer complete control over release boundary conditions. Risk-based strategies require this vital knowledge (Li, et al., 2022)

To improve and evaluate deterministic liquid hydrogen release models, (Liu, et al., 2011) established a novel capability for controlled cryogenic hydrogen releases. The approach that will be used is based on a template that has been used in the past to characterize gaseous hydrogen emissions under high pressure. The suggestion is to include a new, dual-stage heat exchanger in the equipment already in place at the Turbulent Combustion Laboratory. This could lead to the creation of mixed-phase flows by assisting in lowering the supply of gaseous hydrogen flows to the appropriate temperature. A special nozzle will allow the hydrogen to escape. Diagnostics using high-fidelity Rayleigh scatter imaging will be used to quantify the relevant release phenomena. The information gathered from this endeavor will help create release models for hydrogen leaks that occur at low temperatures. This advancement will make risk-informed methods to the safety of liquid hydrogen bulk storage even more feasible. A thorough understanding of release and dispersion characteristics across a wide range of realistic scenarios and environmental conditions is necessary for the development of codes and standards that regulate the bulk storage and transport of liquid hydrogen, particularly in retail fuel station environments.

It is expected that most releases will be highly turbulent and heavily influenced by buoyancy. Additionally, the dispersion of the released hydrogen will be affected by factors such as flashing, multi-phase flows, heat transfer, pool formation, ambient conditions (e.g., temperature, humidity, wind), ground effects, and obstacles/barriers. In instances of spills of liquid hydrogen, the extremely cold temperatures can condense or freeze ambient air, which sets these releases apart from those of liquid natural gas. This can lead to unique hazards that need to be comprehended as well. Currently, there is a limited amount of experimental data available in the literature regarding these circumstances.

To comprehend and assess the hydrogen behaviors pertinent to large-scale accident scenarios, detailed physical models are imperative. Although safety engineering nomograms can aid in identifying immediate hazards, it's crucial to emphasize that complex physical interactions present in most large-scale accident scenarios require detailed behavior models for proper assessment. As an illustration, it's necessary to evaluate the effectiveness of mitigation strategies beyond separation distances. This includes assessing the impact of utilizing strategic barrier walls or determining the most suitable layouts for system components and safety equipment like sensors, process isolation, and flow restriction. This task is challenging to accomplish with process safety charts.

Giannissi et al. (2013) conducted a simulation of HSL liquid hydrogen releases using the ADREA-HF code. They systematically evaluated the impact of factors such as ambient humidity, wind speed/direction, and two-phase slip modeling of condensed water on simulation results. Their results showed that there was a velocity defect between the condensed and frozen water particles and the gas-phase flow in humid air. To effectively duplicate the release required adjusted slip modeling.

A summary of the current state of the science regarding cryogenic releases is given by Zhang, J., (2020), along with a list of areas that require further research. In support of this approach, the study emphasizes the significance of validated, reduced-order models and indicates that a quantitative risk assessment (QRA) could assist in informing fire safety laws. An overview of the scientific literature on cryogenic releases is given in the article, with a focus on the dearth of facts and proven models supporting these theories. Furthermore, the state of the art for reduced-order hydrogen behavior modeling is described, indicating that further development is needed for liquid releases, despite the fact this approach is well-developed for gaseous releases.

The complexities of multi-phase flows and phase change behaviors during cryogenic hydrogen releases pose a significant modeling problem. To solve this difficulty, a reduced-order model is proposed to capture the relevant physics exactly by splitting the stream-wise flow regimes into distinct zones. As the jet has warmed up and is behaving more like a gaseous release, the zones farthest from the release point are predicted to be where the model performs best (Winters et al., 2009).

On the contrary, zones nearer the release point provide greater difficulties due to multi-phase flows and more intricate thermodynamic state modeling. As of now, the models utilized to comprehend pertinent behavior in these domains are predicated on hypotheses that lack complete validation. A new capacity is being created to investigate controlled cryogenic hydrogen releases as a possible remedy. Using a template akin to the one previously employed to investigate high-pressure gaseous hydrogen releases, this capacity can be applied to the improvement and validation of deterministic liquid hydrogen release models.

A homogeneous two-phase critical flow model was developed by Winters et al. (2009, 2011, 2013) to investigate blocked cryogenic hydrogen emissions. The multi-phase components of this model were assumed to be in thermal equilibrium. The model was then updated and varied temperatures for the multi-phase components were allowed for by Travis et al. (2012), building on the work previously done. The homogeneous direct evaluation model (HDE) is the term given to this improved approach. The HDE model was validated by comparison with NASA cryogenic data for oxygen, nitrogen, hydrogen, and methane in liquid and supercritical states under varied stagnation circumstances (Simoneau et al., 1979). After this validation, Li et al. (2012, 2013) used the HDE model as the input boundary conditions and the PHAST software tool to estimate harm-effect distances for cryogenic hydrogen discharges.

Because combustible clouds accumulate and take longer to ignite when hydrogen is released indoors, overpressure risks could arise. It is vital to measure the features of gas build-up since the degree of flammable accumulation and the local concentration of gas determine how severe the ensuing deflagration or detonation will be. It has been noted that buoyant gas releases into enclosures produce consistent accumulation layers at the ceiling based on large-scale fill-box tests (Cleaver et al., 1994; Lowesmith et al., 2009; Ivings et al., 2013; Cariteau et al., 2011; Ekoto et al., 2012; Worster & Huppert, 1983; Kaye & Hunt, 2004). A time-dependent gas accumulation model was created by Lowesmith et al. (2009) to investigate gradual releases of hydrogen/methane mixtures. Their model uses the Kaye et al. fill-box model (Kaye & Hunt, 2004) to compute the layer thickness and concentration over time and integrates a 1D integral buoyant plume model to account for the source. The model also considers the effect of airflow into and out of the enclosure caused by wind. It was discovered that the outcomes of this model agreed with complementary measurements.

13.5 FUTURE WORK

There are considerable gaps in the current understanding and modeling of emissions from liquid hydrogen storage devices, despite substantial study in this area. To accurately predict release boundary conditions, a major issue is to modify current network flow models for cryogenic flows. We need to make models for different types of fluids like liquids and gases. This includes stuff that's not ideal and things in a supercritical state. Doing this, we can understand hydrogen flows better. Having this knowledge is key if we're to learn more about how liquid hydrogen escapes. Safety steps would also come from this understanding. When thinking of a liquid hydrogen leak, there are more things to think of. Things like assessing stickiness and thinking about heat transfer are important. It's also key to remember how temperature might change the flow's volume. Experiments and digital trials could give us needed data on hydrogen. Answering these questions helps us make safety rules that could limit what happens when hydrogen leaks. This will give us a better idea of what these leaks are like.

Professionals must learn more about liquid hydrogen emissions. They need to use understanding and simulations to do this. They also need to make accurate models. These models show what happens when hydrogen touches the surface and collects in pools. Barrier walls help reduce the results of gas releases. But we still need more facts to check how good these walls are with liquid hydrogen. We can make safe plans by creating trustworthy simpler models and checking them. This will help us understand how liquid hydrogen releases act. We can protect people and learn more if we study and check our plans. We need matching facts to see how good walls work at stopping liquid hydrogen. Making and testing solid small-scale models helps. From this, we can put the right safety steps in place. This protects us if liquid hydrogen gets out and lets us see what it does.

13.5.1 SAFETY PROCEDURES

Hydrogen is a flammable gas. It might ignite if it meets sparks or flames. This means folks dealing with it must be cautious. These are some basic safety steps:

- Wearing appropriate personal safety gear is key. This includes items like gloves, safety glasses, and clothes that resist fire.
- You need a well-ventilated area to work with hydrogen. This stops the gas from building up too much.
- Keep hydrogen tanks correctly labeled and tucked away safely. This helps maintain a secure environment.
- We must, as experts, consistently examine our hydrogen gear. This lets us be sure it remains in good condition and doesn't have leaks or possible hitches. It's key for keeping our tasks safe and productive.

Always remember: If you're working with hydrogen, fire risks like candles or gadgets must be on your radar. Don't forget, hydrogen's lighter than air. It can float away if let go outdoors. But, in a tiny closed-off area, danger lurks. It can push out oxygen and you can't breathe. So, make sure there's good air flow. Be careful with hydrogen.

13.6   EFFECT OF HYDROGEN RELEASE ON THE ATMOSPHERE

Hydrogen leaks or vented into the atmosphere can cause environmental harm, therefore, when assessing hydrogen technology's environmental impact, consider the risks associated with hydrogen leaks or venting. It influences Earth's climate system as a temporary indirect greenhouse gas (GHG). For example, independent researches such as Rahn et al. (2003), Derwent et al. (2020), and Paulot et al. (2021) argue that hydrogen is present in the atmosphere for only a short time, from few years before it is completely oxidized by the hydroxyl radical (OH).

Hydrogen oxidation accelerates greenhouse gases' concentration in the troposphere and stratosphere (Derwent et al., 2020; Paulot et al., 2021; Field & Derwent, 2021). Further, there can be a reduction in tropospheric OH due to hydrogen oxidation which leads to tropospheric ozone formation—a GHG—and an increase in water vapor content within the stratosphere.

One of the most significant consequences stems from the mentioned processes. When OH is coupled with methane, it decomposes in the troposphere. A reduction of OH content might lead to methane staying longer in the atmosphere. Additionally, atomic hydrogen that hydrogen oxidation yields will trigger a chain of reactions culminating in the formation of ozone at the troposphere level. As a greenhouse gas, ozone will contribute to heating the atmosphere. Finally, water vapor is yet another by-product of hydrogen oxidation.

It may provide cooling when released in the stratosphere and can accidentally amplify the warming effects. More details and information can be found in the references (Rahn et al., 2003; Derwent et al., 2020; Paulot et al., 2021; Field & Derwent, 2021).

A recent study (Ocko & Hamburg, 2022) looked at different timescales to investigate the impact of hydrogen leaks on the atmosphere. The authors noted that the timeline selected and the estimated leakage rate from hydrogen applications are key factors in this kind of analysis. Therefore, more investigation is required to precisely ascertain how hydrogen emissions affect global warming. Preventing hydrogen leaks is essential from an economic, safety, and environmental standpoint as well.

13.7 CONCLUSION

All things considered, pressurized hydrogen tanks are a tried-and-true technology that has been applied to a variety of situations for many years. When handling and storing pressurized hydrogen, it's crucial to weigh the benefits and drawbacks of this technology and abide by all relevant safety regulations. In conclusion, there are benefits and drawbacks to both pressurized and non-pressurized hydrogen storage systems. The user's needs and the application will determine which storage solution is best. When choosing a hydrogen storage system, it's critical to take all relevant safety precautions into account and handle hydrogen with caution.

REFERENCES

Anoop, C. R.; Singh, R. K.; Kumar, R. R.; Jayalakshmi, M.; Prabhu, T. A.; Tharian, K. T.; Narayana Murty, S. V. S. A review on steels for cryogenic applications. *Materials Performance and Characterization*, 2021, 10, 16–88.

Cariteau, B.; Brinster, J.; Tkatschenko, I. Experiments on the distribution of concentration due to buoyant gas low flow rate release in an enclosure. *International Journal of Hydrogen Energy*, 2011, 36, 2505–12.

Cleaver, R. P.; Marshall, M. R.; Linden, P. F. The buildup of concentration within a single enclosed volume following a release of natural-gas. *Journal of Hazardous Materials*, 1994, 36, 209–26.

Derwent, R. G.; Stevenson, D. S.; Utembe, S. R.; Jenkin, M. E.; Khan, A. H.; Shallcross, D. E. Global modelling studies of hydrogen and its isotopomers using STOCHEM-CRI: likely radiative forcing consequences of a future hydrogen economy. *International Journal of Hydrogen Energy*, 2020, 45, 9211–21.

Duthil, P. Material properties at low temperature. *CAS-CERN Accelerator School Superconducting Accelerator Proceedings*, 2014, 2014, 77–95.

Edeskuty, F. J.; Stewart, W. F. *Safety in the Handling of Cryogenic Fluids*. Springer Science + Business Media, New York, 1996.

Ekoto, I. W.; Houf, W. G.; Evans, G. H.; Merilo, E. G.; Groethe, M. A. Experimental investigation of hydrogen release and ignition from fuel cell powered forklifts in enclosed spaces. *International Journal of Hydrogen Energy*, 2012, 37, 17446–56.

European Commission. *A European Green Deal*. European Commission, European, 2022

European Commission. *REPowerEU: A Plan to Rapidly Reduce Dependence on Russian Fossil Fuels and Fast Forward the Green Transition*. European Commission, European, 2022.

Field, R. A.; Derwent, R. G. Global warming consequences of replacing natural gas with hydrogen in the domestic energy sectors of future low-carbon economies in the United Kingdom and the United States of America. *International Journal of Hydrogen Energy*, 2021, 46, 30190–203.

Giannissi, S. G.; Venetsanos, A. G.; Markatos, N.; Willoughby, D. B.; Royle, M. Simulation of hydrogen dispersion under cryogenic release conditions. *Proceeding of International Conference on Hydrogen Safety*, Brussels, Belgium, 2013.

Granath, B. *Innovative Liquid Hydrogen Storage to Support Space Launch System*. NASA, Washington, DC, 2022.

Grasemann, M.; Kalidindi, H. R. Hydrogen storage technologies for renewable energy and sustainability. *International Journal of Hydrogen Energy*, 2015, 55, 1–1594.

Hussain, M. A.; Islam, A. Hydrogen storage for energy applications: current state-of-the-art. *Renewable and Sustainable Energy Reviews*, 2016, 54, 1087–98.

Ivings, M. J.; Lea, C. J.; Webber, D. M.; Jagger, S. F.; Coldrick, S. A protocol for the evaluation of LNG vapour dispersion models. *Journal of Loss Prevention in the Process*, 2013, 26, 153–63.

Jarvi, T. D.; Jarvi, M. A. Hydrogen storage technologies for fuel cell applications. *International Journal of Hydrogen Energy*, 2018, 34, 4569–74.

Kaye, N. B.; Hunt, G. R. Time-dependent flows in an emptying filling box. *Journal of Fluid Mechanics*, 2004, 520, 135–56.

Krainz, G.; Bartlok, G.; Bodner, P.; Casapicola, P.; Doeller, C.; Hofmeister, F.; Neubacher, E.; Zieger, A. Development of automotive liquid hydrogen storage systems. *AIP Conference Proceedings*, 2004, 710, 35–40.

Laura, P. *Hydrogen Storage Methods*. SynerHy Hydrogen Experts, China, 2022

Li, Z. Y.; Pan, X. M.; Meng, X.; Ma, J. X. Study on the harm effects of releases from liquid hydrogen tank by consequence modeling. *International Journal of Hydrogen Energy*, 2012, 37, 17624–9.

Li, Z.; Pan, X.; Sun, K.; Ma, J. Comparison of the harmful effects of accidental releases: cryo compressed hydrogen versus natural gas. *International Journal of Hydrogen Energy*, 2013, 38, 11174–80.

Li, J. Q.; Li, C. L.; Park, K.; Kwon, J. Y. Investigation on the changes of pressure and temperature in high pressure filling of hydrogen storage tank" Case Studies in Thermal Engineering, 2022 37, 102-143

Lowesmith, B. J.; Hankinson, G.; Spataru, C.; Stobbart, M. Gas build-up in a domestic property following releases of methane/hydrogen mixtures. *International Journal of Hydrogen Energy*, 2009, 34, 5932–9.

Meyers, J. P. Hydrogen fuel cell portable power generators for emergency response and disaster relief, In *2014 IEEE International Conference on Electro/Information Technology (EIT)*, Milwaukee, WI, USA, 2014, 212–9.

Michel, F.; Fieseler, H.; Allidieres, L. Liquid hydrogen technologies for mobile use friedel. In *Proceedings of the 16th World Hydrogen Energy Conference WHEC*, Zaragoza, Spain, 13–16 June 2006, 694–702.

Müller, C.; Fürst, S.; von Klitzing, W. Hydrogen Safety: New Challenges Based on BMW Hydrogen. In *Proceedings of the Second International Conference on Hydrogen Safety*, San Sebastian, Spain, 11–13 September 2007.

NIST. NIST Chemistry WebBook 69, August 2022.

Notardonato, W. U.; Swanger, A. M.; Fesmire, J. E.; Jumper, K. M.; Johnson, W. L.; Tomsik, T. M. Zero boil-off methods for large-scale liquid hydrogen tanks using integrated refrigeration and storage. In *Proceedings of the IOP Conference Series: Materials Science and Engineering, Madison, WI, USA*, Institute of Physics Publishing: Bristol, UK, 2017, 278, 012012.

Ocko, I. B.; Hamburg, S. P. Climate consequences of hydrogen leakage. *Atmospheric Chemistry and Physics, Discussions*, 2022, 22, 9349–9368.

Ogden, J. M.; Lewis, G. J. Fuel cell vehicles: technology, infrastructure, and market readiness. *Annual Review of Environment and Resources*, 2019, 44, 409–437.

Park, W. S.; Yoo, S. W.; Kim, M. H.; Lee, J. M. Strain-rate effects on the mechanical behavior of the AISI 300 series of austenitic stainless steel under cryogenic environments. *Materials and Design*, 2010, 31, 3630–3640.

Paulot, F.; Paynter, D.; Naik, V.; Malyshev, S.; Menzel, R.; Horowitz, L. W. Global modeling of hydrogen using GFDL-AM4.1: sensitivity of soil removal and radiative forcing. *International Journal of Hydrogen Energy*, 2021, 46, 13446–60.

Peschka, W. *Liquid Hydrogen-Fuel of the Future*, 1st ed. Springer, Wien, Austria, 1992.

Pharoah, P. Hydrogen as a fuel for heating in buildings: a review of current developments and potential. *Energy Procedia*, 2014, 61, 1527–1530.

Qiu, Y.; Yang, H.; Tong, L.; Wang, L. Research progress of cryogenic materials for storage and transportation of liquid hydrogen. *Metals*, 2021, 11, 1101.

Rahn, T.; Eiler, J. M.; Boering, K. A.; Wennberg, P. O.; McCarthy, M. C.; Tyler, S.; Schauffler, S.; Donnelly, S.; Atlas, E. Extreme deuterium enrichment in stratospheric hydrogen and the global atmospheric budget of H2. *Nature*, 2003, 424, 918–921.

Sempsrott, D. *Kennedy Plays Critical Role in Large-Scale Liquid Hydrogen Tank Development.* NASA, Washington, DC, 2022.

Simoneau, R. J.; Hendricks, R. C. Two-phase choked flow of cryogenic fluids in converging diverging nozzles. Technical Paper 1484, NASA, 1979.

Song, C. Fuel processing for low-temperature and high-temperature fuel cells: challenges, and opportunities for sustainable development in the 21st century. *Catalysis Today*, 77, 1–2, 17–49, 2003.

Travis, J. R.; Koch, D. P.; Breitung, W. A homogeneous non-equilibrium two-phase critical flow model. *International Journal of Hydrogen Energy*, 2012, 37, 17373–9.

Tzimas, E.; Filiou, C.; Peteves, S. D.; Veyret, J. B. *Hydrogen Storage: State-of-the-Art and Future Perspective*. European Commission: Petten, The Netherlands, 2003.

Ustolin, F.; Paltrinieri, N.; Berto, F. Loss of integrity of hydrogen technologies: a critical review. *International Journal of Hydrogen Energy*, 2020, 45, 23809–40.

Verfondern, K. *Safety Considerations on Liquid Hydrogen*. Forschungszentrum Jülich GmbH: Jülich, Germany, 2008.

Winters, W. S. Modeling Leaks from Liquid Hydrogen Storage Systems. SAND Report 2009-0035, Sandia National Laboratories, 2009.

Winters, W. S.; Houf, W. G. Simulation of small-scale releases from liquid hydrogen storage systems. *International Journal of Hydrogen Energy*, 2011, 36, 3913–21.

Winters, W. S.; Houf, W. G. Simulation of high-pressure liquid hydrogen releases. *International Journal of Hydrogen Energy*, 2013, 38, 8092–9.

Worster, M. G.; Huppert, H. E. Time-dependent density profiles in a filling box. *Journal of Fluid Mechanics*, 1983, 132, 457–66.

Zhang, J.; Dunn, B.; Assary, L. Hydrogen storage for energy applications: methods, challenges, and prospects. *Energy & Environmental Science*, 2020, 13, 8, 2587–610.

Index

Note: **Bold** page numbers refer to tables; *italic* page numbers refer to figures.